From Farm to Field

From Farm to Field

In Their Own Words
Stories of the Battle of the Bulge
From Africa to the War in the Pacific
Kansas Veterans Tell Their Stories

Tad Pritchett

by Tad Pritchett

ISBN-13: 978-0-9843585-7-1

ISBN-10: 0-9843585-7-9

Special Thanks

The Veterans of the Battle of the Bulge Northeast Kansas Chapter

Dee Driver-Pritchett

Dave Murray

Dr. John Gilbert

Mary Kay Rothert

Captain John Monroe, USN Retired

Mary Elizabeth Pritchett

Table of Contents

Foreword		ix
Chapter 1	The Nickel	1
Chapter 2	Uncle Bill	11
Chapter 3	Path to Battle	15
Chapter 4	The Prisoners	19
Chapter 5	The Infantry	51
Chapter 6	The Tankers	189
Chapter 7	The Artillerymen	257
Chapter 8	The Airmen	281
Chapter 9	The Medics	305
Chapter 10	The Mechanics	347
Chapter 11	The Navy	367
Index		389

Foreword

IN THE MID-EIGHTEENTH CENTURY, armies practiced maneuvers on open fields and the event became known as a "Field Day." As the term evolved to become more encompassing it grew to represent an action that was away from Headquarters and in the "field." Over the years, GI's began to embrace it as a description of the battle zone, the area where the enemy was engaged. A person who was "in the field" was in a forward position where the day to day struggle was for survival and human necessities were primitive at best.

In 1940 nearly sixty percent of Kansans lived in rural areas as compared to thiry-six percent in 2009. Rural life was a commonality among all Kansans. If they didn't live on a farm, chances were that they spent significant time visiting family and friends who did.

Farming in the 1930's in Kansas was a desperate struggle for the basics of life. The Depression smothered everyone and farmers were no exception. They had to contend with rainless summers, the dust bowl, credit crunches and losing their farms. They had to be creative in raising their families many taking extra jobs in town or helping out on other farms. They developed a strong sense of pride and it mattered to them what others thought. Teamwork, an overworked modern day cliché, became a fundamental survival skill.

Children, of the 1930's, shared the burdens of farming and learned how to make their own way with few resources. I will never forget the look in Wally Jeffery's eyes when he said, "I never had my own bed until I was in the Army." Social programs were minimal and people had to depend primarily on their own ingenuity and the assistance of others to survive. All they had was their pride and the strength of their word.

The rigors of the farm hardened them in ways that became valuable during the war. They knew how to be innovative taking advantage of anything that might later become useful. They learned how to handle the hardships of living in the field because it was to some extent only an extension of how they grew up. Jim Sharp told me, "It was cold and there was so much snow but it didn't bother me too much. I learned how to deal with it growing up on the farm." Everyone was lean and fit, and strong of will, forged by hard work and a hard life.

Raymond Brown remembered it this way. "After graduation, we were in the middle of the Depression and by the Dust Bowl area. I helped my dad on the farm all the while I was growing up, starting by milking cows when I was six years old and in the field with my own team of horses when I was ten."

From the '30's there emerged an American personality that was unique in American History. The children learned how to sacrifice for what was important and how to handle hardship as they grew into young adults. They learned that they had to trust and help others because they knew at some point the favor would be returned. But maybe most importantly, they learned how to make their own decisions and to be responsible for themselves, always concerned with what others thought of them.

In the 1940's, there emerged a young adult that embodied these unique qualities, a person highly adaptable to the "field." They accepted military life and life in the "field" as another challenge, another normal part of life. And when their job was done and they had created a new world, they went to work creating a new America. What an exceptional group of men and women.

In this book you will meet a few of them who journeyed from "farm to field".

Chapter 1

The Nickel

AS I LISTENED TO EACH of the veterans and their stories I was struck with the "Parable of the Nickel," as my Dad called it. What was the Parable of the Nickel and where in the Bible is the story? It is not a Biblical reference my Dad told me but something about truth just the same. I was 9 years old when I first remember hearing of the Parable of the Nickel and nearly 60 the last time he told it.

My Dad, Lt. William T. Pritchett, was a WWII Navy veteran serving in the Pacific and participating in the occupation of Japan. He started his career as a minister but changed to education before the War started. He wove many of life's lessons from those three braids, the War, religion, and education. A lesson about perspective became the "Parable of the Nickel."

Events, he would say, are like two people looking at a nickel. When you ask, what do you see, they will say, I see a nickel, and when you ask what its worth they both say five cents. But interpretation is a matter of perspective, he would continue. There was always a thoughtful pause at this point and sometimes he would add that history is also a matter of perspective.

But what do they say when you ask them to describe the Nickel from their perspective, he would ask? One would say, it is silver with the head of Thomas Jefferson and the other might say it is silver with Monticello. Both are right but different, he would say, because they both have different perspectives. You wouldn't call one a liar based upon the description of the other until you looked to see what he saw. Each saw the silver as common ground but each had a different view of the Nickel.

Another interesting thing about a Nickel, he would add. It is one of the few coins that can stand up on its own edge even when it's worn.

People's lives and experiences are much the same, he would say, they are all a matter of perspective. But there is common ground in most any disagreement resulting from a person's perspective of an event. Both perspectives seem to be at odds but are really only a matter of where a person might be and what they saw through their eyes. Both agreed that it was a silver Nickel worth five cents but each saw either a head or a building. Perspective, he would say, there are always several versions of the truth.

I was struck by the "Parable of the Nickel" as I made my way to each of the vets homes and listened to their versions of the "Nickel." They were there and saw it all from their perspectives. Who can say if they are right or wrong in the context of history, they saw it from their perspectives and those who record history glean "truth" from others perspectives. They all see the silver five cent piece; some see Jefferson, some Monticello.

For that reason, I have taken each veteran's interview as the truth as they saw it. I have read enough of "George Washington chopping down the cherry tree" stories to know that the truth can only come from those who saw it and lived it. If their stories don't seem to match what others have heard, remember the parable of the Nickel. They saw and remember it from their perspective.

Imagine for a minute you are sitting in your living room listening to music. It is a quiet, cold winter night and you feel safe and secure in your home. From all indications it will be a peaceful night. You close your eyes and for a minute, just a brief minute, you relax, and you sink gently into your easy chair. Without warning the door bursts open, the frigid night air fills the room and you are surrounded by a room filled with armed bandits. There is mass confusion, nowhere to run, no help to call, you are surprised and helpless.

The Battle of the Bulge started much the same way. Many felt that Hitler's Nazi Germany was on the ropes and the war was nearly over. Many were on leave for the Holiday's and green troops were rotated to the front to gain experience during a "quiet time." As Bill Stahl put it, "Morale was high, everybody had the idea that the war was almost over and that we would have a quiet front at least until spring and nobody was too excited about it."

But as John Mock recalls, there were definite signs that something was up. "On December 14th we heard some loud speakers coming from German tanks. With the snow, it was very cold and no wind, everyone could hear the tanks revving up. On the 15th about 4 AM, the tanks started up again."

Carl Shell remembers an odd airplane making nightly runs over the lines. "I did notice one thing, every night while we're staying there a lone airplane, an enemy plane, an old boxcar kind of a plane, seemed to be taking reconnaissance pictures of our position and we named him "Bed Check Charlie."

The strange behavior by the Germans was reported. Nobody at Headquarters seemed too concerned. The Germans were "on the ropes."

Suddenly everyone was surrounded. Confusion replaced the calm December night and The Ardennes Offensive had begun. "I was on guard at that time when the Germans began to attack," Carl said. "Off to my left I saw this machine gun tracer fire and it kind by passed us more or less. But then they came into our positions and I saw the Germans coming across toward me on foot. They were about a hundred yards away."

In addition to surprise, Hitler commissioned a variety of deceptive tactics to confuse the Americans. On December 17 the Germans made their only night parachute jump of the war, targeting an area behind American lines north of Malmedy. Due to a number of unanticipated problems, the troopers were dispersed all over the Belgium-German border giving the impression that there was a large division sized force behind American lines. In truth there were only a few hundred.

In addition to the confusion created by the German paratroopers, Hitler commissioned Otto Skorzeny to organize a battalion of disguised, English-speaking Germans to infiltrate behind the Allied lines. Using American uniforms and vehicles, they infiltrated into the American lines causing havoc and uncertainty. Rumors spread quickly in the confusion and along with the "division size" parachute assault, fear and rumors spread quickly.

Virgil Meyer remembered the chaos. "We had two separate incidents of running into Germans who were dressed up like Americans just as we got to Neiderfelden. We were ahead of the Germans who were marching down the road and were trying to get us to turn south to come down this road rather than to come up this other way."

In addition to the deadly German 88's (88mm artillery), the Americans were subjected to new ways to die. Don Huse remembers the buzz bombs. "One day, when I was on duty, I heard a buzz bomb engine quit. I stood in the open doorway of the van and watched as it dove toward us. It hit an anti-aircraft gun pit about 100 yards from me. I was knocked out from the concussion."

So does Carrol Joy. "I remember in the morning in the early part of June, and it was still dark, and we were pre-flighting, my plane had on its running lights and you would think twice before you got close to that propeller. I saw one coming over, not very high, I could see the exhaust coming out of the back, and I thought, what the devil it that?"

Even the Evacuation hospitals had to worry about the buzz bombs. "The buzz bombs were launched right over the Siegfried Line trying to bomb Liege because this was the hub of everything over there. They went right over us and some hit into top of the hill behind us. The nurses were at their positions when the buzz bombs came right over us. They said that they would just leave their windows open on both side of their building so the bombs could go through. That was a joke then because the bombs sounded so close," remembers Dr. James McConchie.

Of the first hand stories in the book, only a few of the veterans were in the same division. They all experienced the battle from their own perspective. As Carl Shell put it, "It was just survival after that, firing, keeping alive."

Wally Jeffery put it simply. "We all had our jobs to do with different degrees of danger."

The Battle of Bulge was the largest land battle in which the United States participated in World War II. On the morning of December 16, 1944, the Germans launched a surprise offensive through the Ardennes Forrest with the objective of recapturing the supply center at Antwerp, Belgium. The Offensive charged westward pushing the Americans into what was referred to as a "bulge" in the American battle lines. More than a million soldiers including some 600,000 Germans, 500,000 Americans, and 55,000 British battled across and 80 mile front. There were over 200,000 casualties on both sides of the gruesome battle that was fought in one of the worst Belgian winters on record. By January 25, 1945, Americans had pushed the Nazis back to the German border and the Battle was officially over.

The War, however, was not over and it ground on for nearly four more months. Many thought the worst was over until they crossed the Rhine into the German heartland where house to house combat was a norm. There they uncovered the most horrible secrets of the Third Reich and many wondered how mankind could sink so low.

During WWII, more than 227,000 young Kansans entered the military. In January of 2010 veterans from nine NE Kansas counties met to celebrate the 65th anniversary of the Allied victory at the Bulge. Veterans from different divisions and locations gave a diverse perspective of the Battle. Hundreds of family and friends were present to thank them for their sacrifices. They all did their jobs with little understanding of the big picture at the time.

They were an incredibly humble group of men generally feeling that they were only doing their jobs and did nothing special. They still had not gained the insight into their influence on history. They didn't understand that they were not simply witnesses of history; they made history and created America as we know it.

In the January 10th edition of The Topeka Capital Journal, my eyes had been drawn to a front page picture. A WWII veteran named Jim Sharp was holding a red German flag with the large black Swastika in the middle. It looked like another good story about a local veteran who had been able to bring back a Nazi flag from the war but this said something about the Battle of the Bulge. I vaguely knew that the Bulge had taken place in Belgium in 1944 and had actually visited the area around Bastogne fifteen years earlier. Jim and a few other vets were planning a reunion and celebration of the conclusion of the Battle at the American Legion Post in Manhattan, Kansas. This might be interesting, I thought. Maybe I can find someone who was in Uncle Bill's 26th Infantry unit. Bill Driver was from Quenemo, Kansas, and was killed on a hill outside of Hallenberg, Germany. I decided I needed to attend.

It was a bit cold and overcast when I drove from Topeka to Manhattan and was surprised to see the parking lot nearly full. What a turnout I thought! Insecurities made me pause before I opened the car door. Wasn't this the same group of men who had shunned us when we returned from our war in the '60's? I had refused to join the American Legion because of the criticism and the snide comments from some of the Legion Commanders but here I was. The memories still stung. What about that, I thought?

After a few minutes I swallowed hard and headed inside. The conference hall was full of vets, their families and friends. I made my way over to a side table where I could inconspicuously use my video camera. I introduced myself to several people close by.

One couple sitting at the table next to me was Julian Siebert and his wife Barbara from Westmoreland, Kansas. I learned that he was with the 26th Infantry Division and was captured by the Germans on Christmas Day. The other person close by in an Eisenhower jacket and Sergeant stripes was Lenhardt Homier from Wilson, Kansas, and his family. Lenhardt had been in the 2nd Engineers. They were all wonderful people and I began to hear the first of the many captivating, unbelievable stories I would hear over the next year.

The master of ceremonies was a veteran named Jim Sharp, the same person whose picture was in the paper. Jim had been with the 18th Infantry, 1st Infantry Division. Uncle Bill had been with the 26th Infantry, 1st Infantry Division so maybe they were in the same areas I thought. I would have to make it a point to talk to him after the luncheon.

The speakers were very good and I was impressed that the group had been able to attract such high caliber guests. The Consulates from Luxembourg and Belgium told of the sincere appreciation of their people for the veterans in the room. They had faced the loss of freedom firsthand, they said, and understood what it meant to lose it and what it took to regain it.

The agenda allotted time for each of the vets to tell a little bit about their experiences. The reverent silence was deafening as I sat in awe with the others and listened to them tell of personal experiences. I had read about them over the years but now I was listening to them first-hand. Here were the real guys, I thought. All were old but all had a keen memory of their time at the Bulge. I was absolutely fascinated and had to learn more.

At the conclusion, Jim asked for volunteers, anyone who would be willing to work on next year's celebration. Since I wanted to meet him and see what I could find out about Uncle Bill's unit, I wandered up to the stage and introduced myself. He asked if I wanted to help out and I said yes, violating one of the axioms of military service, don't ever volunteer for anything.

Nine of us met at the American Legion the next month and I was appointed Vice Chairman. I offered up the idea that we should try and capture each veteran's unique story and memorialize the stories into a book. They were so unique they needed to be recorded. I volunteered to interview each the vets and edit the stories into a book. The committee thought it was a great idea so with their blessing I started out with my video camera, a stack of maps and a questionnaire.

Soon my GPS became my best friend. Most of the vets lived in areas unknown to me until the GPS led me to their doorsteps. Convenience stores became my lunch room. Many lived on the very farms where they were born and from where they were drafted into the Army. Without exception, they welcomed me warmly into their homes and opened unhealed wounds to tell their stories. They told their stories of horror with the non-humorous smiles that all combat veterans recognize as a necessary mask to facilitate a story. They downplayed their roles and minimized the difficulties they faced. As Jack Gragert said, "It was just a damn war," as though it was an after school event.

Their span of history is astonishing, almost beyond comprehension. "One of my earliest memories was the end of World War I," Herman Westmeyer told me. "I was helping my oldest brother Walter drive cattle from one pasture to another, and suddenly church bells and storm sirens started going off in the town of Farmington. My brother says, 'the war has ended!' I was six years old."

"I spent a year in Vietnam as an employee. I was in charge of all of the entertainment in Vietnam. If you ever went to a USO show, I was in charge of those in 1966 and 1967," said Charles Neale. I had been to several USO shows in Vietnam, what a coincidence and what a connection.

Initially I thought their stories needed to be in the context of the history of the Battle of the Bulge and struggled to learn where each man's unit fought. Soon I realized that history had no bearing or relationship to their experiences. History was the silver five cent piece. In making history, these men require no historical context

Their stories stood on their own, they were the history, the minute to minute actions that weave events into the fabric of history. In combat, as I learned, strategy is some abstract, victimizing view that the combatant never sees. It is irrelevant to you, all you know is what you see and experience. The combatant sees Jefferson or Monticello, not necessarily the silver five cent piece. Let history record strategy, I recorded the personal history as it was lived by the participants, the eyewitnesses.

I had no idea what I was getting into nor could I possibly have fathomed what history I was about to record. I was totally unprepared for the variety of experiences, the warmth of the greetings, and the sincerity of the testimonials. For months I was transfixed on the Battle of the Bulge. Mentally I was whisked back to the dark days of 1944 when the very outcome of the war seemed to teeter on the efforts of these young Americans, many of whom, only a few short months before, had been high school kids. Dreams of desolated cities filled my mind at night and I froze in the snow and ice of dreamland. How different this war was from mine. We used the same equipment, tactics and leadership but it was so different, so devastating to everything is touched.

How could these men return to society and function after experiencing such indescribable horrors and physical hardships? How, I ask myself, how? How could they come home and create the wonderful world where I grew up after the atrocious experience of total war in Europe?

Some of them had participated in the battles of Africa and the invasion of Italy and southern France. Four veterans waded ashore during the D-Day invasion and others, landing later, fought through the Falaise Gap and forced their way across France, into Belgium and eventually the Bulge. Three were captured by the Germans and spent the war in German POW camps. Two helped get airplanes into the air and on target, an effort that helped turn the tide of the Bulge. Some came with their divisions, others replacements for those who had fallen. Still others were engineers, tankers, artillery men, machine gunners, mechanics, and medical personnel. They all suffered terribly from the freezing winter temperatures, snow and no cold weather clothing and the crush of losing buddies.

After the Bulge, they crossed the Roer and Rhine Rivers. Some were part of the 9th Armor Division and after discovering that the Ludendorff Bridge at Remagen was still intact, they scampered across before it crumbled into the Rhine. Later they saw firsthand the horrors of the Nazi concentration camps and POW Stalags. Many went through ruined towns where their German parents and grandparents once called home. After the War, they participated in the occupation of Germany. During the occupation they arrested Nazi criminals and participated in their prosecution. Some trained for the invasion of Japan.

Don Nixon might have had the best occupation duty. After serving with the 82nd Airborne Division, he was assigned the assistant manager for four hotels in Nice, France. "The beach was within walking distance and I went to the beach everyday when my work was done. There were lots of joints that we hit on the way."

Then they came home, I thought to a hero's welcome. Home had changed while they were gone. I was surprised to learn that after some arrived to ticker tape parades in New York City; they found resentment in their home towns from those who thought the returning heroes might take their jobs. Many told stories of the difficulty in finding work, raising families and bottling up the War, never to be discussed again. They would reason in their humble way, who they would talk to? Most everyone had been in the War. They quietly went to work creating the America that we came to know and appreciate.

"I made it a point to try to forget a lot of the War and it works pretty well. When I first got home, any loud noise and I was ready to get under the table," Bob Knight told me.

As Horace High put it, "One of the reasons that a lot of other vets don't want to be interviewed is because they don't want to reveal what they had to do and went through. I put all of that behind me when I left the service."

But the war never left them. I learned of the nightmares that fueled demons in their heads for decades and the continuing health problems resulting from frost bite or war wounds that have plagued them for the rest of their days. In spite of this they became great husbands and fathers and teachers of character, quietly drawing on their experiences in the War. They are proud of their country, proud of their families and proud of what they did. Not many of us can say we saved the world, they can.

Someone asked me what I had learned or what was my biggest surprise? The quick answer is that I met forty Kansas men, some as young as 16, met in an area along the German-Belgian border and along with one million other young men fought America's largest land battle.

But that was a small piece of what I learned. I personally learned much more. For most of my adult life I have felt an outsider to the WII veteran's community as did most other Vietnam vets. I didn't understand why my Father's generation was so uncompromisingly tough, principled to a fault, "my country love it or leave it" but they rejected us. Why?

From what I have learned, it may be this simple. They took the lessons from the Depression and the War, learned from them and moved on. Horace High still suffers from being pinned down on a sheet of ice for days during the Bulge, suffering now from the effects of frost bite. I told my wife after visiting with Horace at the VA Center, "These men are tough guys."

They became survivors during the Depression, learning to be resourceful due to necessity. There was little help so they learned how to survive and think for themselves. They learned to stand on their "own two feet" and abhor those who could or would not.

So, as Jim Sharp may have summed it up to me, people felt frustrated that they couldn't influence the government to change the course of the Vietnam War. They saw where it was going from their own experiences and it seemed to be a runaway train heading for an appointment with disaster. Had they fought their war in vane, was the country they loved destined to relive the history lessons learned in their war and at their expense? What did their experience mean if it was to be ignored?

Returning Vietnam vets were the only thing they could see. We became the symbol of a society that increasingly could "not stand on its own two feet" and a government that disregarded their experiences.

It is a lesson that we should all embrace. These men stood tall for liberty and due to their resourcefulness learned in the Depression, they saved the world from some of history's most notorious criminals. They returned to find others had taken the good jobs. But instead of complaining, they retrained themselves and went about creating a wonderful place to live, where people will succeed if they learn to "stand on their own two feet." Our lesson is that this is how democracy and liberty work.

Here are their stories. We see a silver five cent coin. They see Jefferson or Monticello standing on their own.

Chapter 2

Uncle Bill

This Book is Dedicated to
UNCLE PFC WILLIAM F. "BILL" DRIVER

PFC Bill Driver

Bill Driver, high school.

Place of Birth:
Quenemo, Kansas
Active Duty Date:
July 1944
Unit:
A Company,
1st Battalion, 26th
Infantry, 1st Division
Location:
Hallenberg, Germany
Arrival ETO:
February 1945
Rank: PFC
KIA: March 31, 1945

The 1st Infantry Division entered World War II in North Africa participating in Operation Torch in November 1942. The Division then was part of Operation Husky and Gela in Sicily. On D-Day, June 6, 1944, the "Big Red One" assaulted Omaha Beach. The Division moved on through France to Belgium where it was in the Battle of the Bulge in December.1944 in the Ardennes. On January 15, 1945, the Division again attacked the Siegfried line and occupied the Remagen bridgehead. On April 8, 1945, the Division crossed the Weser river and on to Cheb, Czechoslovakia, where it remained to the end of the War.

Bill was born on February 14, 1923 on a farm close to Quenemo, Kansas, to Roy and Hannah Driver. At Quenemo High School, Bill participated in as many activities excelling in music and basketball. He graduated from Quenemo in 1941.

He headed to Kansas State in the fall of 1941 but an interesting family twist brought him back to the farm in the summer of 1943. Bill's father Roy, a WWI veteran, feared that his son would be drafted if he stayed in school and brought him home in exchange for Bill's younger brother John. Bill would get an agricultural deferment thought Roy but it wasn't meant to be. Bill was drafted in July, 1944.

After completing his basic training at Camp Robinson in Arkansas, he boarded a ship for France and made his way to Weywertz, Belgium where he was a replacement in A Company 26th Infantry 1st Infantry Division. It was February 6, 1945.

On March 31 Bill was killed on a hill in an area south of Winterberg, Germany called Hallenberg Forest fighting the Panzer Lehr Division. The forest is a mountainous area now designated as a popular recreational area utilized for camping, hiking and skiing.

On April 6, 1945, Bill's mother received a letter from his Company Commander, 1st Lieutenant Harold E. Fischer. Bill, he wrote, was buried in a Belgium cemetery. He "was killed by enemy machine gun fire while advancing with his squad against enemy strong holds."

Much could be written about Bill's actions in Germany but nothing can say more than this letter to Bill's mother from his comrades.

Somewhere in Germany
May 2, 1945

Dear Mrs. Driver:

Today we received your letter dated April 18th, in which you inquired of us about our association with your son, William F. Driver, who was affectionately known among us as Bill.

Bill came into our company back about the 4th of

February. At that time we were in a German town near the Roer River, which town we do not now recall the name of. From that time and place he was with us in all our campaigns from west of the Roer river into the heart of Germany, where he met his death. We were with him all along, even at his death and afterward.

Bill was a buddy to us, and we to him, the word "buddy" having a meaning which anyone never in combat cannot understand. We shared dangers together day in and day out, and the officers and Bill's squad leader could always depend upon him without being disappointed. He always did his duty and never shirked any task that was put before him. Bill was a particularly inspiring sort of person, in that as a soldier he was outstanding in always being of a cheerful nature, even when things looked black and discouraging to the rest of us. Now, Mrs. Driver, we are not saying these things in empty praise of Bill just because he was your son, and we are writing to you concerning him; but we are saying these things because they come from our hearts, and we know they are true, as we knew Bill best as a combat soldier and buddy. During our association with Bill, we learned that he was talented in music. When we ran across any horn or trumpet, we would be entertained that night, or as soon as we could get a rest period, by Bill. In hardship or in fun, Bill was always there with us.

Mrs. Driver, you stated in your letter to us that you would appreciate having any of the details about Bill's death. We made an attack on the morning of March 31st, in which Bill was with us, and from which attack he came out all right. But that night, we dug defensive positions on a hill cutting a road, and Bill's position was in the woods on the top of a hill. The Germans attacked us that night about 10:30 p.m., which attack lasted about an hour, and the Germans withdrew. During the last burst of German fire, one shot hit Bill in the edge of the hair in the forehead, and he did not suffer a bit. We carried him down to the road, where he was to be picked up soon.

Every man that knew Bill regretted more than words can express the fact that he was killed, and we in his squad more deeply, because we knew him best. Our sympathy is truly with you, and with all his other relatives and friends.

Signed:

1st Lieut. H. E. Fischer, Commanding
Lt. Joseph H. McGinnis (Co. Exec. Officer)
Lt. George K. Campbell
Pvt. Frank Krygouekl
T/Sgt Charles W. Davidson
Pfc. John W. Weathers
S/Sgt James A. Hutchinson
Sgt. Eugene W. Fenter
Pfc. Charles S. Craig
PFC. Paul E. Coning
Pfc. Jose L. Fuerte
Pvt. Lester C. Dick
S/Sgt Carlton Brown
Sgt. Robert R. Barton

In September 2007, my wife Dee, and I made a journey to the Netherlands American Cemetery and Memorial in Margraten, NL to visit Uncle Bill's final resting place among his "buddies."

Bill's niece, Deetra Driver, in 2007.

Chapter 3

Path To Battle

THE WEATHER COULD HARDLY have been worse as the 175 Navy demolition men pulled away from the USS *Henrico* in thc carly morning darkness. The *Henrico*, an APA, was loaded with the 16th Infantry Regimental Landing Team of the 1st Infantry Division, scheduled to assault the Red Sector of Omaha Beach at 0600.

The ocean swells bounced the small boats like corks as they made their way toward shore. One of the men making their way silently to shore was Seaman John Brooks. His job was to clear away the German beach obstacles and disarm as many mines as the team could find before the assault troops stormed ashore. When he and the others were finished with their mission they were to move off shore, wait until the assault ended and return to the beach to assist with casualties. It was 0200 on June 6, 1944, and before the first wave hit Omaha Beach ninety-one of the original 175 demo-men would become casulaties.

While Seaman John Brooks was disarming mines in the predawn darkness of Omaha Beach, 29th Division Medic Forrest Adams, 90th Division machine gunner Art Cottrell and 327th fighter controller Wally Jeffery were staged on other APA's. They nervously awaited their turn to decend the swinging cargo nets and ride the Higgins boats into man's version of hell. They were the first members of the

Veteran's of the Battle of the Bulge of Northeast Kansas to reach the shores of western France.

By the end of June the 102nd Evacuation Hospital, the 9th Infantry Divisions and 913th Ordnance Company would bring Malcolm Strom, Dale Smith, and Paul Scheid to France and to the war. More units arrived in July as the Allies tenaciously pushed the enemy back toward Germany engaging them at the Falaise Gap. Dick Jepsen, a machine gunner, landed with the 30th Division and Dr. James McConchie came ashore with the University of Kansas Medical School's 77th Medical Evacuation. They moved off the beaches to join the war.

August brought in more and more divisions and units as the First Army pushed the Germans out of Paris. Air and artillery units began to arrive. Carrol Joy landed with the 406th Fighter Group, Don Huse with the 76th FA, and Art Holtman with the XVIII Airborne Corps, all plunging into the war. Paratrooper Frank Rhodes of the 509th Parachute Battalion, made his second combat jump, dropping into southern France as part of Operation Neptune. As the Battle of the Bulge developed, his unit moved north to join the fight with the 82nd Airborne.

Units arriving in September brought Raymond Brown and the 95th Division, Howard Goodwin of the 102nd Division and J. D. Sexton with the 132nd Ordnance Company. They all began their journey toward Berlin, a journey that would make an unexpected soujourn in eastern Belgium.

The human toll was staggaring as the Allies pushed the Germans back to their western borders. Stateside training was cut short to rush much

Queen Mary. *National Archives photo*

needed replacements to the War. They began to arrive in August and September, filling personnel gaps in the infantry units. Don Richards and Virgil Myers both joined the 80th Division in the Saarbruchen area. Bob Pearson reported to the 26th Division in the Montcourt, France, area. In November, Julian Siebert became a member of the 26th Division until he was captured at Christmas fighting toward Bastogne. Replacements were scavenged from college programs to feed the war's ravenous appetite for new young men. Bob Knight joined the 84th Division in November, and in December Lenhardt Homeir was assigned to the 2nd Division along the Belgian-German border. Don, Virgil and Julian would become part of General George Patton's Third Army as they pushed north in December toward the besieged Bastogne. Bob and Lenhardt would be pushed to the north side of the bulge with General Courtney Hodges' the First Army.

Units continued to pour into France and Belgium from England. In October Kenny Luigs and Carl Shell arrived with the 9th Armored Division and headed to the Luxemburg-German border. Harold O'Malley's 335 FA BN headed west of Metz with the 87th Division. All would be part of Patton's Third Army.

The 106th Division arrived in mid-December and infantrymen Bill Stahl and John Mock took up defensive positions on the Schnee Eifel. Within days they would become prisoners and spend the rest of the war in camps. Soon Carl Shell's tank company moved from the Luxemburg area to support the 106th.

Carl White and Harlan Henry arrived in early December. Carl was assigned to the 749th Tank Battalion and Harlan eventually was assinged to the 774th Tank Battalion. By the end of December Herman Westmeyer and the 11th Armored Divison, Elmer Blankenhagen and the 666th FA Battalion, Charles Neale and the 17th Airborne, Ivan Woellhof and the 238th General Hospital and Art Holtman's XVII Airborne Corps had moved into the Belgian area, innocently unaware of what was happening.

When the Nazi's struck the Ardennes on the morning of December 16, 1944, Jack Gragert, Don Nixon, Jim Sharp and Horace High were still enroute to Europe. As the American lines bulged westward, the divisions parted around the semi circle, holding back the onslaught.

Together they pushed east flattening the bulging line and by January 26, 1945, they were ready to head on to Berlin. The Nazis still had over three months of life left and the young men still had horrors that awaited them in Germany.

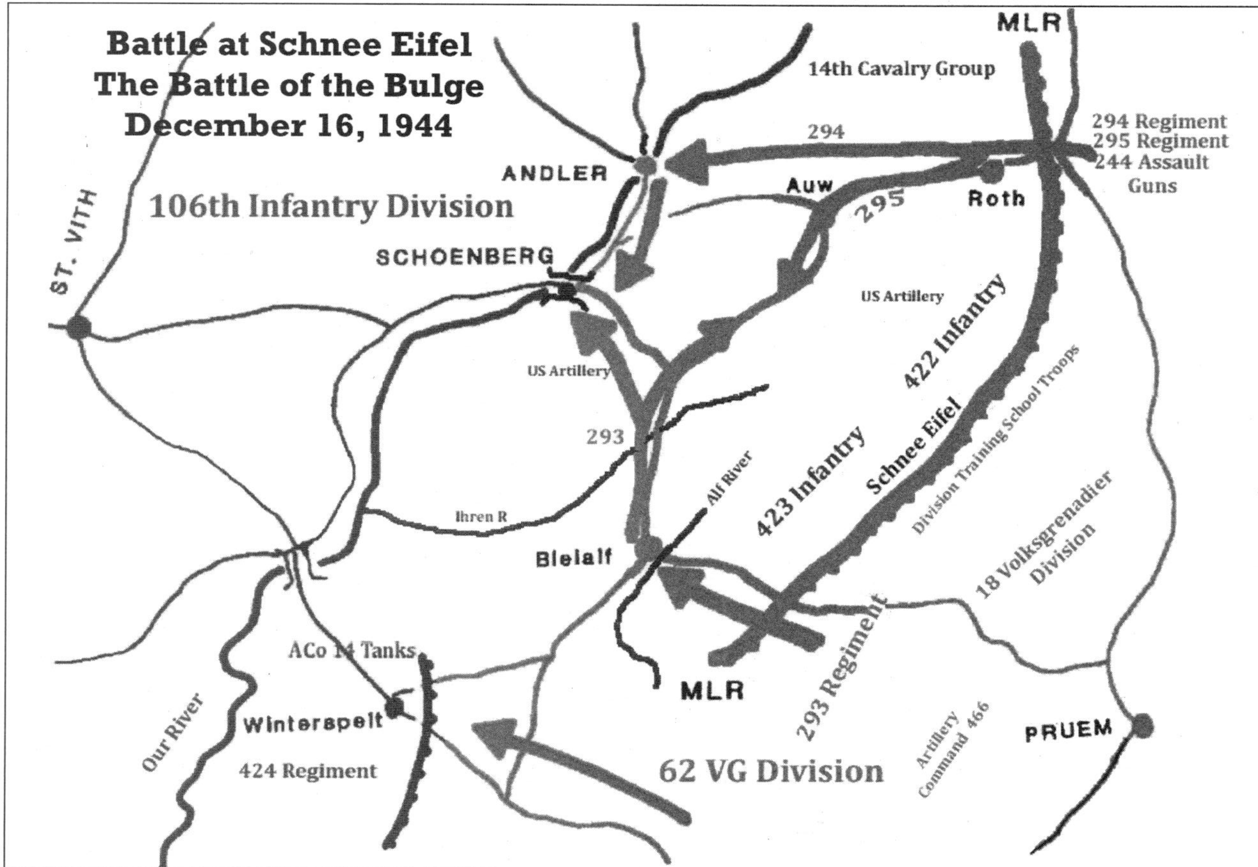
Battle at Schnee Eifel
The Battle of the Bulge
December 16, 1944
106th Infantry Division
14th Cavalry Group
MLR
294 Regiment
295 Regiment
244 Assault
Guns
294
ANDLER
Auw
295
Roth
SCHOENBERG
ST. VITH
US Artillery
422 Infantry
US Artillery
293
423 Infantry
Schnee Eifel
Division Training School Troops
Alf River
18 Volksgrenadier
Division
Ihren R
Bleialf
MLR
293 Regiment
ACo 14 Tanks
Winterspelt
424 Regiment
Our River
62 VG Division
Artillery
Command 466
PRUEM

Chapter 4

The Prisoners

THE EIFEL PLATEAU CONSISTS of three important ridge lines along the Belgian-German border. The middle ridge running about 26 km from southwest to northeast is named the Schnee (German for "snow") Eifel protruding about 185 meters above the 513 meter valley floors between the ridges, and in 1944 was the keystone connecting the Siegfried Line.

The Ardennes Forest was a dense, foreboding forest thought to be unsuitable for an armored attack. It was mountainous maintaining broken terrain, steep hills, deep and narrow valleys with a great number of small streams. It was covered by fir trees. Into this terrain, covered by snow and shrouded in fog and mist, the German Armies rolled into the American lines.

The Ardennes front was thought to be a quiet area by the Allied high command. It seemed to be an ideal time for providing rest for some of the experienced divisions and an area of operation suitable for the newly arriving divisions to gain experience before the spring.

Julian Siebert was assigned as a replacement to the 328th Infantry in the 26th Infantry Division, the Yankee Division, in November, 1944. The division had arrived at the Cherbourg Peninsula on September 7, 1944, and had run its first combat operations on October 7, 1944.

In taking Sarreguemines, France, they had overcome tough resistance and had sometimes been engaged in house to house fighting.

Julian's his unit was moved to Metz for a well deserved rest at the completion of the operation. He had no premonition that his part of the war would end in less than a week as he and his buddies relaxed, trying to shake off the nagging fatigue of combat. They had reached the Maginot fortifications along the French-German border.

His rest was cut short when on December 19 and 20 he and his company were moved in trucks to an area north of Luxembourg. He was captured on Christmas Day and spent the rest of the War in a German POW camp.

Bill Stahl and John Mock had barely arrived when General Von Rundstedt launched what would be later known as the Battle of the Bulge. They had trained with the 422th Infantry in the 106th Infantry Division at Camp Atterbury, Indiana and had arrived in the St. Vith area on December 10, 1944.

Secretly, Hitler began to secretly assemble 250,000 men, over 1,000 armored vehicles and 200 guns in October 1944 in the Eiffel area. The objective was to drive toward Antwerp, Belgium to capture the port and cut off supplies to the Allies and forcing peace talks. Resources were short for the German Army and the plan required capturing American supplies of gas for the tanks and food for the troops. The operation was called "Unternehmen Wacht am Rhein" or "Operation Watch on the Rhine." Americans would call it the Ardennes-Alsace campaign or the Ardennes Offensive.

At 0530 the enemy began to lay down a deadly barrage starting north, onto the 14th Cavalry position, then walking down the front and through the 106th positions. Trees were splintered, the noise deafening beyond believe. The boys of the Golden Lions hunkered in their fox holes, enduring and dreading.

The surprise attacks came along a slightly bulging forested ridge of the Schnee Eifel approximately 12 miles east of St. Vith. Through a tenacious and gallant defense put up by the 422nd and 423rd Regiments, the Germans encircled and cut off their avenues of egress, separating them from the rest of the division in the vicinity of Schonberg. Quickly regrouping for a counterattack, they were blocked by German tanks and infantry and surrounded. Soon Bill and John were taken prisoner.

All three of these brave men experienced the horrors and hardships of the Battle of the Bulge compounded by months of captivity, their whereabouts mostly a mystery to their families.

Here are their stories.

John, November 1945

John Mock

Place of Birth:
Eureka, Kansas
Active Duty Date:
1943
Unit:
3rd Plt., L Company,
3rd Battalion,
422nd Infantry,
106th Division
Location:
Auw, Germany
Arrival ETO:
December 10, 1944
POW:
December 22, 1944
Rank: Corporal

JOHN MOCK

The 106th Infantry Division was activated on March 15, 1943, at Fort Jackson, South Carolina, and was transferred to Camp Atterbury, Indiana, on March 30, 1944. It was the last of the 66 Infantry Divisions that were activated during WWII. It was continually stripped of its trained manpower for service in other Divisions or as replacements sent to replacement depots. Some were brought into the 106th at the last minute from other training programs in the Army such as the Air Corps where the need was not as acute as it was in the infantry. As a result, when the 106th shipped to France on November 10, 1944, a large proportion of the men were under trained and unit integrity was poor.

After a brief training period in England, the 106th landed at Le Havre, France. They were trucked to the area around St. Vith, Belgium arriving on December 11, 1944, and replacing the 2nd Infantry Division. The Division was spread thin along the snowy ridges of the Schnee Eifel as the Infantry Regiments moved into old German bunkers.

After only five days in the line, on the morning of December 16, 1944, at 0530, over 8,000 German artillery pieces blasted the American lines in the Ardennes. During the next three days, the German armor and infantry were able to quickly and stealthily surround the troops of the 422nd and 423rd Infantry Regiments. The men of the 106th fought valiantly. But as they slowly ran out of food and ammunition, they lost their ability to fight on and were ordered to surrender. The Division lost 8,663 casualties. Over 7,000 men became POW's for the remainder of the War. It was a difficult defeat but the 106th was able to delay the German's timetable for the assault, a delay that they were never able to make up during the Battle.

The 424th Infantry was south of the Schnee Eifel and were able to avoid the encirclement. They were assigned to the 7th Armored Division and participated in the offensive at Manhay and the retaking of St.Vith. They went on into Germany and were at Mayen, Germany, at the end of the War.

Here is his story. . . .

I was born on September 18, 1925. I was born in the oilfields since my dad was working there. We had oil on our place and were also farming.

I was drafted in December 1943. I volunteered for the air Corps, but failed the eye test; I went back home and waited to be drafted. My basic training started at Camp Fannin in Texas where I was put in communications. I did well enough on the tests so that I didn't have to go to the infantry then.

From there, I went to Fort Campbell, Kentucky, with the 20th Armored Division. Each squad had a half track and we didn't walk anywhere, we rode. We would go out into the boondocks and have a class, and then we rode back to camp in a half track.

I started out in communications when stationed at Camp Atterbury located south of Indianapolis outside of Edinburg, Indiana. I was assigned to the third platoon, L Company, 422nd Infantry, 106th division. The 106th was formed at Camp Atterbury and at this point I was now an infantryman. I had a hard time learning the Morse code.

As we were building up the division, we had part of the Air Corps, because they had too many people. We got men from everywhere and it took a while to get them straightened out. We trained at Camp Atterbury for about three months and from there we were transferred to Camp Myles Standish, Massachusetts for about 10 days. During that time we were going through all types of drills, such as lifejackets and lifeboat use.

From there, we were transferred to New York Harbor where we boarded the ship Aquitania for overseas. The ship was fast enough that we were not in a convoy. On the third night out, we picked up a submarine. We had a 7-inch gun aboard, a 20mm and a 40mm on both sides of the ship. If the sub surfaced, we could usually see it. It left a big wake that we could see. (The wake looked like a big wave on each side of the ship.) We out ran it. The ship was running wide open and vibrating all over.

The next day, we slowed down and conditions on the ship grew calmer and more settled down. Our ship was a British four stacker. It was a big ship, and had the biggest rats I've ever seen in my life. We ate British food and it was terrible! Hardtack, hard bread and pickles. We didn't see much of the British crew. We had bunks six high with just enough room to get into the bunk. I missed one meal because of an upset stomach but after that, no problem.

We provided the gun crew on deck A. I was on the 20mm and 40mm was next to me. The 40mm had two guns mounted together. The 20mm was like a big machine gun with a large ammunition drum. To cock it a cable had to be put on it and then had to be tipped in position so it could fire. It could not be cocked by hand. It had a large spring and was automatic.

We had to do lifeboat drill. Duty was eight hours on the antiaircraft gun, and then eight hours off to sleep and then back on gun crew. Because the galley was so far down to the bottom of the ship to eat and get back to bed, we would soon be awakened up again. It seemed like we had no sleep at all. Then every third day, we would get the day off.

We landed at Glasgow, Scotland. We came in and the *Queen Mary* or *Queen Elizabeth* was just leaving as we passed her. We had a destroyer on each side of us and escorts all the way into the port. From there we went on a train to Stow on the Wool (a town) in England. We walked all over the countryside. The old farmer's had stonewalls built up on the property lines that had been in place for hundreds of years.

From there we went to Southampton and boarded a ship. Here we were ordered to board an LCI that was waiting for us and unloaded us on Omaha Beach. Here we saw the concrete houses where every window and every door was a gun port. The whole beach was that way. We walked for couple of hours until we came to some open pastures. We are told to pitch our tents. It rained every night, and it was a mess trying to keep dry. The daytime was nice but it was rainy at night. We were waiting for the Red Ball express, a truck convoy, to take us up to the front. We were there about a week when they finally came and got us. This was around December 1, 1944.

We rode the trucks to the Belgian-German border, got off and met the 2nd Division. We are able to talk to them for a few minutes before they had to move out. Our duty was to patrol and we had people stationed on the front lines watching the Germans. You had to get them relieved before it got light in the morning and after it got dark in the evening. At one place we put a piece of canvas going up and down beside the side of the road. Then we were pretty well invisible.

On December 14, we heard some loud speakers coming from German tanks. With the snow, it was very cold and no wind, everyone could hear the tanks revving up. On the 15th about 4 AM, the tanks started up again. On the 16th here they came, but they bypassed us. There wasn't any road by us, so they just went right by us. We are along the Siegfried line where all the pillboxes were. Col. Thompson had his headquarters in a pillbox, and even had telephone wires running into the back door.

The Americans had little wooden huts built, and you had to crawl down into them. The engineers had built little log cabins and each squad had one of them. The kitchen had one below ground level. We had a railing with wire and cardboard is a bed.

We were on the top of the hill in concrete bunkers on the backside of the Hill, because they were looking towards France. Our section was about 4 or 5 miles and had a little road around it. We went on trails and the road when we went out on patrol and it was all on foot. For supplies, we had one 6 x 6 truck. The 106th Division was spread out as far as 28 miles.

The Germans came into the little town Auw, behind us in the Valley, and I think the 81st Engineers were there also. The Germans took them prisoners. We came down out of the Hill and chased them back about a mile. It started to get dark and we were ordered to make a lot of noise. So we started to make noises, banging on trees and things we could find. As soon as it got dark, we put a dead man on a stretcher and moved out. We went back to the little town where our 105 was. The gun crew had gotten caught. There was a stone wall around the barn. The 105 got right in behind the rock wall and was pretty well protected. We dug foxholes around this area and all around the edges of the town.

About 11 PM that night, the trees where we had made all the noise were hit by German artillery fire, and they made toothpicks out of them. They were up on a hill, and they did not know what we were doing. We waited a while but the Germans figured out where we were, or someone of the town told him. They were dropping rounds outside all of the buildings, but they didn't seem to want to hit the buildings in the town. The next morning there were two small tanks that came around the hill. The 105 got both of them. They were good, because it was about a half a mile shot. The next day there was a convoy that came out of the trees into an open field. There were two trucks and the 105 got them. Everything behind them was a team and wagons and that was their convoy. The105 also blew up one of the team and wagons stopping the convoy.

The next day there was a tank coming toward us. He just came up over the hill to get the 105. We didn't hear him until he started shooting. I was out of my foxhole but sitting close by. You had to use your rifle as a crutch to slide yourself in but my feet did not make it. A shell hit 5 feet away.

You couldn't hear anything. You feel pressure, your ears ring and you are stunned. He can't move or do anything for a while. In time, we could hear the shells going off. We were in front of the 105 and they were shooting at the tank.

When I got back into my foxhole, I found a piece of metal that had gone through my shoe. It didn't bleed too much, but the rest of the day, I just sat and started shaking and couldn't help it. I told Lieutenant Christiansen there was something wrong. I'm shaking but I am not cold. The sun was shining. He told me to sleep in the hay mound that night where it was nice and warm. I did, was warm but still shaking. Sometime in the night an old sheep came up there. He was noisy and snoring! The next morning I was over it.

Then the Captain said to go take Schonberg before the Germans got there, but they were already there and they started shooting at us. Then they started hitting us with tree busts and it made a lot more shrapnel. So L Company asked, what are we going to do?

Lieutenant Christiansen said we're going to head south and get back to the American lines. In the daytime we hid and tried to find some evergreens and put the limbs over the snow to help us get warm. Then during the night, we would go single file through the snow. We saw a little town on our left and we could hear the Germans and saw the lights. It was night and they shot up flares and started shooting at us. The whole company was eventually surrounded. Lieutenant Christiansen was leading and he got shot in half, then they hit Captain Spudola and others. Another boy and I were about half way down the line and we got on our hands and knees and with our arms through the slings of our rifles. We were going through there fast and we were going to get out of there. All the ones that were hit were carrying on quite a bit and crying for help. We looked at each other and said, "We can't leave them," so we went back and stayed with them. We were then taken prisoner near Auw.

The Germans put us in an old school room. They had straw around the outside walls for us to lie on. The next day the Germans came in and asked if anyone was wounded. Eight or nine of us got up but I did not want to leave. The Sergeant said, "You better go and get that foot taken care of or you are going to be in big trouble so about 10 of us went to the first aid station. One guy had shrapnel in both legs and we ended up carrying him piggyback. They cleaned my foot up and put black coal tar salve on it, the same kind of salve my grandmother used back home in the '20s — "black coal tar salve." They wrapped it in crêpe paper bandages, because Germany didn't have cotton fields to make regular bandages. I didn't get any infection in it. Our feet and legs were also frozen.

The Germans didn't know what to do with us. We were wounded, and one day we stayed in the hospital. Then on Christmas Eve we stopped in a German first aid station. It was a house and barn with a big enclosed hallway between them. There was straw on the ground and we got a bowl of soup and a slice of bread. The Germans ate the same thing.

About 11 PM three German officers came in with a bottle of schnapps. We took a drink and they gave the wounded Germans a drink. They started singing "Silent Night" in German. We sang it in English. It sounds the same.

The boy who had shrapnel and his legs had to be carried piggyback. We told the guard about him. The guard said he would take him to the hospital. The next day we asked how he was doing, and we were told he died. You don't die overnight from something like that, so we don't know what happened to him.

We went from town to town along the Rhine River. We went through Koblenz five times and we saw that big church with spires each time.

A lot of times we were on regular passenger trains. One time one of the Germans got really upset because he had lost his family in a bombing. We had to go back to another car. Generally, the German guards were usually good to us. They didn't abuse us, and they didn't want to walk either. One day we were riding in a truck, and it started missing. So they pulled over. There is a big stove hanging on the side of the truck. They opened the top, put in woodchips and sealed it up. It had screws in it. Then they lit a fire under the bottom one. There was a tube that ran from it into the truck tank on the top of the cab. They were making wood alcohol. They ran the stove for about a half-hour until he got enough alcohol, then started the truck and away we went.

When we crossed the Rhine River, they had little flat bottomed barges. They had no motors, but they had an anchor up in the middle of the river with a cable. They would use the rudder to go from side to side carried by the current. They could get a team and wagon, a Volkswagen, or some cows across on the barges. This is how they got across the Rhine because most of the bridges were blown up. When crossing bridges, there was usually only walking traffic. We walked around bomb craters in the middle of the bridges.

About January 18, we ended up in Stalag XIIA, Limburg Germany. I spent the rest of the war there. First thing in the morning we have roll call and they would count us. If we were short, we would stand there while they went in to check the barracks. Some of the men couldn't even get up and walk anymore. They would count them and then us again to verify everyone was present. We would get a bowl of grain like paste that they had roasted, and then we would get a loaf of bread to a six man squad.

BLACK BREAD RECIPE

Former prisoners of war of Nazi German may be interested in this recipe for World War II Black Bread. This recipe comes from the official record from the Food Providing Ministry published as Top Secret Berlin 24.X1-1941 from the Director of Ministry Herr Mansfield and Herr Moritz. It was agreed that the best mixture to bake black bread was:

50% bruised rye grain
20% sliced sugar beets
20% tree flour (saw dust)
10% minced leaves and straw

From our own experiences with black bread, we also saw bits of glass and sand. Someone was cheating on the recipe!

This bread recipe could have some protein in it. When the grain goes into the grinder, there could be some insects and maybe a mouse or two.

You sliced it so each man would get the same, and we took turns at picking first. At 4 PM we got a bowl of soup. It was cooked and the bugs floating on top because they had been cooked, and they didn't bother you any. We also got a small piece of cheese once a week. When we first got there, we got one cigarette a day. In February, we got two cigarettes a week, and in March, we got no more cigarettes. They were regular cigarettes.

One man from Tennessee would trade his bread rations for a cigarette. He smoked his cigarette under his blanket so we could get every bit of smoke into his lungs. He finally died.

You could tell the ones that smoked because the thumbs and fingers would be black. They would put out the last little bit of cigarette and put it in their shirt pocket. When they had enough pieces, they would make another cigarette. We got one Red Cross package for 10 men and we got one only once. I know more were received, because we unloaded them off the boxcar and put them in a warehouse. We spent one day flattening tin cans and that was the only work I did. There were thousands of men in XIIA. We just sat around the rest of the time. We picked lice and talked about food. It started out with breakfast talking about pancakes with marshmallows and sorghum, then a big steak and banana cream pie. Banana cream pie was a favorite of everyone. Each person would talk about what they wished for.

Each day we watched the bombers go over. Some would get shot down and we would count parachutes but we never saw a full crew get out. We would only see three or four at time because the plane would start spinning before all the crew could get out the door.

I was there about a week when someone came into the barracks and asked if anyone was from Kansas. I said I was from Eureka and he said "that is where I'm from." It was Mr. Dunlap from Dunlap School near Hamilton, Kansas. We used to go there for track meets and things. I had never seen him before. Later I was moved to another part of the barracks and I never saw him again. He made it home also. If you had a "J" on your dog tag when you were captured, you are taken up to the castle and never seen again. Jewish soldiers were segregated.

By March 21 the Americans were getting close. One night they put us in a boxcar when the Black Widow bombers were flying at night by radar. The air raid sirens were going off all the time and we had to go back to camp. The next night we got back on the train. We would go for a while and stop for a while. The next day a P-47 and P-38 strafed the engine. They didn't get the boxcars but they got the engine so there we sat on the tracks. We were outside Bergholms, Germany. We were locked in the boxcars for about a week. After three days we got something to eat and drank water out of the little stream and then we were back to the boxcars for another three days.

On the 27th, a train crew came along and unlocked all the doors. The guards were gone. We broke into their boxcar and found bread and canned goods. I didn't get any the bread. I was too late, but I got a can of peas and carrots to eat.

There was an overpass over the railroad and it was a main street of Bergholms. The safest place we could get was under that overpass. It was covered with rock, but we lay down. It was the first time we had been able lay down for a week. There were 50 of us in the boxcar. The cars are only 21 feet long and 8 feet wide, a Cadillac is about the same size. The first POW's sat down all around the sides of the car. The next group in sat down back to back down through the center of the car. We were just crammed in and could not lie down.

There was a big can by the door that was the toilet and the guards would empty it every day. That night we heard some machine gun fire and a

grenade. Then we didn't hear anything else. The Americans had just bypassed us. The two POW's who could walk okay went out and got hold of the 99th Division. The rest of us were under the overpass popping cooties. We would take a louse and roll him around and when we dropped him he would walk off. You had to put him on your thumb and put him under your other thumb and pop him.

C Company, 393rd Infantry, 99th Division liberated us on March 28, 1945. Here they came. We were just skin and bones. They looked like the fattest GI's we ever seen. After our tears stopped, and we got our voices back, we started talking to them. When they saw us, they said they had seen a lot of dead bodies but it had never seen anything like us. One of the GI's was Bill Tumblin who sent me a picture of the rescue.

That night we stayed in a house. I had a canteen cup, the German cup that is about twice as big as ours, and I had a square piece of blanket I used to put over my head to keep my ears warm. I also had an old worn-out razor that would not cut anymore, all in the pillowcase. A chaplain with the 99th division held a religious service on the side of the underpass.

I put an article in the 99th Division paper and talked to a lot of people who remembered the train at Borgholms. I found one guy in New Jersey, who also was on that train. The men were from all parts of the different divisions. That evening the 99th Division came in with a weapons carrier. The whole back end was filled with fresh baked bread. It was great! There were 277 of us on this train. The next morning ambulances started taking the ones that could not walk. Each ambulance could only take four at a time. It was late that night, before they took me.

We rode a long time but soon we stopped at a field hospital. There were a bunch of big tents with rows of cots. Two of us went to a tent and there was a nurse sitting at a desk. She just looked at us. We were walking skeletons, dirty and long whiskers and just skin and bones. I had on a German overcoat with the left pocket torn out.

She did not say anything. We started speaking English and told her we were Americans and had been prisoners of war. She started to cry. And then she jumped up and took the man with me to another tent. She came back and took me to a cot. Then she left and came back with a bucket of warm water, soap, wash rag, towel, razor, a metal mirror and a new pair of pajamas.

I was to put my clothes in the middle of the floor because they were full of lice. After I got cleaned up, I shaved and was in pajamas, she came back with a coffee cup of sliced peaches. You cannot believe how good they tasted! We talked for a long time. We were the first prisoners of war she had seen. She said there would be an ambulance plane in first thing in the morning to take us to the hospital in Reims, France.

We got on the plane. There was one stretcher on the floor and then they were stacked up one on top of the other and on both sides of the plane, nothing but stretchers. At Rheims, we were put on the third floor of an old school house that had been made into a hospital and we were assigned to bed.

One evening meal was served by two German POWs, and we didn't go for that and complained. We didn't see them again! French workers served our food and they gave us bars of soap and clean pajamas. We went to the nurse told her we wanted G.I. soap because we were loaded with lice. We took shower after shower until got rid of lice.

We suffer from malnutrition, and I had trouble with my ears. They have been ringing for 65 years, and I can't hear out of my left ear. My feet had been frozen when we were walking in the snow. We had to cross a stream one night, and it had ice on top. So we were walking in ice water and our feet could never really get dried out. When I got wounded in the foot, I had to march without a boot.

In Reims for treatment, my feet were put in ice cold water and then warm water. Then every day, they made the water a little colder, and that did more for my feet and circulation than anything else.

When we first arrived at the hospital we had to eat in bed. We were wearing red clothes and had to stay in our room. We could see a chow line down below us in the yard and they had on blue clothes. After five days we got ourselves blue clothes (not issued to us.) The reds were served first. After finishing eating we would go in the bathroom and change into the blues, go down the back stairs, and get in line again. We would take jello, fruit, pies and cakes all the good stuff because we were starved. One day the nurse caught us. She said, "We are not serving you in bed anymore." While in blues you could go back for seconds and have as much as you wanted as long as you cleaned up your tray. We started to gain weight then.

I was not sent back to the states until May 25, 1945. They wouldn't or couldn't send us home since we're walking skeletons! I weighed 105 pounds and was 6 feet tall, and I could touch my finger and thumb when I put them around my arm.

From Reims we went to Paris where a guide showed us some of the sites. We were there four days and issued new class A uniforms.

From Paris, we boarded an army DC-6 four engine plane. We landed next in the Azores for fuel just for a few hours. There was a restaurant there but it was closed but the lady running it asked if we were hungry. Well a little bit, we told her, so she opened it back up and made sandwiches and served pie. She wouldn't take any money when she found out we were POWs.

From there we flew on to Newfoundland, Long Island, Detroit, Chicago, Des Moines, and landed in Kansas City. Each man was sent to a hospital close to their home. The Army picked up three of us and took us to Winter General Hospital in Topeka, Kansas.

Of course my family was there. My family only knew I was missing in action and they first found out that I had been in a prisoner of war camp when I arrived at the Rheims hospital. There they took our names and addresses of our folks and sent a telegrams that we were okay. It had to be a shocker for the family but a happy one. I was discharged October, 1945.

When I was discharged from the Army, there were not many employment opportunities in my home community of Hamilton and Eureka, Kansas. I went to work in Hobbs, New Mexico. Everyone there was short of help. I worked overseas for oil companies for 19 years, so didn't marry.

My job was chief mechanic for drilling rigs. I worked on offshore rigs, land rigs and helicopter rigs.

I taught local mechanics to do maintenance and overhaul of drilling equipment in the countries where I worked. We drilled wells in Brazil, Colombia, Venezuela, Singapore, Indonesia, Iran, Saudi Arabia and Kuwait.

After the war each state received a 40 and 8 box car from France refurbished for display. The one for Kansas is located at the American Legion Post parking lot in Hays, Kansas and has been converted into a small museum.

I retired in 1984, married in 1993 to a retired schoolteacher and World War II Wave veteran. We now make our home in Eureka, Kansas, and have been very active in our community, especially veteran's organizations. I am presently Commander of the Air Capital Chapter Ex-POW's in Wichita. I am also an active member in the Disabled American Veterans in Eureka and past Commander in Emporia. In addition, I am an officer in the Purple Heart, VFW Military Funeral Commander, and former member of the Lions Club. My hobby is show cars.

John was awarded the Purple Heart, Combat Infantry Badge, ETO Ribbon with 3 Battle Stars, American Defense Service Ribbon, and the Good Conduct Medal.

PFC Julian Siebert

Julian Siebert

Place of Birth:
Westmoreland, Kansas
Active Duty Date:
May 4, 1944
Unit:
F Company, 328th
Infantry, 26th Division
Division Location:
Eschdorf,
Luxembourg
POW:
December 25, 1944
Arrival ETO:
November 1944
Rank: PFC

JULIAN SIEBERT

The 26th Infantry Division was called the Yankee Division because it was made up of units from New England National Guard units. The Division was shipped directly from the U.S. to France and landed on Utah Beach on September 7, 1944. In October they were in defensive positions in the Slonnes-Moncourt sector. They entered Saarefuemines on December 3. They were resting in Metz when General Patton was called into the Battle of the Bulge moving north to Luxembourg. They captured Arsdorf on Christmas Day, regrouped and continued the assault taking Grumelscheid crossing the Clerf River on January 24, 1945. On January 29, they were shifted to defensive positions in the Saarlautern area. Advancing through Germany, they moved into Austria and Czechoslovakia liberating the Gusen concentration camp as the war ended.

Here is his story. . . .

For years I just could hardly talk about the war and once in a great while, I'd get to feeling like I wanted to talk a little bit but generally I didn't want to. And even now, sometimes it makes me kind of nervous to do much talking. Anymore now it seems like a dream.

I was born on November 28, 1923, and raised in this house here near Westmoreland, Kansas. In fact, there used to be an old house out next to our house where my dad was born in 1890. There were eight of us kids.

I was drafted and left home on May 4, 1944. I first went to Fort Leavenworth where I was inducted into the Army the following day. From there I went to Camp Hood, Texas, for seventeen weeks of Infantry basic training. It was terribly hot during basic training. Following a 10 day furlough in route at home, I went to Fort Meade, Maryland, for preparation to go overseas.

On October 22, I boarded ship from Camp Shanks, New York, for the trip to England, and changed to a British ship to cross the English Channel for France. About the third day out of New York I got seasick. They say that the first day, when you first get seasick you're afraid you're going to die. And about the second or third day, you're afraid you won't and that's just about the truth. Oh, you get sick.

I don't remember the name of the ship or anything but it was just a ship. You had living quarters, different levels, there were living quarters on all of them. And I can remember the ship swaying up and down, I started to walk up the stairs one day and they say you can fall up stairs and I did. The first step I went to take that old ship went down and I fell up there.

I was assigned to the F Company, 2nd Battalion, 328th Infantry Regiment 26th Division, as an Infantry replacement. The division patch is a YD and they were called the Yankee Division because it was made up of "Yankees" from the northeast.

The 3rd Battalion was in this place, Dieuze, and on November 22nd the 2nd Battalion captured Munster. I don't think it was a very big place and you never know sometimes we might not have got right into the town either. I remember at least seeing the sign to it.

The reason I remember this place is because it was my first operation and I remember a good friend from St. Mary's, Clyde Aubert. Clyde and I left for the Army the same day and trained in Texas together. We left on the same ship and were both assigned to the 328th Regiment. Clyde was killed on our first attack, my first experience with losing a close friend.

A lot of times I never had any idea where I was. Once in a while you'd maybe see a sign pointing towards this town or something and that's all you knew about it. You had no idea where it was anyway once you saw the sign. You couldn't tell what country many of the cities were in by the name because the borders had changed so many times; there were German names in each one of the countries and that kind of thing.

I remember Saarbrucken. I was there during November and it was muddy and sloppy and then it kept getting colder and colder and finally everything froze up and it just kept getting colder.

We were around Munster on November 28, my 21st birthday. The night before that we were dug in at the edge of a small town. Our squad went on a night reconnaissance patrol. The next morning we learned there was going to be mass in the church just behind us, so several of us went to Mass.

That afternoon the artillery was falling on us, and when we moved it would follow us. Later on a sniper started firing on us from the church steeple. That told us that he was directing the artillery. Between the artillery and the sniper, it got so hot I looked for cover and I found an old foxhole that was full of water. I broke the ice on top with the butt of my rifle and jumped in up to my shoulders in the icy water. That night the sniper tried to escape, but he didn't make it.

Once we took refuge in a school house to get in out of the cold. I saw a book lying on a table. I open it up and on the inside cover was written the name "Nicholas Siebert," the family name was the same as mine.

Around December 15 or so we went to Metz to be on rest and I don't know just how long we were there but I don't think we had more than maybe two nights. We could sleep finally and I wanted to take a shower. They had a little thing about four foot square tacked up outside and it had cracks so wide you were barely out of sight. There was a bucket hanging up above with holes punched in it. So you went outside and it was very cold to start with. You went out and got a bucket of cold water out of a well, went in the shower, took your clothes off, dumped the water in the bucket overhead and that's how you took a shower.

It was cold but it was refreshing. It was the first time I had a chance to take a shower in over a month probably and I had some soap. When we got our rations and whatnot, sometimes it would have a little soap with it. In Metz we stayed in a big building, no heat in it and slept on a rough brick floor and boy, that was really great to be able to lie down and to sleep all night for once. All you had was just a blanket to lie down on and you pulled the other end over yourself. That's the way you slept.

We just had our regular boots and we had heavy clothes. It was our regular uniforms and then we had an overcoat and a jacket so we were dressed pretty warm, but still it was cold.

We didn't dig foxholes every night because we were pretty much on the move all the time. Sometimes you might find an old foxhole or something that you could jump in if things got a little bit hot. Sometimes you might be able to get inside a building for a while.

There was one night that I was standing guard looking out of a window and that's where I spent most of the night. The next morning there was an artillery shell that hit during the night right outside the window but it was a dud. If that thing had not been a dud, I would have been blown sky high.

There were a lot of German artillery shells that were duds. I think maybe the reason why is because they might have been using prisoners to help build them and maybe they would do something that they would forget on purpose.

I carried the M1 and then for a while I carried the anti-tank rifle grenades. There was a little launcher that went on the end of the rifle and you put a blank shell in it. They didn't do much to a tank unless you could get it tangled up in the track or something. We used them more like we would use a hand grenade when it was too far away for you to throw a hand grenade.

We would eat whatever we happened to have, C rations, K rations or maybe even a D bar (Hershey's chocolate.) Once in a while we would find a barn and there would be a cow or two there. I was the only one that knew

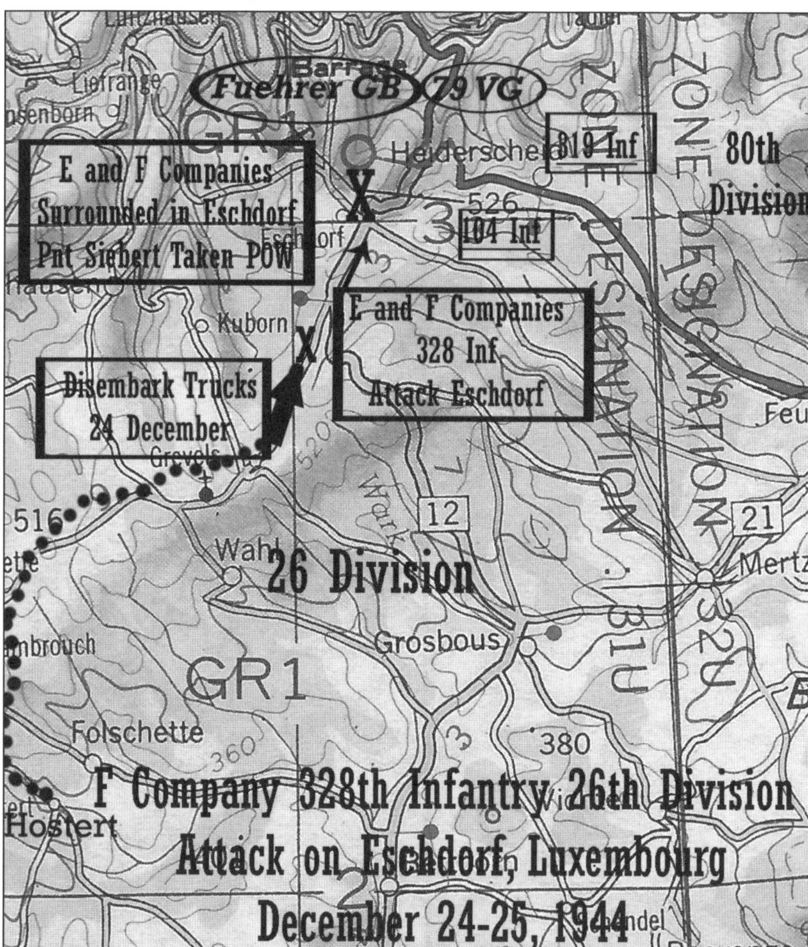

how to milk it so that was the job I got, to milk that cow and we'd shave off some of the D bar in it and heat it up and have something hot to drink once in a while.

I don't know that I ever did get any mail while I was over there. It never caught up with us. In fact, when I wrote home there was only one letter while I was a prisoner that finally came here and that was the only indication from me that my folks had that I was alive. Our mail carrier from Westmoreland returned from his route and found that letter had come in the afternoon while he was delivering the mail. He saw it and he made a trip down here to bring it to my folks specially.

From Metz we traveled up north of Luxembourg. On about December 23th, the whole regiment moved to Arlon, Belgium, and this was the beginning of the Battle of the Bulge for us. Our first objective was Eschdorf and it was also my last.

You didn't know for sure who you were bumping into when you saw somebody because the Germans were wearing some of our uniforms and they could speak very good English. There was lots of snow on the ground and the Germans would sometimes wear white sheets over their uniforms so it was hard to see them at night.

I think it was the first night that we went out and three of us were standing guard on the road. Everything was dark and suddenly one of our ambulances came down the road. The first guy yelled halt but the ambulance diver paid no attention and kept coming. The second guy also challenged the driver and he kept coming so I had him in my sights ready to shoot. After the third challenge, the driver slammed on his brakes and another second I would have pulled the trigger. He claimed he'd been out so long that he didn't have the password.

Is he telling the truth or isn't he? Is he American or is he German? We just took him up to headquarters and let them deal with him. Maybe he was an American but why would an ambulance driver be so far out that he'd never get the password? It didn't quite make sense to us. It was difficult to figure out who the enemy was.

On Christmas Day I wasn't with them anymore because I was captured early in the morning. One of my first thoughts was how do I stay alive.

It was only my squad that went in to the town and all I saw was one building and I don't remember whether there were any more buildings. We crossed a ditch and went up toward the building and they open fired on us. We were laying flat on the ground and I remember shells going off close in front of us. The exploding shells ripped the pack from my back and scraped my back up a little bit, and there was a guy right beside of me that had just joined us the day before and he was killed.

Then we went inside this building. It was a rock barn-type building with living quarters in it because that's the way a lot of the homes were built. We went into a room and there was a hallway to get into the room. We were able to hold them back for a while but we couldn't hold them back forever. They would try to get in and we would knock them back out as fast as we could. It was dark and soon they found a way to get a grenade into the house. One of them found one little hole high in the wall and they got a hand grenade in that little hole. We couldn't find the grenade in the dark and we knew when it was time for it to go off so we hit the floor in the corners of the room when it went off. My ears still ring.

Soon we saw a tank outside and he swung around toward us pointing his 88 at us. We knew that was it. The tank got us to come out. When we saw that the tank was there, we just laid our guns down and walked out with our hands up.

A couple of the Germans took us back a ways by foot and were getting ready to shoot us when a vehicle drove up and an officer got out and talked to them and they gave up shooting us then. But we were sure that that's what they were getting ready to do. In fact the guy right beside me says, "Well looks like this is it." I said, "Sure looks like it."

But you know that really didn't scare me. I knew it was going to happen so I figured it was just going happen and there was not much I could do about it. You just had to get over being nervous. I probably have blanked some of that out.

My parents had no idea what happened to me. They received the following telegram from the War Department:

> Mrs. Aurelia R Siebert
> Westmoreland Kansas
> January 31, 1945
>
> The Secretary of War wishes me to express his deep regrets that your son Private First Class Julian A Siebert has been reported missing in action since twenty six December in Luxembourg. If further details or other information are received you will be properly notified.

My older brother Nate was in the 9th Armored. He was up there in the vicinity of Luxembourg and Belgium. When he found out I was missing, he was able to get in touch with one of the officers. The Officer told him that he was almost sure that I was probably taken prisoner. This was all they knew but at least they knew I was alright.

They made us walk for days with little to eat or drink. They continued to collect prisoners as we went along and we walked to a railhead with one single boxcar parked on the siding.

PFC Julian Siebert

They had a boxcar full of prisoners and there was only room enough for each person to stand up, nobody could sit. But they didn't move the boxcar, it sat and we stayed in it for several days. Nobody could lie down, people began to get sick and it was miserable beyond belief.

One night our planes began bombing and strafing close to the railroad and we could just feel like any minute that some pilot was going to see that little boxcar out there in that railroad yard and blow it up but they didn't. What we did learn was that people could sleep standing up but that still didn't give us much rest.

From there they took us to Stalag XIIA somewhere near Limburg, Germany. We were questioned by the guards and then they gave us identification tags and took our photos. The picture that they took of me was a surprise to me. After I got home, a letter came addressed to me and all that was in it was that picture, not who sent it or anything. I have no idea. I still have my German dog tags with my prisoner number on it and of course I wore my Army dog tags too all the while. I wore them both around my neck.

I can't remember seeing the outside of Stalag 12A at all and I have no idea what it looked like.

After a few days maybe, we moved to a work camp. I don't' know where that was but my POW records in the National Archives records it as Stalag 2A at Neubrandenburg, Mecklenburg. The camp was bare bones, no facilities for bathing or washing and food was just terrible, consisting mostly of soup. It is funny now to look back on it but everyone tried to make the best of a really bad situation and we tried to laugh in spite of what was happening. When we got our soup, one of the guys said, first if you see a worm in your soup, you will probably throw the soup out. Then if you see a worm in your soup, you will pick it out and eat the soup. After that, if you find a worm in your soup and he tries to crawl out, you will knock him back in. The worms seemed to add the only flavor to the bland soup.

I was part of a work crew of 30 other POW's who worked in a forest cutting pine trees. It looked like we were cutting firewood. They had kind of a barracks type thing for us to live in. It was fenced in and we went out and cut trees down, sawed them up in meter lengths.

No mail ever got to us but we were allowed to write. They just gave us a form called a "Kriegsgefangenenpost" which meant "POW post" and it was sent home from the Stalag. The letter finally came home and that is how they learned that I was a alright.

Here is what I wrote:

> February 25, 1945
>
> Dear Folks,
>
> I am going to write you a few lines to let you know I am still alright. I wish that I could get a letter from you to let me know how things at home are going. I suppose you have been worrying about me. I am fine. I don't know anything to write so I will close for now. I hope you are alright and not working too hard.
>
> Love Julian

Of course, I was desperately homesick but didn't want to worry my folks.

It was strange but one night a German guard came to us and said if things got too bad we should break out of the barracks we were using. He gave us an ax to break down the doors and said it was for our own

Julian's Dog Tags. Note the XIIA on the POW tag on left.

protection. Then, the next day we learned that the Russians were getting close and we were heading to the American lines.

We walked for just that one day. We left in the morning, we walked all day, and it was in the evening by the time we got to the American lines. I don't know where we were but we must have been quite a ways east from Limburg. We walked all day and I don't know what direction we went or anything. The Russians were firing on the other side of town as we went out the other side. The Russians were so close behind us that we could hear them coming. And it wasn't a very big town either.

I think it seemed as though the Americans were just sitting there waiting for the Russians to come. They weren't going ahead and doing any fighting then. We were getting close to the American lines when the guards turned their guns over to us and said, "We are your prisoners now."

We were almost to the American lines. This was May 2, 1945. The war was almost over. I remember the American flag somewhere after we got back across our lines and I have never seen anything so beautiful.

On June 23, 1945, my mother received the following telegram:

> Mrs. Aurelia R Siebert
> Westmoreland Kansas
>
> The Chief of 'staff of the army directs me to inform you your son PFC Siebert, Julian A is being returned to the United States within the near future and will be given an opportunity to communicate with you upon arrival.

This was the first that they knew for sure that I was still alive and coming home.

When I came home, I didn't have a chance to call until I got to Manhattan. When I got in, I called home. I had an aunt and her family that lived in Manhattan and I went over to their place. My cousin had been killed in the war, on Guam. It was sure great to be home.

I met my wife, Barbara through my younger sister. They went to school together. We kind of knew each but I was older. When she turned twenty, we started dating and got married. We have five children. Four of them live in the area and one is an aerospace engineer in Houston working for NASA. He has four kids, three triplet boys. He was in Iraq in Desert Storm.

The '60's was hard times especially farming. I don't know how it was for somebody that had a job but it was pretty hard times. I had a family and I thought how in the world am I ever going to send the kids to college with what I'm making? I knew I wouldn't be able to do it. The mail route at Westmoreland came open and there were probably thirty of us that took the exam. I thought I probably wouldn't have a chance but I was picked in the top three finalist and they were going to choose who would be the next mail carrier out of that top three.

I was really interested in the job. They set up a democratic committee out of the townships that the route served. So I went around to each one of the committee members and talked to them. When they voted, only one of the committee didn't vote for me and I got the mail route. That was in 1966.

My folks were still living here at the farm and my mom wanted to move to town so my dad just sold me the farm and they moved to town. What he sold me the farm for wouldn't buy any portion of it anymore. But he sold it to me for what he figured was fair. He didn't want to charge me too much but he wanted to be fair to the other kids so I paid him what we thought was going price for it.

And things got better then it seems like. At first the mail route was short enough that I was home shortly after noon and I had most of the afternoon to work the farm. After a few years, they made the route bigger. I would find myself out there working at 11:00 o'clock at night a lot. Finally they were going to make it bigger again and I said that's it. So I rented the ground out to one of the neighbors. I kept my cows and I still have the descendants of those cows.

What I remember most about the Battle of the Bulge? After you were discharged from the Army, you thought it ought to all be over. But it wasn't. I had nightmares for 30 years afterwards. I would wake up in the middle of the night just scared to death, I'd be yelling and wake Barbara up. What I was dreaming about, I have no idea. I couldn't remember any of the dreams. They seemed to be getting a little worse and then all at once they quit just like you shut a switch off and I never had another one. That was almost worse than the war. That's what I remember most.

Julian was awarded the Combat Infantry Badge, ETO Ribbon with 3 Battle Stars, American Defense Service Ribbon, and the Good Conduct Medal.

Bill Stahl
106th Infantry Division

Bill Stahl, 2009

Place of Birth:
Junction City, Kansas
Active Duty Date:
July 1943
Unit:
K Company,
422nd Infantry,
106th Division
Location:
East of St. Vith
Arrival ETO:
December 10, 1944
POW captured
December 19, 1944
Rank: Corporal

BILL STAHL

The Schnee Eifel formed a slight curve and the 422nd Infantry 106th Infantry Division was positioned between the 14th Cavalry Group on the north and the 423rd Infantry to the south. The bulge of the curve put the 422nd on the eastern most position. 424th was on the division's right flank next to the 28th division. During the onslaught, the 14th Cavalry Group was forced to withdraw, exposing the 422nd left flank. The resulting wedge allowed the oncoming Germans an opportunity to advance around the exposed flank.

Here is his story. . . .

I grew up right here in Junction City. My Dad was a railroader with the Union Pacific. My family was Pennsylvania Dutch who came to the middle of Pennsylvania in 1749. My family was Mennonites and they came from Berne, Switzerland into Lancaster, Pennsylvania and then to Kansas. They were not Mennonites by the time they got to Kansas.

I entered the Army in July 1943 right out of high School. My whole high school graduating class was drafted. We all went down and all but one passed. We knew that we were going and couldn't wait to get in and after we got in couldn't wait to get out.

As a machine gun helper, I got to carry all the ammunition it seemed like. I finally made Corporal a few months before I was discharged. The 106th was formed at Ft. Jackson and sent to Camp Atterbury in Indiana. All of a sudden in the spring of 1944 they took everyone except a cadre and shipped them over to Europe as replacements. So we were guys that came in to replace those people. A lot of them were ASTP (Army Specialized Training Program, mostly college students,) there were some guys who were plucked out of pilots training. They said we were the smartest division in the whole Army but not smart enough to get away. The ASTP men were not too happy to be assigned to an Infantry Division. I was scheduled to go into that program but they cut it off. They were supposed to send us to college for 3 or 4 years.

They took draftees and filled in for those who were sent as replacements. My basic training was in anti-aircraft and one day 3 or 4 of us got mad and went to the First Sergeant and said we wanted to be in the infantry. He said, don't worry about it, the whole outfit will be in the infantry in the morning.

I had six weeks of infantry training at Camp Carson, Colorado, I guess it is Fort Carson now. The only thing I remember about that is that at 4 PM everyday it rained for about 10 minutes. From there we went to Camp Atterbury on March 28, 1944. We had a group of Polish boys from Chicago who had never been outside of Chicago. By the time we got to Atterbury, we had them convinced that we would be attacked by Indians and buffalo, and they were plastered against the windows afraid of being attacked. Atterbury is about 40 miles south of Indianapolis. We were there about 6 weeks, and then went to Ft. Kilmer, then to Miles Standish on October 10, and onto a boat in Boston on November 10 for Europe. At Miles Standish we were there about 4 days and we loaded onto a boat at night. Then everything was secret. We couldn't call home or write. Now everyone is on TV.

We went to Scotland, took a train to Stow on the Wall in England. Our whole battalion was in Quonset huts. I tried to find it on a subsequent trip to England but couldn't find it. We were there for 30 days then took a train to South Hampton and after 2 days we landed at Le Havre on December 6. It was cold, a drizzling rain fell. The Red Ball Express took us in open trucks to Belgium and we relieved the 2nd Infantry Division when we got to the front.

I was crossing the English Channel and as you got on to the boat they took our duffle bags and when I got to the other side mine and another guy's was missing. I went to the First Sergeant and he said I had better find it or I would be digging slit trenches. Foxholes are for one or two people. Slit trenches are 30 or 40 feet long like a trench. It was a latrine. I had taken out my combat gear so when you got to the front lines they took them back anyway. It might have had a shaving kit I needed but you didn't need them on the front lines. I have read different stories about these things but you just took out what you needed or thought you needed. When I got on the front lines you had a pretty good choice of things. Like on one patrol I took a Thompson machine gun. I shot it and it went straight up in the air. They weigh a ton.

We arrived at St. Vith the night of December 10 and were assigned to the front the next day. The road to Belgium was terrible, rough, pitted by heavy traffic and cratered by gunfire. We went through bombed out villages and were greeted by smiling town's people flashing the V for victory.

We replaced the 2nd Division, and made contact with the 28th Division on our south flank and the 99th on the north. We were on the eastern most part of the curved front next to the 14th Cavalry Group, attached to the 106th on our left.

We were on what was called then the Eifel plateau along the Belgian-German border. The middle ridge running about 26 km from southwest to northeast was named the Schnee (Snow) Eifel and it tied the Siegfried Line together. The Ardennes is a dense, dense forest and we took positions on the eastern slope of the Schnee occupying abandoned German positions. We began to run patrols immediately.

We were about 12 miles east of St. Vith. The Germans went around us with tanks and infantry before we knew it because our lines were so thin and we got separated from the rest of the division. We were around Schonberg.

The 2nd wanted to get out of there, you take this gun, you take this and that and we stepped into their shoes. But they had been there before. Morale was high, everybody had the idea that the war was almost over and that we would have a quiet front at least until spring and nobody was too excited about it.

They sent out patrols every day. And guys like me, they were 'volunteered,' you know like you, you and you. I went out a couple of times and it wasn't all the guys from our company, just different guys who wanted to get out there on patrol. We were 18 and 19 years old so we were gung ho to go. We went down in a gully between two mountains and thought nobody could see us. There was snow on the mountains up to our waists and didn't realize the whole world could see you.

There was a river or kind of an indenture. There was a Black outfit between the 99th and the 106th. At the 106th Division convention there were several guys who were in this unit. They were in the artillery.

At 0530 the enemy began an artillery barrage starting north, onto the 14th Cavalry positions, then walked down the front and through our positions. Trees were splintered, the noise deafening beyond belief. We were deep in our fox holes.

We were kind of on the side of a mountain and we ran patrols around the mountains, about 300 yards or so. They were not huge mountains but they were mountains. At 5:30 AM on December 16, we were right in that initial blast. I don't think anyone got hit but they sure scared us. They led with artillery followed up by infantry. At that time they didn't come very fast. In fact the next morning there was talk of tanks coming through and you could hear them coming. They put me with a machine gun to stop a tank. Thank goodness they didn't show up. The first day or two we didn't get much action but on the 18th they began to pull us out of the line and we were to fight our way out. We were told we were surrounded. We didn't know it at the time.

There was a lot of snow and patches of forests. Maybe half way up a mountain there would be a forest and half way down behind us there would be forest but once you got back up behind us it was not so much. We were headed toward Schonberg, pushed back from the front, headed toward St. Vith. It was to be our destination. The 422nd and 423rd were headed toward Schonberg and we were told that we should take Schonberg but when we got there we were completely surrounded. They had Schonberg under their control.

The German's were trying to take St. Vith and its network of roads. German tanks and infantry moved in and around us.

We were surrounded but we kept fighting. We did not know what was going on. We withdrew by foot, no panic. We were told we had a problem and had to fight our way out. Well by the time we got to where we were captured there was no food, no ammunition, no nothing. I can vaguely remember where I was captured. That afternoon I was on top of a hill and I remember looking down the hill at the Germans who were fighting toward us. I must have had my arms around a tree and an 88 went off close and I got shrapnel in both hands. My son keeps telling me to go back and find that tree. I was captured on December 19 about 4 PM. I was probably injured about 3:30 but in those times unless you lost an arm or leg it didn't make any difference.

Nobody panicked and nobody threw down their arms to quit. I never saw anyone run. I keep reading in books that artillery pulled up their guns and escaped but I never saw anything like that. Our General, General Jones, apparently had a heart attack and lost his nerve. We knew nothing of that. I was under Sergeant Braden who was about 5 years older but looked like an old man and he took care of his own. There were roads but they were running the wrong way. I remember we crossed a four lane road like I-70 autobahn. We had never seen anything like that. The Germans were well ahead of us in transportation. I was moving across fields and going from one point to the next. That was about it.

When I heard the last clip click from my M-1, I felt bewilderment. When we were told we had to surrender it was like the end of the world, it was degrading and something you couldn't comprehend. We were ordered to surrender, it was not voluntarily. And I think all of the rest of the guys were in the same boat. Nobody panicked and nobody wanted to do anything that would hurt them like getting shot in the back. One of our last radio messages was, "Can you get some ammunition through?"

We were cutoff, no food and out of ammo. The word came to us that we were to surrender but we didn't want to surrender. I was really ashamed of it for about 40 or 50 years though. What is important is what I know to be true. I wondered what my parents were going to think, even after I was a prisoner. My thinking was I know where I am and what I am going through. I wonder what my folks are thinking. My Dad was 49 years old, he had sandy hair when I left and it was pure white when I came home. We wrote letters once a month, sometimes twice a month. It was very slow and of course if the train got bombed they were gone. When they notified your parents, they just got a telegram notifying them that I was missing, not like it is today, much better now.

We were transferred to 40 and 8 box cars with no food and no toilets. I was sent to Stalag IVB in Milberg. It was one of the bigger camps, about 450 km east of where I was captured and after 30 days, because I was a PFC, I was sent to IVD in Toga, about 215 km further east. I was moved to D because it was for privates and PFC's which all of us were at that time. A lot went out on farms. There were sixty of us that went onto IVD and of course no medicine, no doctors, no nothing. A great big husky boy from Nebraska, a big guy, very friendly, he got something in his arm, some kind of poisoning, blood poisoning and that redness started up his arm and in about three days he was gone. I can't even remember his name.

IVD was a work camp. There are two big machines there and we loaded logs into them and did everything to get them to churn and kind of make a pulp out of it. And we were always told to do everything to foul them up and within a half hour we would have them crossways or something, the warning lights would blow.

Of course, there were always the air raids and we never hurried back from the air raids. Planes were launched by Americans. Almost every night there was some sort of an air raid but apparently it was mostly English bombers. When there was an air raid siren we're supposed to go outside and go down to steps into a cave type deal. One time two or three of us decided we would just get under the bunk and after a week of that, here came a German guard screaming and yelling to get out of there. That night a bomb hit a German gun and the bomb went up into the air and hit the fence. It wiped out the whole side of the barracks and if he hadn't have gotten us out of there we would have had some dead bodies. So we probably owe some German our lives and a thank you.

I don't remember how close together Stalag B and D were. They must have been close together, I just cannot tell. Apparently IVD was a group of different camps of people who went out. A friend here in Junction City worked on the farm in Poland and he apparently had a pretty good deal there. He was fed and things like that. He had a good place to sleep and that was not the same with us.

The guards were just regular German Army and they were mean to us. Several years ago we planned a trip to Italy and I did not get trip insurance. I had not been in the hospital since I had my tonsils out in 1935 so we didn't buy it. The day before the trip was to go I ended up in the hospital with pneumonia. The doctor came to me and asked if I had had an operation on my nose and said no, but he said you have a perfect hole right up here. I remembered that a German guard got mad at me and hit me with the butt of his rifle and that is the first I knew that I had anything wrong.

I remember when I was released. Apparently toward the end of the war the POWs were placed out on the road and marched generally toward the American lines because the Germans were afraid of the Russians. The guards kept saying that we have orders from Hitler to kill all the POWs when the war is over. We had been out on a march about three days and woke up one morning and they were gone. The 69th Infantry Division caught up with us. I don't remember exactly what happened but we probably had 200 to 300 guys in that line so they took us under their control and that was it. There's a guy in Manhattan named Uttermueller who is in the 69th.

They put us on trucks and sent us to a collection depot. I remember that we started to eat. We would eat something you would eat every day and we just threw it right back up. They said that you're going to do that for another 4 to 6 days. You have to have powdered milk and it took that long for our stomach settled down. Our normal diet consisted of the little soup in the morning, probably at noon you had two potatoes and then you generally split a loaf of bread which is about half sawdust among eight or nine guys and that was it. We worked six days a week 12 hours a day, that's not much food.

I'm not sure what date I was released. The day before we were released, one of the guards had said that the president had just died on the 12th of April so I think it may have been either the 13th or 14th of April of '45. Of course we didn't know whether to believe them or not being a Republican.

We were not released immediately from active duty. We had to serve out some time because we were on a point system then. Eventually we were flown from where we were in Germany to Le Havre. After about two weeks we were put onto a ship and sailed into choppy seas. I did not notify my family and as soon as we hit the United States we were put on a train to Leavenworth and turned loose. I took a bus to Junction City. I got off the bus in the middle of the night. Of course, we didn't lock our doors in those days and I just walked into the house lay down on the couch. My mother came down the stairs in the morning and found me on the couch. I probably shouldn't have done that. They thought I was still in the camp. There was no way to get in touch with them.

We had a 60 day furlough. In fact, when I was home the war in Europe was over and we got in cars we and we went up and down the street honking and yelling. After that they sent us to Hot Springs, Arkansas. We had two weeks of rehab there which meant all you had to do was to lay in bed and sleep. From there I was sent to Camp Lawson in Seattle, Washington, to a desk job and that's where I saw my first female Officer. All we did was keep books of people coming in from the South Pacific. They were assigned to barracks.

But in November, 1945 I was out and by the middle of January 1946 I started at Washburn going summers and winters. I finished six years of college and law school in 4-1/2 years.

This is my second marriage. I had two children from my first marriage, a boy and girl. Mary Lou was teaching elementary school when we met and eventually retired after 32 years of teaching. We also have a daughter, Marcia.

One of the things I remember most about the Battle of the Bulge was the surprise. It was the biggest thing because we had been told it was a quiet area and we would get some training and eventually we would be in battle two or three months down the way and wham. Within less than a week we are right in middle of it. That day and the afternoon I was captured were both most outstanding points in the Battle of the Bulge.

Bill was awarded the Purple Heart, ETO Ribbon with 3 Battle Stars, American Defense Service Ribbon, and Good Conduct Medal.

Chapter 5

The Infantry

WHOEVER SAID "Infantry, Queen of Battle" obviously had a desk job. Dick Jepsen summed it up like this: "It was all an adventure, every single day but it was a very unpleasant adventure. It was always uncomfortable."

No one is closer to the enemy than the infantryman. They exist only to "close with and destroy the enemy." It is a vocation like no other where boredom begets inattention and combat becomes the ultimate high. Survival is animalistic and random death is as quick as any on the Serengeti plains. Not only is it a "very unpleasant adventure every single day," it is an adventure where an instant of inattentive behavior can result in immediate, violent death or traumatic injury.

National Archives photo

"I recall when I first entered combat. The Battle of the Bulge was during a pretty severe winter. I remember dirty, muddy fields, and ungodly weather. People were freezing in the muddy, wet, soggy ground," Bob Pearson remembered about the Battle. He was a replacement in the 26th Division. "All I had were the socks that I put on in England. I had to wrap my feet because they were freezing. I know a guy who used to play a professional instrument of some kind and he couldn't play again because he had his fingers frozen."

But in spite of its inconveniences and dangers, its horrors beyond comprehension and memories indelibly etched into one's psychic for the remainder of their years, combat is an addictive companion. Throughout life, it begs to be resurrected at the most inconvenient times, manifesting itself as a simmering subterranean volcano in dreams and flashbacks.

As the need for new troops grew, training cycles in the States became shorter. The new replacements had to learn quickly on the job if they were to survive. The first few weeks were critical. If they could live for just a few short weeks of combat they greatly improved their chances.

For replacements, combat required constant focus on new details foreign to everyday civilian life. They had to learn to read warning signs, anticipate danger and react without thinking, instinctively registering suitable locations for protective cover or routes for escape. Scenic terrain no longer was a source of beauty but a perpetual topographical puzzle rendering avenues of approach by the enemy. Slowly, as they gained experience, they learned to separate emotions from the harsh realities of combat and survival became everything.

These skills were experiential, coming mostly by trial and error and learning from other survivors. The consequences of "error" were catastrophic and seldom was there a second chance for relearning. Experiences registered in the fiber of their muscles, teaching them to react on a "hunch" rather than thought.

The finality of the steep learning curve drove casualties quickly. Men came and went before anyone even knew them. "There were people who came to our platoon and I never did know their names. They lost their lives before I ever got acquainted with them," said Raymond Brown.

Young men were scooped up from as many stateside programs as possible and put into infantry training programs because of the high rate of casualties. Bob Knight was an ASTP (Army Specialized Training Program). It was the largest educational program in US history, sending over 200,000 young soldiers to college. The program was suddenly terminated when it became apparent that manpower demands in the field were more important. Divisions that once were cohesive, homogeneous units were either stripped of experienced personnel or inundated with men pulled from specialty training.

"I was sent to the 84th Infantry Division at Camp Claiborne, Louisiana, along with a whole lot more of ASTP guys, said Bob. "Here we were a bunch of college guys in with this old regular Army outfit. It was pretty hard to get accepted. We were not the regular guy that they were used to and that was a difficult situation at first."

There was little time for acclimation when a new soldier joined their unit. Don Richards was a new Lieutenant in the 80th Division. "I was a platoon leader and that first night that I checked into the adjutant at the 80th Infantry Division, I can remember we were under 88mm gunfire from the Germans." Don was wounded in an attack along the Mosselle River returning to the 80th Division just in time for the Battle of the Bulge. "In nine months of combat our division had 11,000 casualties. That's quite a turnover."

To help fill the personnel gap, Airborne Units performed as Infantry after the combat jumps at Market Garden. Frank Rhodes was assigned to the 504th Parachute Battalion that started the war in Africa unattached and eventually became under the control of the 82nd Airborne Division. "We were in France when it started. It was so darn cold, I froze my feet the second time. We were supposed to jump across the Rhine but we were only there about 10 days when the Bulge broke out," said Frank.

In addition to the miseries of war, as in previous wars, the Infantry was usually the first to experience the new weapons of war. The 82nd Airborne had been in the war since it first landed in Africa in 1943 and Don Nixon joined the 505th PIR in July 1944. In addition to the buzz bombs and V-2 rockets, the Germans used another new technology, a noisy aircraft with no propeller. He saw the new German jets first hand.

"I remember General (James) Gavin, 82nd Division Commander. He was looking around at a whole bunch of us standing around and it was the first jet I ever saw, they had them before we did. And he said, 'all that and not a shot fired.' I actually saw the jet; it went right in front of us. I did not know they had jets, it was an odd airplane, no propellers."

Horace High was a replacement with the 75th Infantry Division and summed up the horror of the war for the Infantry in a few words. "We stacked the frozen dead men five this way and five that way until we got them high enough for protection and the bodies would freeze. We were weak, we couldn't pick up too much. I got down to 125 pounds."

Combat bonds, virtually welds participant's souls, facing death, as they share the inconveniences, accomplishing the impossible and becoming one in spirit for the remainder of their years. They are combat vets.

Such was the war for these brave men. Here are their stories.

RAYMOND BROWN

Place of Birth:
Olpe, Kansas
Active Duty Date:
July 1942
Unit:
Company F of the 379th Infantry, 95th Division
Location:
Saarlautern
Arrival ETO:
September 1944
Rank:
Lieutenant

The 95th Infantry Division was reorganized in 1942 and after extensive training landed in France in August 1944. The Division was called the "Ironmen of Metz" because of its tenacious defense of Metz during October 1944. During the Bulge they were in the Saarlautern area defending against the southern German offensive designed to pull resources away from the Bulge.

Here is his story. . . .

I was born April 23, 1915, on a farm three miles south of Olpe, Kansas. My parents were Mathew G. Brown and Emma K. Trier Brown. We lived on a farm; we were diversified farmers. We had cattle, crops, hogs, geese, sheep and chickens of course. The whole bit. When I was six years old, I started to school at St. Joseph's school in Olpe. I went there for twelve years and I had Franciscan nuns as my teachers the entire time. After graduation, we were in the middle of the Depression and by the Dust Bowl area. I helped with my dad on the farm all the while I was growing up, starting by milking cows when I was six years old and in the field with my own team of horses when I was ten. We marketed some cattle and hogs locally. There was a packing house south of Emporia at that time, Morgan Packing Company. We marketed our hogs there some; otherwise we would ship them to the Kansas City market with our cattle.

The twenties were good years but the crash of 1929 changed the whole complex. I was 21 years old in 1936. As I said earlier, there were no jobs to be had. There was no work. The WPA (Works Progress Administration) started at about that time. My parents were too proud to participate in that. Only heads of families qualified.

I started farming with my dad again. He lent me horses and his machinery. Then in 1938, I rented another farm on my own and with his equipment I farmed that. In 1939 my brother was due to move off of the farm where he had been renting. He rented a river bottom farm on the Cottonwood River southwest of Emporia on Lockerman Road. They got me a deferment from the draft. I didn't know until I got the notice that I had been deferred because I was on the farm.

This was in October of 1941, the third to last working day in October I'm not going to give you the exact date. But I was in the car department of the Sante Fe as a laborer for 50 cents an hour, which was a good job at that time. In two weeks I got my first pay check for three days work. I immediately went to the store and bought a new pair of work shoes because my others were pretty well used. Things were going well for me in the railroad. I was getting promotions. I was moving along. In July 1942, I was notified by the draft that my expiration date of deferment had lapsed. I was called up for the draft, which I knew would happen if I left the farm.

Our family didn't pay much attention to foreign affairs and what was going on in the world. We were concerned about survival during the Depression. The *Emporia Gazette* was the only paper that we got. Maybe we got the Kansas daily *Drover's Telegram* which was a stock paper. We had no electricity, so we had no running water and no radio and no lights. We had kerosene or coal oil lamps on the farm. Rural electrification came through while I was in the service.

Pearl Harbor was on December 7, 1941. December the 7th was a church day in our parish. I heard this after church. The townspeople who had a radio heard it on the radio. I thought where is Pearl Harbor? Never heard of Pearl Harbor. Generally our family had the same thoughts. Otherwise, it was something that happened. It was something terrible along with the Depression and the drought and so forth.

By going through the Depression, the hard work, the long days on the farm, the dust bowl era and the anxiety of not having enough money to do what you wished to do and going in debt deeper and deeper. Pearl Harbor was a minor problem probably.

I was called by the draft board and sent for induction to Fort Leavenworth, Kansas, which was a distribution fort actually. I was assigned to the infantry. My thought was, since I'd worked with a bunch of men and lived in bunk cars in western Kansas in 1937, I didn't like the idea of living with a bunch of men in a barracks. But I decided then I would take whatever the Army could give me or whatever I could get out of the Army, if that was honorable. Take what came. I didn't pursue a transfer or really volunteer for much. I said I'll just jump in. I'll do what they want me to do and go from there.

They sent me from Fort Leavenworth to Camp Swift, Texas, which was near Bastrop, Texas. With pass privileges we enjoyed Austin. I did my basic training at Camp Swift and was assigned to the 95th Infantry Division. It was just being re-activated at that time all new men except for the cadre and so forth. Training was not easy because it seemed to me like some of the things you did were physically hard, very hard. Were they physically necessary? What you had to learn was that they were necessary. The adjustment had to be made there. The adjustment is bad. I'd never been away from home and to tell you the truth, I got homesick. I wrote letters when I could. At Camp Swift we had thirteen weeks of basic training. We were not allowed to go to town through those thirteen weeks. So it was all on the base. We had very little leisure time for one thing. If nothing else they'd call us out and we'd police the area for cigarette butts, anything just to keep us busy.

I have every respect for all the military and especially the infantry. I'm close to the infantry, but I respect all the branches. One thing I noticed down there, there were all kinds of people, all varieties of life. We had illiterates. We had lawyers. We had business people. We had rich people who worked for good companies if they had a job. Many of them had no job when they came. Some were fat, some were skinny. The Army made those who were skinny put on weight. The fat ones took weight off pretty fast, pretty quickly.

After that, it was late in the summer or early fall. Shortly they transferred our Division to Fort Sam Houston, Texas, which is just outside of San Antonio. We had good quarters, very good quarters, and access to San Antonio when we were on leave. From there we maneuvered or marched to outlying tent cities for further advanced infantry training. The 95th Division was in the process of being re-created then. (*The 95th Div. was created, however briefly, in WWI.)*.

The thing that stands out in my mind, we were to go to this camp city, I don't know if it's necessary to give you the names, but we started at midnight as a platoon, as a company, walking for 35 miles to this Camp Cibolo, a tent camp. This was unusual. Some people didn't make the march. They just didn't fit the land. They didn't have the physical ability to do that.

After playing war in Camp Cibolo with a lot of wet mud and so forth in June of 1943, they took us to the Louisiana Maneuver Area, where we practiced war, fighting the enemy and fighting another division which was our enemy. We were the blue and they were the red. We were going without food; sleeping under a raincoat, (if we got any sleep), on the ground, wet most of the time, if not from sweat then from the heavy dew at night as you tried to get some rest. You fought off the ticks and the mosquitoes and tried to get by on a quart of water a day and took brutal punishment.

This is in my memory anyway. I don't mean to exaggerate, but that is what my memory says. You know after sixty days or whatever maneuvering in Louisiana they took us to Camp Coxcomb on the California desert near the Salton Sea where we did desert maneuvers. They were preparing us for whatever theater of operation wherever we might be sent, whether it is the Pacific or the mountains of Italy or North Africa. We maneuvered there until February of 1944. That's where we are now.

We did our normal, I guess, three or four months, maneuvers there. Then we were shipped to Indiantown Gap, Pennsylvania. That was the name of the camp, near Harrisburg. We got oriented there and we got replacements. At Indiantown Gap we reorganized, got settled in our barracks, and we received a group of ASTP men (Army Specialized Training Program), and our battalion formed a cadre. I was one of the sergeants on the cadre who gave these people infantry basic training. I remember I was elected to give the bayonet training, which took some physical efforts and some meanness. After that they took us by truck to West Virginia mountain maneuvers where we forded rivers, rappelled mountain cliffs, slept out on the ground, chased bears out of our camp. We helped build bridges across the river and got wet and cold and dodged rattlesnakes. This lasted probably from February until June 1944. We went back from mountain maneuvers in West Virginia to Indiantown Gap where we prepared for our embarkation center, which was Boston, Massachusetts. This was in the fall of 1944. After being in camp there for several weeks, we finally boarded the ship, which was the SS America, a former luxury liner. I remember my bunk was number five, stacked five high. I was in the top bunk. It was cold wherever we were. I think there were around 12,000 of us on the ship.

They had sent an advance guard to Europe to lay out our landing, where we would go there, so that cut down quite a bit of it. But the trip across the Atlantic took us nine days, and every thirteen minutes the ship would change course. This created a zigzag line across the ocean, supposedly to keep a submarine from zeroing in or sighting in on us. We landed at Liverpool, England. I don't know the date exactly, but it was in the fall of '44, August, I guess.

It was late summer. From there we went to Camp Miles Standish in England and waited for the crossing to France. So we crossed the Channel in infantry landing craft and landed on Omaha Beach. It must have been about late August. We landed at D-Day plus 100. That was scary. That was the real thing. Time and dates meant nothing to me.

From there it was kind of foggy for awhile, but we billeted in Normandy in an apple orchard. We lived in our pup tents while part of our unit volunteered to run the Red Ball Freight Line, it was called, truck drivers. The front was moving so fast, they were running out of supplies. These people drove day and night, two people in a truck to relay the supplies up to the forces ahead. After this was done, they moved us on.

I was with Company F of the 379th Infantry, 95th Division, 3rd Army under General Patton. I remember my first day of combat. This was early in the morning. They moved us into holding positions, replacing another unit, and the day that we met our baptism of fire was in late October. Our first battle was to try and capture or annihilate a huge, huge fort which we could almost see from our position. Other units had tried to enter it, but failed. This fort was self contained. It had its own electrical system, its own kitchen. We found out later, of course, and it was manned by 80 to 90 men at different times. It had concrete alleys running out from the fort where they could fire and this is what happened when we attacked the fort. It was early morning, and the Germans threw a flare. We started shooting; they were shooting. We were shooting, and we kept moving and they eased their fire. They ran back into their fort. I was a squad leader at that time and when we moved forward, my officer was wounded, and my platoon leader sergeant was killed. His platoon guide was killed, the other two squad leaders were killed and I was in command, unbeknown, unasked for command, because I was the ranking person from that platoon. We lost a number of others, but we did bypass the fort and reached our objective on what we always thought was a hill. But when we reached this hill, we had lost contact on the left and the right. There was nothing to do but wait for replacements or for orders or whatever else to do. We were headed toward a big American objective which was Metz.

From our so-called isolated hill we had no contact left or right, so we waited. By that time, of course, we were wet, cold, and snow on the ground. Our feet were wet. Our bodies were soaked. Our clothes were soaked, and we had no communication except our artillery observer, so we dug holes, dug in, and the artillery observer called and dropped artillery around our position which made us feel a little safer. If anybody stirred at night he was dead. A German came into our area, and he didn't make it out. We had orders to shoot anything that moved. Otherwise, we waited, and after two days we finally got contact from the flanks and moved on into Metz across the Moselle River and into the city of Metz itself.

From Metz we moved on toward Saarlautern. We went through small towns up that way, farther away up that way, until we got to Saarlautern, now called Saarbrucken. It's a bigger city. I remember waiting outside the city while our bombers softened up the city before we entered. We had a lot of resistance there until we got through Saarlautern and secured it. It is on the Saar River. We stopped at the Saar River while one of our units secured the bridge across the Saar which was a big, big thing for the whole 3rd Army because it left a passageway to the next move. The next morning early, we attacked the suburb town of Saarlautern Roden. This was across a flood plain, at least 1,000 yards where the earth was absolutely flat with shell holes with water and ice on them. We would jump into these shell holes to catch our breath a bit. We did marching fire as we went across that and got into Saarlautern Roden. We encountered strong resistance there also and lots of confusion and so forth.

The casualties were very, very high. I would say at least 50%. In fact, the company, our division, our regiment had over 100% casualties either killed or wounded through the entire war. So we got lots of replacements and this was another thing. There were people who came to our platoon and I never did know their names. They lost their lives before I ever got acquainted with them. Anyhow, back to Roden, we captured quite a group there. I think it was 78 Germans, we captured them in one of their fortresses and from there we tried to take the city which was house to house. We went through the walls, the ceilings, the basements from house to house. If you were to go outside the house, you were dead. Eventually, with the artillery, our machine guns, and mortars, in two days we had captured this small place with a lot of small forts around it. A thing of note, when you're in the infantry, if you stay close to the enemy it really, in a way, is safer than some other place because artillery is going over you both ways, from the enemy and ours also. This is just a note. From there on we went forward up through Nancy, and, I'm quite sure, to Trier. From there I know we went through Hamm.

I was commissioned an officer because of the high casualties. Because in this process thus far, two or three months, whatever, in fact the 95th Division was in the line for 105 days without a break. This means we had contact with the enemy, [though] there were lulls in the fighting, so we did have some rest time along in there. But, going back to the casualties, we would get a replacement officer, and he'd be either killed or wounded. I don't know, I'm kind of like the Energizer bunny, I guess. I just kept going and going and finally the Company commander asked, "Brownie, how'd you like to be an officer?" I said, "If I'm going to get killed, I imagine my folks would be a little more proud of me if I was an officer." So I was awarded the battlefield commission of 2nd lieutenant and later on a month or so, I got the promotion.

From there we were up near Luxembourg and very thin in strength. The Battle of the Bulge was in progress, and we were slated to go to the Bulge, but we were too far under strength. We didn't advance from our holding position until later on when the Bulge was reversed; then we went through the Battle of the Bulge battlegrounds and on into Germany. This would be on the west side of the Rhine. We got, I guess, near Wessen, and from there they trucked us across the Rhine to Munster, Germany. From there we fought back toward the Rhine in what was called the Ruhr Pocket. This was the climax because we went day and night. In seven days, I got something like 6-1/2 hours sleep because we did not stop. We'd take a unit or town or small area and go on to the next.

We took a lot of prisoners in the Ruhr Pocket which was created by pincer movements by 3rd Army units. You would bypass, and so these had to be annihilated. From there is where our war ended right then when we got reached the Rhine. I think that was pretty much where our war ended.

I was shot through the rear end with small arms fire. That got me a Purple Heart. Going back to Metz, when we got there we were almost out of ammunition. They brought up ammunition, and I was opening a box of grenades that had a steel band around it and when I cut the band, it flew and cut my forefinger on my left hand, and peeled enough of the skin of my knuckle away. The medic happened to be handy and he patched it up, put sulfa powder on it, and a splint, cut the finger out of my glove, and I went on with my one finger sticking out in the air. Otherwise, there was the one small arms fire through the rear end which hit no bones. They just ran a patch through the wound and bandaged on each side and I went on. That's the only wound I got really, never missed a minute with our Company or Division.

I was not afraid, but there were a number of times, many times, when I was confused, really really confused. I'll admit that I called for help from upstairs too, several times during these battles, real battles, bad times. Going back to the Dust Bowl, the farm days, no money it made us hard, we were tough, we were ready for stuff like this. Even the kids from the city, they had no jobs, they were in soup lines and they knew what it was to suffer. So I think we were the right people at that time to be in that situation, and I have no regrets except for the loss of my friends. That's the only regret I have. It did wonders for me in my life to witness this. Nobody else could ever experience what I did.

I came back on a Liberty Ship they called it, a small one. I had good quarters coming back. I got to Olpe on the 4th of July, 1945. We were one of the later divisions committed over there in Europe, so we came home early to go to the South Pacific. After thirty-day recuperation we went to Camp Shelby, Mississippi, in preparation to go to the Pacific. Otherwise, those who had gone ahead of us stayed there for occupation service and got home later.

We were in Mississippi at Camp Shelby, Mississippi, when they dropped the bomb over there. We still were slated to go as occupation people, but McArthur cut the occupation quota.

When they dropped the bomb and the war in the Pacific ended, I felt mostly jubilation. We could see, well we've got a chance to live, you know. That was my thought. We weren't going to have to fight our way into Japan. Everybody was ready to go home and be people, instead of just animals.

Our original opinion was that General Patton was a rascal. He was an egotist and just didn't care how many people he lost to do what he wanted to accomplish. We found out later that he was the right general to follow because the Germans were so afraid of his 3rd Army. That made it a lot easier, I think, for the 3rd Army in a way than for some of the other armies. We gained respect for him as you served under him.

My war experience made a tremendous difference in my life. I learned so much in the service. I learned to communicate with other people, to realize their infirmities and in the social part; I gained a lot of social ability, I think, and confidence. You know, I'm just a different person, I'm sure, than I would have been if I had stayed at home on the farm. I tolerate other people's errors more. I really think I do that all the time. Everybody has a right to their own character, their own way of doing things. The military is a great thing. I think it's a little too soft right now for some of us, some of the military people. I'm thinking about the National Guard. I know some National Guards who just are not organized or whatever. I don't think I should be saying this, but anyhow that's my view.

The family is my life, definitely my life. I married M. Janet Rossillon in May of 1944 before we embarked for Europe. I was anxious to get home. I hoped to be a father, a dad, and a husband. We had three daughters after I got home. I had my job as a carman with the Santa Fe Railroad testing air brakes on the cars. I got promoted to the top of the craft. This enabled us to send our three daughters to St. Mary College in

Leavenworth, where they got their various degrees. They married. Now I have seven granddaughters, and six of those are married. I have two great granddaughters and one great grandson. So I have a lot of women in my life. They try to control it, but I enjoy it. They are my life; there's no question about that.

One thing that sticks in my mind very, very strongly is, and we've mentioned going into the town of Saarlautern, I had a lieutenant then, Lt. Mullins, and I was the platoon sergeant. He said, "Brown, you take the squad down this street, and I'll take the other two squads around the other street and we'll meet at the intersection." I said, "Well, okay, Moon." It was only maybe three blocks or so to the intersection. We could see down that way. One of the squad leaders, a sergeant, was there, and I said, "Okay, Sack, this is your baby. You take off."

I was going to be second in following him, because he had people behind me, a pretty near full squad which was twelve men. He just stood there, and he had the awfulest look on his face. I said, "Okay, let's go." I started in the lead, and in just a short while there was movement on the intersection, and before I knew it, I can still see this red streak. The Germans had set up a 40mm anti-aircraft gun for flat fire. A shell hit the tree right behind me. Sack was right near, and it just knocked me on my back. I found myself on my back in the street and you couldn't see anything what with all the dust and the smoke and everything. I thought it had killed everybody because Sack was lying there, and he had no head. That sticks in my mind, and you know that's just one of the extreme instances.

Another time we were going down one of the streets, I think maybe it was in the same town I'm not sure which town. A mortar hit in the street and part of the squad platoon on the other side of the street were over there walking in single file. There was a big explosion and this kid fell to the sidewalk and he tried to get up. Shrapnel had cut his throat and his jugular vein. He was trying to breathe, which was why he was struggling, and in about thirty seconds he just fell down and just died in his own blood. That's why I'm so against wars. Just stuff like that. You see your own people. I can take all that, but you don't forget those, you know. One forms a lasting opinion.

This is a Flint Hills Oral History Project World War II Veteran's Series interview, and Mr. Brown's personal recollections taken by Loren Pennington, April 8, 2006.

Art and General Patton on the Rhine.

Art in 2010.

Place of Birth:
Corning, Kansas
Active Duty Date:
January 7, 1942
Unit:
M Co. 357th Infantry,
90th Division
Location:
Bastogne
Arrival ETO:
June 6, 1944
Rank:
T Sergeant

ART COTTRELL

Art Cottrell waited below deck of the Navy Attack Transport (APA) listening to the air attack. He was an "embarked troop" staged for the assault and he listened intently to the planes taking in the airborne troopers. It was June 6, D-Day, and Art headed down a cargo net and into a Higgins boat. He was beginning his part of General Dwight Eisenhower's Great Crusade. He and the 90th Infantry Division would fight their way through France, to the Battle of the Bulge, and end the war in Czechoslovakia.

Here is his story. . . .

I was in the first draft for Nemaha County. I was not the first one drafted but I was in the first draft. I had been going out with my wife for about a year and a half when I got drafted. She went to welding school in Topeka and was a certified welder. She worked in Kansas City welding together frames for fighter planes while the war was going on. She was a real Rosy the Riveter.

I got these two medals while I was in Germany. The medal that has the 2 is for World War II. You didn't get that unless you were in the War. I was qualified as a number one machine gunner, A gunner as they called it, on an air cooled 30 caliber machine gun.

I got home a week before Christmas of 1945, and we were married January 6, 1946. We were married two weeks after I returned home. She said let's not wait so we didn't.

I owned my own lumber business 25 years. In 1958 the business was owned by C. E. Friends out of Lawrence. They wanted to sell it so I went to the bank for help. My brother went in with me. We bought the business but then had to stock it because the stock was very low. We had a railhead into town so I could bring my stock in by rail.

People were good to me, I lost a few accounts but it all went pretty well. Then there were people who didn't pay their bills, guess they thought I didn't need it. I had two full time men working for me. I sold out in 1983. The current owners gutted it out and put in a grocery store and restaurant.

The War seems like a dream now. I wasn't brought up to take a life and now I have to live with that and it hurts. I am glad my folks didn't know it. My dad died while I was in the service, he died at 51 years old with a burst appendix.

I was drafted and got my basic at Camp Barkley, Texas, across from Dyess Air Force Base. It was rugged training. Our trainers were a cadre of 20 year career men and if you looked at them cross-eyed you ended up cleaning the barracks or on KP. They were pretty strict. Right before the war was over, we came back and were being trained by a cadre of men who didn't know what we had been through and they were also tough. They didn't know what we had been through, Normandy Beachhead and all. That cadre didn't give a darn about what we had been though, they were tough.

They were training us for Japan. After Roosevelt died and Truman dropped that bomb, if he hadn't done that we would have been headed for that after we had gone through this other war in Europe.

We trained in Texas and then went to the desert in California, the whole 90th Division, so we could train for Africa. Then they changed their minds and they told us we were going to Europe. They transferred us to Fort Dix, New Jersey. We went from the desert where it was hot to New Jersey where it was 10 below zero and snow on the ground, and we had no underwear.

Once we had a general's inspection with all of our gear out next to our jeeps. Right before my jeep was to be inspected, I noticed some grease on the engine so I grabbed my friend Lee Lonsinger's shirt that he had out for inspection and wiped the grease. He didn't see me take it and afterwards I tossed it into the river. He got in trouble for being one shirt short and I got complimented for a clean engine. He never said a word about it then.

Then we went to a secret camp, Camp Kilmer, a staging area for Europe and part of the New York Port of Embarkation. It was camouflaged, the trees were painted like buildings, and they sprayed them to look like buildings and the wooden building were spray painted a camouflage.

We left in March, and they took us by small boats like tug boats to a large ship, the *Dominion Monarch*, a British ship. It used to be a luxury liner and it was moored at Pier 89. The ship was so large we got the whole 90th Division on it, can you believe that! We were loaded at night and we walked single file across a long gang way carrying our gear. There was a swimming pool at each end of the ship but they drained those and that is where we put our duffle bags.

We were in alphabetical order and my name was "C" and it was on the bottom. It took us thirteen days and 14 nights and we were in a convoy. We landed in Liverpool but everything was blacked out because the Germans had been shooting rockets over there. They had that and South Hampton flat and they were starting to bomb London.

We started unloading and got on trucks, no lights on the trucks and went south into the country and dug foxholes in the rocks. It is rocky over there. After two or three days they marched us back and we reloaded and went to Scotland for about a week, then on to Ireland and then back to Liverpool. They did that for espionage. They didn't want spies to see the large number of troops.

We were supposed to hit the beach on the 4th of June but it got so foggy Eisenhower postponed it. We were loaded on the ships. We got across the Channel on the *Miss Clara Barton*, it had a painting of a naked lady on the side. It was a liberty ship. The whole channel was full of ships; you will never see that again. A lot of them sank too.

I heard the planes taking the 101st and 82nd Airborne flying overhead. It sounded like thousands of them.

On June 6, D-Day, we went over the side and down cargo nets and into the Higgins boats. A guy stepped on my hand. We were going down as fast as we could and my hand was on a rung and he stepped on me. People were dropping equipment into the water as they came down.

There were three men on each machine gunner. I was first gunner, there was a second gunner and then the third gunner, the ammunitions carrier. You could hand pick the third gunner if you wanted to so I said let's get a big, strong guy. The first guy was a big guy, a Swede from North Dakota. He would go back and get the ammo, come back with belts of ammo around him and he had an extra barrel to carry. If you got trigger happy the barrel would warp, which we did a few times. We trained on water cooled guns but we threw those away. They were way too heavy.

Our Higgins boat hit a small sand bar and we had to wade ashore when the ramp went down. It was dark, about 4 AM in the morning but we could hear what was happening. The Germans were sitting up o the ridge shooting and the artillery blasted a lot.

After we took Utah Beach, in about three days we had a foot hold on top of the ridge. St. Lo was our first objective and we had those darn hedge rows. We lost a lot of men in the hedge rows.

We fought in the Falaise Gap battle in July 1944, where the Germans were almost surrounded. It was tough. We went on through France. We had the Moselle-Saar River crossing in September. The Germans flooded the river by breaking the dam and it was nearly a mile wide. We were in the water when a 20 foot wall of water came down on us. It was awful. We lost a lot of equipment and men, drowned because the water came up fast and they had on heavy packs on their backs. There was no warning, just a sea of water came down on us. We crossed eventually over a railroad bridge.

I remember crossing the Rhine. It was flat on our side and there were hills and trees on the other side. The Germans were hiding in there. We crossed on Bailey Bridges where they hooked them together. They had a power boat on the end, pushing against the current to hold it until they could get another piece out. It was night before they got it built across, the current was swift. Along came Bed Check Charley, a Junker 89 we called him. He circled around, you could hear him coming and he dropped a 500 pounder right in the middle of it. That happened three times before we finally got across. It took so long and Patton was antsy, he wanted to get it done now. We got across there and we had quite a few battles after that.

In combat it is all a blur, you didn't know what day it was, nobody knew nor did they care. Our third gunner was a signal man too and he would get a signal on where we were going next, go to the left flank, right, withdraw, he got the signals and we did what he said. We had a liaison up front who knew what was going on and it is a good thing we did. We were de-motorized when we hit the beach. We trained as motorized.

When the Battle of the Bulge came in December, we were supposed to get hot meals brought up to us at Christmas. We were regrouping about 10 miles from the Bulge. It hadn't started yet. I saw a jeep come to us, flags waving on it and I said oh oh, there is something up. The rider said trucks are coming, the battle started, Germans are making a stand, they are coming to either going to kill us or take us prisoners.

That is how I found out about the Battle. In 15 to 20 minutes, here came a whole fleet of trucks, stake sides. We were 10 miles from where the brass wanted us to be. We loaded on trucks, standing up, no room to sit, they just kept packing us on and we had our full field packs on. I had the machine gun, the second had the tripod and the third gunner had the ammo. We got up there and it was dark. The trucks had the cat eyes, the blackout driving lights, on.

We got into snow and it was cold. We got as far as they would take us at the point of demarcation, we were still about a mile away. I jumped out in the dark and hit my knee on the ground, broke a bone off. We had an Italian Doctor, Dr. Diminski, who was in the next truck. The whole medical team went with us. He got out his truck and I told him I hurt my knee. He looked at it and said yes, you knocked a bone off the side but you will be alright. It won't hurt you in combat but don't let it get into the joint or it will lock your knee up.

That was the only time I rode during the Battle was the 10 miles to the Battle. It was still dark and we had to walk another mile to get to the designated part. They tried to surround us. We headed for Bastogne and we hit them straight on, very hard. It took us quite a while to take Bastogne. Several weeks at least and it was January before we got the Bulge pushed back.

We took Normandy and the hedgerows, liberated the French by pushing out the Germans, then on to Reims, and Luxembourg. The Consulate from Luxembourg at the meeting celebrating the end of the Battle of the Bulge in the Manhattan, Kansas, came over and shook my hand and said he wouldn't be there if it were not for us. He was 14 years old at the time when his mother told him to hide because the Americans will shoot anything that walks.

I remember there was an old fellow sitting on his porch crying. He had geese in the front yard, pecking away for food. I hollered at him, he was close by. He spoke no English. I asked him in German what the geese were and he said 'goose' in German.

In France, when we liberated it, I never saw a Frenchman in uniform. I saw men the age to be in uniform. The women came running down to us with candy or an apple with appreciation. They couldn't speak English. Coming back they didn't say anything, it was a different story.

The battle for Bastogne was bigger than I thought it was going to be and I first realized the size of the battle when we ran into our first Germans. Our artillery pulled up behind us and they started blowing the place apart. We started to enter and they withdrew us about a mile to get out of the line of fire. They were blowing everything up, you couldn't believe it. It was the darndest of all the battles I was in. I never saw anything like it. I bet I lost so many buddies there. They used to be like brothers, better than brothers. One guy said if you don't make it don't worry, I will tell everyone about you back home and I said you don't have to, I'm going home. You train with them, sleep in the same barracks, you do things you don't want to do, you get really tight with your acquaintance.

When the weather cleared and the American planes were flying again. I thought the war was over. I was never so glad in my life to see air support.

When I got home, in the Company I was with, there were seven of us who came back from Europe. My wife and I went to 5 of their funerals. We went to Abilene, Texas, Sioux City, Iowa, Binkleman, Nebraska, and Kansas City, Missouri. We went to Sioux City to visit one boy because his wife wrote and told us he had bad cancer. His name was Lee Lonsinger, the guy whose shirt I used for cleaning my engine. When we were ready to leave he said I want to give you something. He went out to the garage and came back with a brand new rod and reel and he said I want you to fish with this for me. He also told me that he knew I was the one who took his shirt.

Out of the 200 men I trained with only 7 came back from the war. After the hedge rows at Normandy we only had 47 men left in our company. Then we lost a lot more at the Saar River and then more at Bastogne.

I didn't count the replacements. Some of them were good soldiers and some hadn't even shaved yet but they filled in.

The Siegfried Line was made up of cement pillars; they had them in there for years. They had expected this. On the main road they knew which road we were going to try to take. They went into the ground and they were tapered and about 5 feet tall. They were zig zagged and they were called dragon's teeth. The tanks couldn't even pull them out, we had to have them bulldozed out. A tank would come in with a big blade on the front and they couldn't even budge them or break them off.

Then at the Falaise Gap, that was a terrible place. It was a low area and they let us get balled up in there and then here they came. We were surrounded, we didn't know it, and we were in a valley. They caught us there. I don't know how any human could ever get through it. Sometimes I feel like this all happened to someone else.

The first time I saw General Patton was at the hedge rows. He had on those white gloves that he always wore. He had a briefcase under his arm, a flimsy thing. To me he was the best General we had. He wanted to get things done and didn't pussyfoot around. One time we were walking down a bayou, 12 to 14 feet wide, and trees on both sides. We were in columns of four. He would stop a bit and say, soldier do you have a clean pair of socks on, how long have you had those socks on. When was the last time you had a good meal or when did you have any sleep. But then, if he saw someone messing around he would get on them something crazy. If he saw someone hiding he would ask them if they were afraid and told them to get out and start shooting somebody. He was something else.

I drove for him for two weeks. Patton had two drivers, a primary and a back up. If something happened to one the other would take over. Sometime before the Rhine a sniper got his first driver and the second one was in the field hospital with an appendix operation. In those days you weren't out in a few days, you were in for a few weeks.

My Captain, Captain Waldo or Oliver, we lost so many Captains I can't remember who it was. With battlefield commissions, they would promote First Lieutenants to Captains and give them a company real quick. So my company commander, Captain Waldo, volunteered me to drive for General Patton. I really didn't have any choice. He told me he was volunteering me to drive for the General.

So they took me back to where he was and there was his jeep and his garb. They got the jeep with all the stuff in the back; you couldn't believe all the stuff he wanted to take with him. Anyway I drove him for two weeks and that is where I got my Bronze Star. We were pretty close to the front lines and I said we are in the wrong territory with this jeep. All I could see was foot soldiers everywhere and I knew where the fire was coming from up front. The Germans were in there neck to neck. He jumped out of the jeep and I watched him pointing his finger and shaking it toward the enemy, motioning some soldiers toward something. He had that briefcase under his arm and we turned around to start back.

He started yelling that he had told those guys to get out of there, they are in enemy territory. I told them to get the heck back, retreat a little. We got back about a half a mile. He took his briefcase out from under his arm and he started looking around and he said "Where in the Dickens are my movement papers?"

I said, "I don't know General." He said I must have lost them up there. I asked him if he had it zipped up and he said, not that old thing, you can't zip it up. So we got up there and here came the Captain. The Captains and up are usually back behind the lines and the Second and First Lieutenants are up there with you because they are lower on the totem pole.

He told Waldo, I lost my dad gum papers showing the next moves and where we are going to move, left or right. That is the whole darn battlefield. I have got to have those papers. I had them on a little clip board. The

Captain volunteered me to get out of the jeep and go find the papers. Boy! Like a dummy, I was twenty-three years old and had no sense so I started back to where we came from. We had gone over a rail road track so I ran and crawled as close to the ground as I could get back toward the tracks. When I got to where we crossed the rail road tracks, the Germans were sweeping the road and tracks with machine gun fire.

I lucked out and there was a tree about 2 feet in diameter and I got just about a yard from the track and hid there until the shooting let up. I couldn't see the Germans, they were hidden in trees and things. So when it let up I looked over the tree and there were the papers laying just on the over side of the track. They hung on the track sideways and I could see the edge. I waited and waited and I thought, boy, they know I am here and they are just waiting for me to make a move so I waited some more, it seemed like five hours but it was only 10 minutes.

I didn't have any weapon with me, what good would that have done for me. So I ran along the side, grabbed those papers and zig zagged back and made it back.

He said, "Well you made it."

I said, "Yea, I did. General Patton, why don't you put it in there and zip it up?" And he did.

So we got back into the jeep and he said, I want to go back to echelon something that I didn't understand, take a right here and we finally found a little trail. We finally got to F Company and he told the Captain what he wanted to tell him, he gave orders all the time. We didn't come over here to pussyfoot around. They are our enemy and we are going to take them as an enemy. He didn't care what he said.

We got back into the jeep and he said, "Where you from?"

"Corning, sir," I answered.

"Kansas," he asked?

"Yes sir, Kansas."

"Farm boy?"

"Yes sir grew up on a farm."

"What are you going to do when you get back home?"

"Well, if I get home, I will probably get back to farming." He was good like that, always talking to his men.

We heard that he had gotten killed in a car accident. When we were going down the Autobahn they had the sign up already. He and his jeep driver both died. I don't know what happened, if a tire blew or what but they went down a steep embankment and turned over. The jeep was gone of course when we got there.

I am a strong Legionnaire man and I belong up here in Corning. We had 80 or members so but many have died off. We were having an open house and it might be a woman, or a young boy, and they ask you the craziest questions. Our Legion had a program at the schools and my wife and I would talk about my experiences in the War.

One day twenty years after the war ended, a guy from California, a photographer, came to the house and asked me if I was Arthur Cottrell. He had a picture of the 1st Sergeant, the Company Commander and Patton on the Rhine River. He showed me a picture General Patton and me and General Patton was urinating in the Rhine River. He was on his way to Missouri to visit his brother and wanted to give me the picture on his way through.

The thing I remember most about the Battle of the Bulge was the amount of German soldiers heading for us. I hoped the good Lord was going to watch out for me. I also remember clearly the severe cold and snow. I remember jumping off that truck and breaking a bone in my leg and I thought, what am I going to do now?

My grandson has a project to do in school and he wrote this about me. Very nice isn't it?

My Grandfather's Footsteps

By: Wes Burdiek

Amazement of the roar of the plane is short-lived
Overwhelming sense of pride
and anticipation fly along beside me
I cannot seem to grasp the difference
between my landing and His . . .
Just to attempt to compare our different
emotions is simply impossible
With me traveling for baseball . . . and He for war . . .
First breath of German air sends a chill of pride down my spine
Indisputably knowing because of Him
I have the chance of my sixteen year-old lifetime
Incredible sense of connection
between the two of us as I walk in Frankfurt . . .
The very same town he and
his battalion liberated in World War Two
Retracing my Grandfather's footsteps allowed for me to see
what an impact he had on an entire county . . .
an entire world . . .

For his service, Art was awarded a Bronze Star, ETO Ribbon with 3 Battle Stars, American Defense Service Ribbon, and Good Conduct Medal.

Author and Harry Perkins at a German bunker at Utah Beach. *Author photo*

Utah Beach Monument
Photo by Author

Mons, Belgium 1945

Howard now.

Place of Birth:
Emporia, Kansas
Active Duty Date:
July 19, 1943
Unit:
D Battery,
548th AAA
AW Battery with
the 102nd Infantry
Division
Location:
Limnich, Germany
Arrival ETO:
September 30, 1944
Rank:
T/5

HOWARD GOODWIN

The 102nd Division was activated in 1942 and landed at Cherbourg in September 1944. By the end of November it was along the Wurm River in Germany positioning for an assault on the Roer River. They held the northern most American position next to the 29th Infantry Division during the Bulge. While crossing Germany, the 102nd Division discovered the massacre at Gardelegen and by War's end met the Russians at Berlin.

Here is his story. . . .

I was born at Newman Hospital in Emporia, Kansas, March 31, 1924. My father was named Byron K. Goodwin. My mother's name was Ada May Fry Goodwin. My father was a printer in Emporia, and he printed labels and cost books and things like. My mother was a practical nurse at St. Mary's Hospital for years. We were poor, but we didn't realize it because everybody else was poor but you didn't go hungry.

I started out in grade school at the old Century School from kindergarten through the sixth grade, and to Lowther Junior High School, seventh through the ninth, and then to Emporia Senior High School from the tenth through twelfth. I graduated from Emporia High School on May, 1943.

I remember December 7, 1941, the Japanese attacked Pearl Harbor. Everybody was kind of scared; a lot of commotion going around. I was downtown on Commercial Street with a buddy. I think there was anger, we were mad because they did that.

I got my draft notice in February, 1943. My birthday is March, and I had to go to the draft board, and they let me go on to finish school in May. I got called up in July of 1943. First we went to Leavenworth, and that's where we got inducted. From Leavenworth some of us went to Texas and some of us went to Camp Haan, California.

It was kind of infantry training; the articles of war were read to us and all that kind of stuff. We exercised, and after so long we went out to Camp Irwin with our field artillery, our anti-aircraft guns. First time I was ever out of Kansas was when I went to California. It was bad because I was really homesick.

I think our training was at least sixteen weeks. The first outfit I was in was the 833rd Anti-Aircraft, and they said that they had too many anti-aircraft units. At Camp Irwin in the Mojave Desert, an anti-aircraft outfit, the 548th was headed for Europe. So they let anybody from E-5, corporal, and the PFC and private, buck private, if they wanted to, could volunteer for overseas duty to go with the 548th. That's what we did.

I thought the training was rough, going through the infiltration course and under the barbed wire with machine guns firing over you. And everything around when you're going through training, it says "Kill or be killed." It was good training.

I got married on November 6, 1943. I was homesick, so my mother and my girlfriend that I went to school with decided they'd better come out and see me. So they came out to Riverside, California, and on November 6, 1943, it was on a Saturday, we got married at noon. I married Mary Alice Geiger.

After we were married she and my mother came back. Back in those days, a young girl like that wouldn't dare travel by herself, so my mother went with her, and then they came back together. Then my wife, Mary Alice, and another wife, Bob Chamberlain's from Madison, came out together after basic training. I got to go home after your basic training for fourteen days.

I was on a train coming through Emporia, and it stopped there to get some exercise on the platform. I saw a girl who worked with my wife at the Santa Fe. They had guards on the train and wouldn't let you off unless you got off with guards to drill. I couldn't get off the train to call her. The girl ran in and called my wife at home and then my wife would tell her what to say, and then she'd run back out, and it was just back and forth for about thirty or forty-five minutes. We called it long distance. And then we went up to Camp Shank, New Jersey. It was your embarkation point for overseas in Europe.

Our trip over was kind of fast, faster than it was going through the English Channel. Going over, we missed our convoy at the port in New Jersey, and on September 24, 1944, they put us on the *Queen Elizabeth*. Three days out, we passed our convoy that we were supposed to be on that we missed, and before we got over there, we'd out run a couple of submarines, and we landed at Glasgow, Scotland.

We went from Scotland down to Leeds, England, and we had everything, all of our guns and everything was 40mm and was in cosmolene; that was the stuff to keep it from rusting, to keep it dry. We had to use a lot of hot water.

We went down from Leeds through London to Southampton and then we went across the English Channel to France with our guns and stuff, so it wasn't a real small boat and we landed at Omaha Beach. And then we got off at Omaha Beach. It took us six days to cross the ocean but it took us ten days to cross the English Channel because it was so rough and bad weather. And then we got off the beach and we pulled up to a little town I can't remember the name of it now; there was an apple orchard around, and we had to clean our equipment again.

Somewhere along the way, while we were going to Paris we got orders. We were assigned to the 102nd Infantry Division, the 9th Army. For a little while after we got up there in the north of Germany, we were assigned for about two or three months under Field Marshall Montgomery. I didn't like Montgomery at all. Montgomery couldn't get his air force, his lend-lease airplanes, they were our airplanes, but he couldn't get them organized. He got them over too far; he was strafing our own men.

My brother, Kelly, was in Company B [137th Infantry, 35th Division], and I think they were in Patton's 3rd Army. During the Battle of the Bulge they moved them up to help out Field Marshall Montgomery. I was able to visit my brother once. I went AWOL.

We bivouacked outside of Paris, and we went up through France, clear up into a little country, Luxembourg, and that's where we got strafed. The first enemy fire we had was at Luxembourg. We were strafed by German ME109s and FW190s as we were going up. And then we finally arrived at Setterich, Germany. It was just a small town over in Germany. All the farmers' farmland was outside town but the families lived in little places together in the town. And then we were at Setterich for quite a few months because by the time we got up there the Germans broke up some dams and flooded the Ruhr and the Rhine, and so we couldn't get across. And then the bridges were blown up too. When we were at Setterich, we got a lot of mortars on us and 88s from the German guns, but not too many at that time.

We had near my gun section, one of these American guns they called a Long Tom, and it would fire once a day. And it was a big shell. And this master sergeant, he's in charge of it, and he'd pull up pretty close to our outfit, and he'd fire. And they claimed it went clear into Berlin, this shell. It was called a Long Tom, and when it went off, the ground shook just like an earthquake.

But then, after he fired, they'd pick up that gun and move it quite a ways away. The Germans had these JU88s with cameras and stuff, and they'd come over looking for that big Long Tom gun.

But this guy was at our guns section five or six times, and he was really friendly. He'd ask guys names, and he'd ask if you had any relatives over there, you know, brothers. And I told him I had a brother in Company B in the 35th Infantry Division, and I didn't know for sure where he was. He said, "If I ever find out, I'll come let you know." And here one day he came back in his jeep, just him and his driver. He said he found my brother. He told me where he was. To this day, I don't know how I got up there to see my brother without getting lost.

I just stayed that day. I went up there, started out in the morning on a bicycle. We didn't have rubber tires. The MPs, the Company B MPs, stopped me, 35th Infantry Division MPs, and made me get off. They were afraid I was going to wreck a truck or something, and then they tried to get me a ride with the commander and all kinds of things going up there. They finally got me in an ambulance. And so I rode with an ambulance driver and his helper. And they said, "Oh no, we know where he is." And as we got talking, wherever this town was, I said he was in the personnel section. They said, "Well, you don't want to go where we were going to take you! They're behind the lines quite a ways in Holland, somewhere in the Netherlands."

So they got me there, and I found my brother. And I had to get back because I went AWOL. I asked my captain if he'd let me have a jeep and a driver to take me up there, and he said, "No. We're on alert to move at any time. If you're not here, you're gone, you're AWOL." So my chief of section, my sergeant, he got together a couple of other sergeants on the phone, and they said, "Let him go."

Every morning they'd call in, and the chief of section said everyone present and accounted for. He said I was accounted for, but I wasn't there. And so I told them I had to get back, so my brother's captain of the personnel section, he said, "Kelly, why don't you go back with Howard and stay for a day or two."

He came back with me, and going through that night, I don't know why we didn't get shot; I mean we were stupid. But my brother hid through the night. He didn't like the guns going off, and the white phosphorous bombs going off and just like daylight. It was bad. He wanted to get back to his outfit. He left the next day.

As the Bulge turned around, we headed into Germany around Linnich. There's another town right close to it named Roerdorf, and the bridge was blown up on the Ruhr River there. The engineers tried to get bridges, pontoon bridges across, but it was too swift. They didn't get them across for quite awhile. And I was up there by Liege. One day, a couple of young kids come up the bank, right outside Liege there where we had our gun position, 40mm. We thought they were SS or something but they were prisoners of the Germans.

They were Hollanders. They lived at Heerlen, Holland, and they were coalminers. The Germans took a lot of coal miners prisoners, political prisoners they called them and made them work for them. Here we got a red alert that the airplanes were coming in. We were interrogating these two kids, and this one kid said his name was I said, "Was du namen?" He said, "Namen ist Helmut, Helmut de Jong." And I said, "Nazi?" He said, "Bosch? Nein, nein Bosch. Nein Bosch. Hollander. Hollander." I went running to the gun to get my position on the 40mm.

And my steel helmet liner had broken, the strap, and it fell off. And I got on the line to raise and lower the gun. And then all of a sudden I heard a "Halt! Halt! Halt!" but there wasn't any machine gun going. Just about that same time, we're hearing something up ahead, and I turned around, and here was that kid, that Helmut de Jong, and he had run and got my helmet.

He put it back on my head, and it wasn't very long after that, he started running back to the jeep. They were going to take those two boys into the command post. Something hit my helmet and put a dent in it. I don't know whether it was a rock or whether a bomb went off. I never did find out what it was, but it knocked my helmet off, and whatever it was, if I hadn't had my helmet on, it probably would have gone through my head.

I think he saved my life. And so then, right after that, we didn't know what happened to these two boys, never did find out. So right after that on January 26, 1945, we went on a special mission. The general took us on a special mission right outside Linnich to see if we could use our gun section, three 40mms, as anti-personnel guns against ground forces.

So they were in a big building, two or three story building, and just as they were lowering the gun, I guess the gunner said were all so tired said let her go, or let her drop. And I didn't hear it, so I caught all the weight and I couldn't get back up. I wasn't caught under the gun. It just pulled me down and my back gave out. And so they went ahead and laid me on the floor up where the officers were seeing where the shells were hitting. And then they had a Piper Cub radio back where the shells were hitting. He was over German lines.

They finally knocked out the German command post, and they took me back, or somebody did, to the first aid station at Heerlen, Holland. We had our first aid station there. They had hot and cold running water. They had me taped up from the tailbone clear up to my chest, and I took hot showers about every hour or so. Finally my captain came and said that they wanted me back up there, needed me back up there, I guess he put it.

I got back to my outfit sometime in February. I don't remember dates or anything, but anyway, the engineers had gotten the pontoon bridge across the Ruhr, so we got the guns over. There were four M-51s that went across the bridge and set up east of the Ruhr River and then there were four on this side west of the river. A lot of German planes came over and tried to knock out those two pontoon bridges.

We started protecting them. Twice in about two or three days, there were just almost constantly the tracer bullets from the 40mm M51s going up. Because we could hear German planes and see them at times, and they were dropping bombs, and not a one hit either one of those bridges, and there were two times we got credited for knocking bombs out, exploding them before they hit the bridge. We hit the bomb with anti-aircraft guns and that's what you call luck.

I think I was lucky all through the war, more than some of my buddies. Then we went on from Liege towards the Ruhr River. This is when I got the Purple Heart for that but I didn't know it at the time.

Sometime in February, we were already across the Ruhr, or they went on across the Ruhr, and I was back at the medical station. And then at Kerfield, on the west side of the Rhine River this was in March. I got back about the end of February, and then March 1, 1945, I'd just gotten back a few days from this injury, my chief of section got hit with a machine gun from an airplane, one that he was fighting up there, and it went down and took out a testicle. And he went clear to England with his injury.

March 5, four days later, there was a big bomb hit and a lot of airplanes came over. They think it was a big bomb; I don't know for sure what kind of a shell it was, but it hit right close to Gun Section 3, my gun section. It hit with so much force it dug a big hole in the ground and took me and threw me up over the gun. The drop hit me with so much power it pushed me under the wheel. I was under the gun. The first thing I remember, none of the fifteen men in the gun section heard anything coming in the air or anything but if it's coming right at you, you don't hear it.

I just remember turning to the right and a big puff of black smoke, and that's all I remember. And the next thing I knew was somebody said, "I'm all right. Better look at Goodwin." And I thought, "Look at Goodwin?" And I tried to move and I couldn't because I was stuffed under that wheel. And so they said, "Just wait a minute. We'll get you out." So they had outriggers they had put up there and dug me out.

Whatever went off gave me a concussion so bad for months afterwards it felt like my ears were pressed together, and oh, the ringing in my head. Anyway, right away an ambulance was there, and our doctor, he gave us too much morphine, I think. There were four of us that got hit. He said, "Take your clothes off, take your jackets off and your shirts." It was cold, but we did. I was hit in the right shoulder, and another soldier was hit in front of the right shoulder. He was jumping in a foxhole, and the blood when he came around, he lost control of his arm and the blood was just flowing.

The captain said, "Take your pants off." We didn't want to because it was cold, but we did. And this one kid was hit on the right side of his arm; besides his leg looked like it was just cut half-way through clear down to the bone, and the skin was open. It wasn't bleeding really that much. I thought it would be. And then he went into shock when he saw that. He didn't even know he was hit there on his leg. So then finally we were operated on at the field hospital, and I went into shock just as soon as I went into that hospital because there was so many wounded a lot of them with legs gone and arms gone, a lot wounded a lot worse than I was. But I went into shock, and I remember them putting the blankets, army blankets, just a bunch of them, grabbing and putting them on me because I went into shock. And then I found out that they gave me the new drug, sodium pentothal. That was a new drug that was first put out to interrogate prisoners.

Truth serum is what they called it, but they found out it worked good to put you out. That morphine the doctor gave us didn't work too well with the sodium pentothal, and for the next three days, I was in and out of it. They said they were going to fly us back to Paris, but the weather was so bad that they couldn't. So they put us on 40x8 trains. I don't remember much of the ride because I was in and out of it so much. But we pulled into Paris, and I thought we had been captured. I didn't see any Americans except the ones that were wounded, but Germans, Germans, and here they were the prisoners of war and they were using them as litter bearers. Boy that gives you a thrill when you think you're captured because all you could see were Germans.

When we got back to our outfit, and my captain said, just as I was getting out of the jeep, "Goodwin, I want to congratulate you. You're the only one in Dog Battery that got two Purple Hearts." And I said, "No, I've only got one." And I showed him the one I got in Paris. And oh, he was mad. He got the personnel section and wanted to know where my first Purple Heart was. The personnel section man had it somewhere with all the literature. I think he got relieved right then. And then the captain was really mad, because he had to take that and send it back to the Paris hospital and then write a thing where the Purple Heart that the Paris hospital had awarded me had to be rescinded and made an oak leaf cluster, because during World War II, you could only receive one medal and then a cluster for each one after that.

When I went back, January 26th, to this coal mine area, I was on my cot up there reading the *Stars and Stripes*. Here comes this kid just covered, totally black. His lips were red, his tongue was red, and his eyes were white. And he was just jabbering. And I said, "No compris. Nichts verstehen. Didn't understand." And he said, "Ja, ja, ein moment. Helmut, Helmut de Jong. Ein moment." He went down and took a shower and here was that kid. And before I got relief to go back to my outfit in February, I got a pass, and I visited with Helmut and did quite a few things with him. I visited his family and then up until about two or three years ago we wrote. I kept in touch with him. My wife and I went over to Germany twice after the war, and both times we saw him and his wife and family, his brother. So we kept track of him. I figure he saved my life.

I thought I had a million dollar wound; everyone said we were going home this time but there I was back on the front again. So we went from Hanover then, after we got back to our outfit, and we went up to Bismarck, and clear over to Stendal, Germany, and that's right on the Elbe River. And that's where the Russians came across the river and met us. We met them, the Russians, at Stendal and very shortly, we had the end of the war.

We weren't there very long after the Russians and the Americans met, and then we came back down; we came to a town called Gardelegen. And as we pulled in, here was a big barn, and there were 1,016 political prisoners that the Germans had cut holes in the roof and thrown gasoline in on them and set them on fire. And oh, it stunk for days after that.

It was one of the big war crimes of the war. So we were at Gardelegen for quite a while. For three or four days after we were there, it was still smoldering and stunk so bad. Some of them saw something move in there. They said, "Well, somebody's moving in there." Come to find out, five or six, maybe more than that, I can't remember how many it was, had lived in there because if they did get out through the wooden doors and stuff or dug out, the Germans had machine guns around. They mowed them down before they left town. They [the Germans] left town by the time we got there. So we had to take the ones who had lived in that mess. They said there were 1,016. There were so many in that barn, they couldn't even all lay down at one time. They just had straw in there. And so we put them up in a big house. Oh, it was bad. Then we were there quite a while, quite a few months. That's when we started guard duty and pass and looking for SS and stragglers.

The war is over and we were on occupation duty. German occupation, it was called. They made everybody except little tiny babies and real old people get out and dig graves for all these people in Gardelegen. And I guess they have a huge, beautiful thing there, and they've got to keep it green at all times and cleaned up.

I thought the Germans were pretty damn mean when I saw that. I mean, it was so stinking; it would make you cry to see them like that.

I felt hatred mostly for the SS. The regular German army was just like we were. They were drafted. And they either had to fight or the SS would kill them. They were up in front all the time.

We felt differently about the Nazi's and the rest of the Germans. And this Gardelegen, town of Gardelegen, it was Nazi clear through and they were trying to get to these ones and kill them, so we had to have armed guards. The ones that escaped death in the barn, the SS were still trying to kill them because they knew. They knew what they were going to tell us.

While we were on occupation duty in Germany, we just talked to the German civilian population, you couldn't fraternize with them. That was against the rules. If you got caught fraternizing, you're liable to get Court Martialed. Maybe shot. Some outfits, not my outfit, but some of them did pretty bad things.

It was sometime after the second time I got hit. Sometime between Wesel and I can't remember where it was, but anyway we had just pulled into position. This was after we crossed the Ruhr. I guess it was right between the Ruhr and the Rhine somewhere. We pulled in position and down at a crossroad, our captain was trying to get our attention, and we thought he was getting shot at or something because he was firing a gun down there ahead of you.

We'd already pulled into our position. And he gave the sign to start your engine, come on the double, and so chief of section said, "He's wanting us down there at that corner." We got down there and he said, "Just line your guns up, your 40mm, you're so much distance away, your M51 machine guns." We had four machine guns, and the bazooka man, and then the next four on one side of the road.

They're out ahead of that. And this road that we were on both sides of, it went down and we'd already escorted the infantry boys down there along that other road. And so that was trouble; the captain said we got orders to stay here until the last man. The infantry is going to have to retreat." American infantry is going to retreat back through your line because tanks, quite a few German tanks, were coming. And the weather was so bad, there hadn't been any planes out and didn't have any air cover.

Anyway, after we got all set up there, you could see the German tanks coming, Tiger Tanks. They were quite a distance away. And you could just hear guys praying. We were scared. It was cold; you were shaking because it was so cold. And you were just scared to death. And here everybody's praying and the sky just opened up like, I'd say like a funnel, and way up there, way up there, you could see blue sky. And just as soon as the sky opened up, here were the American airplanes coming down through there, and they started strafing and dive bombing those tanks and knocked out a lot of them. And the tanks turned off, turned and went back. I think the Lord was working with us. The Lord was with us. He heard us praying. I never heard that outfit pray; they weren't really a religious outfit, but they were praying. This was my tightest spot in the war. But I was scared quite a bit of the time. Boy, I'll tell you, tanks fighting each other, and you're pretty close to them. Let me tell you, it is bad. You know how it is.

A lot of them got hit, a lot of them got killed, but for the aircraft we counted eighteen at one time, FW190s strafing from different positions and stuff. When the ME-109s and the FW190s come down at your gun section, they're coming at a motion that goes back and forth; they're coming down to you and they're strafing you. And time after time, there wouldn't be a single person hit, but gas cans would have holes in them, tires would be blown out, and stuff would be damaged so bad, but none of the fifteen men in that whole gun section were hit, not a soul hit; now that was something.

We came back through Cologne, and then we ended up at Mons, Belgium. They had a prison there, and we pulled guard duty sometimes. And they had a big warehouse that we were protecting from stuff getting stolen because there was a lot of black market going on. And we were at Mons, Belgium, for quite awhile.

In the Black Forest, we were looking for Nazis, Hitler youth, SS, and people of this type. Going through the forest, it was kind of hard to see. They gave these crickets, because you couldn't see each other on each side of you. They'd go along the road and drop a man off, every so often. We were searching for people and we found quite a few Hitler youth. They had a lot of guns but no ammunition. But they all had little Hitler knives we called them, little Hitler knives or daggers. And they had one or two different groups. We caught three or four groups, and they had one or two women cooking for them.

We were in Mons, Belgium, when they came out with the points system. If you were married you had so many points, so much time in the service, each medal you received had so many points. I was married and had two Purple Hearts. That put most of my outfit out so I had to leave them and they put me in the 80th Amphibious outfit to come home. So I was with them for a while, and then we moved from Mons back down to Cherbourg, and I can't remember if we were at Camp Lucky Strike or Camp Chesterfield. We were at one of those.

I think I came back on a Liberty Ship or a cargo ship. We had to climb down a big ladder on the side. Boy, I was sicker than a dog. We went through one of the worst storms they'd ever had. You know a ship in a storm can only go over a certain amount where the keel comes up and tries to come out of the water. And we were past that for three days and we didn't know it, though, until we got back to New York. They wouldn't let us unload because of the longshoreman's contract. We couldn't get off the boat. Then the captain got on and told us about it, and he said the load had shifted and it was just ready to go on over. Flip the keel out of the water. It took us a long time. I got discharged January 31, 1946.

They talked me into staying in the Reserves. When you went to get out Camp Chaffee, Arkansas, is where I got out, where I got my discharge. And they were saying, "You're in for the duration, for ten years, and they can call you back anytime if something happens. And it's better to be in something you like than in the infantry that you don't like." So I signed up for the Reserves and never did get called. I finally got discharged January 30, 1949.

Well, I'm alive. I'm lucky. I was just in the wrong places a couple of times. I've had a lot of surgeries with my back problem. I came home to find out my back had three ruptured disks. I suffered for four years after I got out of the service. I had it checked out they gave me a fifty-fifty chance of walking again. And they took hip bone off my hip, and when they opened up my back, those disks came right out. And I have two metallic screws holding that hip bone onto my back.

Then my arm, see, I was on the job training when I first got out, and they found jobs for me. But they were jobs that didn't pay too well; they were jobs like woodworking and carpentry, and things would get rough and I'd get laid off. And then I can't remember the guy's name here in Emporia, would find me another job. They were just working on temporary jobs. This was through the G.I. Bill.

But I couldn't do much with the G.I. Bill because the first five years I was out, I had bad dreams. I choked my wife. The stress of war continued. They didn't have what they've got now, you know, these things, to help this stress stuff.

I went back to the hospital for my back quite a few times. They told me my back problems were all in my head; there wasn't anything wrong with my back. I think they told me this because they didn't want to operate on me. They blamed my back problems on the stress of war.

I took the Civil Service test and finally Earl Gadberry at the post office called and wanted to know if I wanted to get on with the post office. So I got on there and that was my career. I was a window clerk most of the time.

Howard was awarded two Purple Hearts, ETO Ribbon with 3 Battle Stars, American Defense Service Ribbon, Good Conduct Medal.

This is a Flint Hills Oral History Project World War II Veteran's Series interview, and Mr. Goodwin's personal recollections taken by Loren Pennington, May 19, 2006.

Horace in WWII.

Horace in 2010.

Place of Birth:
Ashley County, Arkansas
Active Duty Date:
August 1, 1944
Unit:
D Company, 2nd Battalion, 290th Infantry, 75th Infantry Division
Location:
Garonne, BE
Arrival ETO:
January 12, 1945
Rank:
T-Corporal, BAR

HORACE HIGH

The 75th Infantry Division was organized at Fort Lenard Wood, Missouri, in 1943. The Division landed in France on December 13, 1944, and was rushed to Belgium at the start of the Ardennes Offensive. After the Battle of the Bulge, they participated in the operation liberating Colmar and proceeded into Germany ending the war in Braumbauer.

Here is his story. . . .

One of the reasons that a lot of other vets don't want to be interviewed is because they don't want to reveal what they had to do and what they went through. I put all of that behind me when I left the service. I started to school the third day after I was discharged. I got 6 and half years of school in and retired after 42 years of teaching in Louisiana and Arkansas. I grew up in Arkansas and was the 15th child in our family. I had 7 brothers and 7 sisters.

None of my other brothers went to the war; they were 10 and 12 years older than me. I was born on June 11, 1921. I grew up on a farm in Ashley County, Arkansas. I started out teaching chemistry, physics and high school science. The juniors and seniors knew they were going to graduate and wouldn't study so I went to the principal and told him I wanted to resign. He said we're not going to let you resign. I have a fifth grade class you can teach and you only need 6 hours to certify for it and I taught 20 years of it. They gave me 10% more when I taught special education and I taught it for 12 years. Special education is really special and

I got really close to the kids. After that I retired. I had to have my knees replaced in 1992 and 1994 because of war injuries.

I had to raise my two children by myself because my wife died in 1960 and I was left with a 13 year old girl and 7 year old boy. I never had one minute of trouble with either of my two children. We would go to school in the winter and travel in the summer. Judy is my oldest, retired from Washburn University. My son, John, is a retired accountant and lives outside of St. Louis, Missouri.

I taught 42 years each. When I got out of high school, my dad wanted to give me the farm and all of the equipment but I wanted to get a job that paid me each week. I just didn't want any part of farming. My sisters and I picked 10,000 pounds of dried peas before we started picking cotton and I had enough of farming.

I went from Arkansas to California and worked putting up airplane canopies in an airplane factory. It was my job right out of high school in 1940. I was getting deferments and they were drafting men with four children they gave me one more six month deferment. It was my fourth one and then I was drafted.

I was drafted on August 1, 1944. It was right after the D-Day landing in June. I had 17 weeks of very intensive training. It was intensive and they were trying to cull people out. I was being trained to replace a person who had either been wounded or killed. We got out of training on December 14, 1944, and we had 10 days at home. On December 24th, in the afternoon, I stepped on a train in Little Rock, Arkansas, about 4 PM, I left my mother and dad and also my wife.

My mother and dad and my wife were there when I had to leave. When I stepped on the train one of the boys I had met in training, Jesse Matlock from Harrison, Arkansas, was crying. He left his wife with a set of twins and they both were in the hospital with pneumonia. Jesse and I met in training and are still friends to this day.

The train was going to Baltimore. It was December 24, Christmas Eve, and we went through a little town in Kentucky around lunch time and we were starved to death because there was no food on the train. The town's people had cooked turkey, dressing and everything. They gave us pitchers of lemonade and put it all on the train. I have never been so thrilled in my life. The train stopped and it looked like thousands of people came to the train with a gift package for each soldier. We couldn't eat it all so we saved it for supper. The American people are the most charitable people anywhere. We didn't have any food and we left our parents. We didn't think of bringing anything with us. It was so welcome. When we ate it for supper, we licked the bones clean, nothing left. We didn't get any more food until we got to Baltimore.

We shipped out from Baltimore at night. I have never seen so much ice in my life. It rained on the porches and the guy in front of me slipped and fell and broke his leg. He was going to ship out with us and he said, this is a million dollar leg, even though it hurts, I don't have to go to the war.

We sailed out of Baltimore about midnight on January 1, 1945, on a Liberty ship and there was one ship after another. We boarded and went directly to our bunks to stash our gear, and they got us up for supper. The bunks were six high and I thought I had better get a top bunk so I wouldn't get baptized. But I got seasick. The sea was rough and dark so that nobody could see as we slipped out of the harbor. I got deathly sick, I felt like my feet were coming up in my mouth. The doctor gave me some purple medicine that coated my stomach and I didn't get sick anymore. We landed at Le Havre, France.

We went to a central point and boarded a French train of 40 and 8 boxcars that were joined together with chain links. They said they were for 40 men or 8 horses. As we rolled along you would hear the constant clanging of the chains as they sped up and slowed down. There were different sizes of rails in Europe, extra narrow, standard as we have it here and wide. As we went on up into Belgium, we would sit down and hold hands to keep from turning a flip when the train slowed or sped up. After the war, we were coming out of Germany into France and they had the same kind of trains.

When we got to Belgium at a collection point, they blew the whistle and we got off. The snow was up to my chest and I wondered what a guy from Arkansas was doing there. We had no winter gear. We had combat boots, OD trousers, shirt and OD jacket but not an overcoat. We were not equipped for that kind of weather but we were losing the war up there. After we got there men started arriving who had only three weeks training.

I was assigned to D Company, 2nd Battalion, 290th Infantry, the 75th Division as the BAR man and carried that until I got calluses on my shoulders. I had a leather strap for a slide and shot the BAR from the hip. When it was hot from shooting, I put it in the snow to cool and watched the steam come off the barrel. Once I saw three boys filling their water cooled, 30 caliber machine gun with urine because they had no water.

I got to the Division on January 12, 1945. We were in two or three different places during the Battle of the Bulge. On January 13, the 75th Division was along the Siegfried Line. I can remember some of the cities. We were in Goronne, Vielsam, Neuville, Burtonville, and Poteau in Belgium.

I never saw so much snow in my life. We were not equipped for that kind of snow and winter. Some people found overcoats and others had blankets but I did not. You would see very few Belgians out in the weather because the weather was so terrible and when you did it was always the man. The wind would go through you; it seemed to be blowing fifty miles an hour.

Once during the Battle of the Bulge, we were pinned down on a sheet of ice. Germans were shooting at us and the fire wouldn't let up. We couldn't move and all we could do was just lie there, pinned down. I lay there so long I finally passed out from lack of sleep, lying the whole time on one shoulder. After a while the ice froze my shoulder and I when we got free I had to see a medic.

The 75th Division was called the "diaper division" because age wise it was the youngest division in Europe. There was such a big influx of boys with only three weeks training and they were very young. The 75th Division had only been in Belgium since December 20th.

When we arrived, we didn't have any time to get used to anything, they told us there are the Germans, go get them. They didn't stutter one minute about it. There were some foxholes but they were full of snow. We were directed to go down about thirty feet to a foxhole. There was a dead American boy lying next to a BAR. He had just been killed. They made me the BAR man to take his place.

I was told to pick up the BAR and the ammo and use it. I weighed 137 pounds and I carried the BAR with tins of ammunition around me. We walked over to the Platoon Sergeant and he said we had better get behind this hanging sheet because the Germans were coming and the sheet provided some concealment. Some Sergeants came through and told us to kill anything that moved. Eight Germans, two deep coming toward me out of the woods apparently lost and I mowed them down with the BAR.

I was given an 18 year old boy named Franklin Moore from Kansas as my assistant. He was bigger than I was and I told him, son, do exactly what I tell you to do. I will tell you time and one time only or you will lose your life.

My Sergeant was from Kentucky. He was an excellent sergeant but he was completely illiterate. He had never seen the inside of a school. His grandfather was a bootlegger, his daddy was a bootlegger and he was a bootlegger and they all had fine homes, very wealthy. The law didn't bother them because they knew they would have to go on welfare if they were arrested. His name was Charles Meyers. We were in Germany and I found him back behind a piano trying to figure out what his wife had said to him in a letter. I asked him if I could read it to him. He said yes. I asked if he could write and he said no but would I write her a letter from him. After that, I wrote letters to his wife from him and read her letters to him.

We got a letter back in about two weeks from his wife thanking me, she found out more about Charles in this one letter than all of the rest. He was one of the sweetest men you would ever meet in your life. He saved my life many a time.

On January 17, 1945, we attacked a town called Petite Their which was a few miles west of St. Vith. We captured that town, Patteaus, and Neuville. We took so many prisoners, hundreds of prisoners. The Germans used everything on us, 88's, mortars, everything. Our Division captured Aldringen and that ended the Bulge. The Germans were in houses and had to be dug out like rats.

I had my assistant, Moore with me; I couldn't breathe without him close by. During the last of February, we took a sponge bath in our helmets, and then we walked to the kitchen to eat. On the way we passed a kitchen which was a buggy shed in front of a castle that we used for refuge. The kitchen was full of officers and Sergeants but not our Sergeant. We heard a shell coming and we ran toward the castle. The shell just made a mess of the shed and killed everyone in it.

Once Moore and I were running from one place to another following Germans and they didn't want to be followed. They were trying to get back into Germany so we shot them. It felt like freedom. We jumped into a foxhole and there was a dead German in it already. We didn't want to dig a new foxhole but we were afraid that the fleeing Germans had booby trapped the body so I put a rope around his neck. Moore took the end of the rope to pull him out. Nothing went off; they must not have had enough time to set up a booby trap. We had killed him and his buddy had jumped out and ran. We laid him right on the side of the foxhole and we could hear bullets hitting him. That is how close we came to getting killed.

The last day of February, Sergeant Meyers came to us and he said, give Moore your gun and ammunition High, he is going on patrol and you are going to write a letter for me. Before he left, Moore asked if he could use my pen so he could write a letter to his parents. When he was done he gave me back a pen but it turned out to be his pen.

At midnight the patrol came right by me and got into a boat and went across the Meuse River in Holland. There was a machine gun set up on the bank right in front of them and they went directly into it. The Germans waited until they were right on the levy and began firing. All but one was killed in the boat. One jumped out of the boat and got away and returned to our group. The four dead Americans drifted down into some willows and the next day the Canadians saw the boat and recovered the bodies. One of the guys killed was from Holton, Kansas, Franklin Moore, my assistant. I just missed it and felt divine intervention.

I had intensive training in Arkansas. The Sergeant told me that when I have hand to hand combat with a bayonet, he wanted to show me how to deflect a bayonet that is coming toward me so I could kill him instead. When I heard hand to hand my hair stood up on end but sure enough it happened. I believe that was an act of God.

The training sergeant was a short guy, 5 feet 6 or so and he was an angel to me, don't know why he took to me but he did. He showed me exactly how to deflect a bayonet.

Sure enough our sergeant told us to fix bayonets, we got to get them. It was daylight and a group of Germans were charging my position. You would hear screams here and screams there from both sides but I didn't want to scream. We were charged by the Germans, a group of about two dozen coming at us and we had to get out of the foxholes and fight them.

There were Americans who had bayonets in them. You see the whole picture in 5 seconds. The German was coming at me and I diverted his blow and stabbed him like I was trained. You could hear the screams of both the charging Germans and the Americans.

We got in there and we fought like tigers. At this point I had put down the BAR and had an M-1 because it had a bayonet. You live a whole lifetime in 5 to 10 seconds. I believe it was nothing but my mother's prayers that got me through and if you have ever talked to God you talked to him then. Like one Sergeant said, you are either a Christian or a hypocrite and you had better get on the right side.

About 20 feet away I saw the German coming. He whirled around to get me and I went at him. I deflected his bayonet; apparently he didn't have the training that I had. I deflected his bayonet and put mine into him. As just as soon as he fell, I put my foot on his chest and pulled it out. The muscles will hold it in; it all was exactly like the Sergeant told me it was. I have been in many close places and it is something I have put aside.

There was a guy walking around with mittens or glove inserts. I didn't have anything like that; I kept my hands under my arm pits. No hair grows on my legs from the knees down because of the frost bite. My feet were so numb it was like walking on Tom Walkers. I didn't realize my feet and hands were frozen. This guy froze to death in a foxhole. He was just walking around and the next day he was frozen to death. It took about 15 minutes for people to freeze after they lose consciousness. So I told him, "old boy, you won't need these anymore," and I took the mittens. They were warm but my hands were already frozen. Too bad I couldn't find any good boots.

We stacked the frozen dead men five this way and five that way until we got them high enough for protection and the bodies would freeze. We were weak, we couldn't pick up too much. I got down to 125 pounds. Then what we called "sabers," a rat type animal with a long mouth came out of the wooded area and they began consuming the dead men. They would go for the private parts first we would hear them all night eating them, crunching on the frozen flesh. They would smell and take those teeth and rip the trousers off and all around them there was money, bill folds, watches and Belgian money would fall out and I didn't want any of it. It was gruesome. These were American boys and their possessions were everywhere. We were so far advanced, the graves registration guys were not around there yet. It was the most terrible thing I have ever witnessed.

At the Siegfried Line the Germans had trenches and they were in those trenches. We were on tanks going about 25 miles an hour and the tanks ate the trenches up. Those Germans didn't shoot one time because the tanks would have turned and shot right down the trench, they were straight trenches and they were scared to death. We were right behind their lines and that was when I used the 50 caliber machine gun on the top of the tank.

We got on our tanks and went through the Siegfried Line and we scared the Germans to death. One night there was a German who was going to harass us all night. He was behind a tree, a sniper. I didn't even have a gun and had to use the machine gun. I was sitting on a tank by a 50 caliber machine gun and I let loose into the tree. A big slab of tree five feet long sliced off the other side of the tree and hit him and killed him.

I came very close to getting killed during this time. Our squad was patrolling toward German lines and they opened up on us with machine guns. The bullets were just cracking over my head and around me when one hit my helmet and knocked me down. I really thought I was dead but I jumped right up and felt like I had more energy.

After the Battle of the Bulge, we went to southern France to the Colmar battle area. We then moved back up to Holland and then into Germany. We had pockets of Germans everywhere. They knew at this point that they had lost the war but they wanted to antagonize us so it lasted about 6 more weeks.

We were so far ahead of the food we were starving and we had to steal food. We would go into German houses to find food. You know the cord in the trousers; we would tie it into like a pouch and go into the German cellars and get jars of meat, canned beef, and put it in our trousers. Later, we would put it under our jackets and it would melt the grease. The Captain told us that anyone with German food would get court marshaled but hunger took over.

When we were crossing the Rhine River, a German plane came right down on us and they didn't have a chance. There was too much fire power. I saw one river where two German trains were heading toward a bridge in opposite directions and a plane bombed it. Both trains had gone into the river. I didn't see it happened but my two friends did. I saw the bridge down and the trains were in the water below.

We crossed the Rhine River the 6th of March and I will never forget that date. We were crossing it in motor boats. It had warmed up and the snow had melted and the river was up, very wide. We got out to about the middle of the river when the boat ran out of gas. The driver yelled at the next boat that we were out of gas. We had to hold both boats together and got to the other side using the power of one boat motor.

It was night and we all stepped in each other's footsteps when we left the boat. The next morning they removed thirty-two land mines that we had stepped over. That night we followed a wire across the river. A guy from New Orleans right behind me was cutting up. He didn't step in the tracks and a leg blew past me. I picked up his rifle and threw him over my shoulder and stepped in the tracks to carry him back. It blew his sock up into his severed knee and he didn't bleed a drop. He said he didn't feel anything but he said I knew my leg is gone. I just carried him out on my back and a medic took him from my back.

We came up on a prison camp and my assistant and I crawled through a big culvert, killed the guard and blew the lock off the gate of a prison camp. Frenchmen, Russians and I don't know who all were in there. They came marching out and a great big Russian woman, very tall, grabbed me and kissed me in the mouth. She was at least a foot taller than me. They were so happy they were out.

You could hardly go in because of the stench; a person could hardly live in there. There was an 18 month old girl who couldn't even walk because of malnutrition. Some men got her and carried her to our hospital and began feeding her and really took good care of her.

I saw a stack of human bones as large as any gymnasium. They were all Jews. I don't remember where that was. I was 23 at the time. They were just stacked up there, high. They had loaders and were stacking them up. You couldn't even imagine bones like that. Later, we came to another place where they had stripped the clothing off people. They could sell the clothing and would sell their houses too. In this field, there were men out there shot, and butt naked. They were living only hours before we got there. They were dead and stripped naked. It was terrible.

Toward the end of the war, we were surrounded and detained by German SS. One night a patrol of SS troopers came into our lines to take prisoners and in their attempt they encircled ten of us and we became their prisoners. They kept us prisoners for about two weeks. They didn't have anything for us to eat and I seemed to have gained their trust so they let me go out to the road and pick and eat raw turnips which were all we had. I also had a bag of lifesavers in my gas mask that I got on the boat over. That is the way I would eat.

The war was practically over and they wanted to harass us, we thought we were just going to be detained and released. Two of our guys said they were leaving and I said maybe you better stick around.

The Germans were SS Troopers and they were mean. They were the ones who had 150 German ladies on the top of a mountain and were breeding Hitler's super race. They picked out the Norseman head and had to be at least six feet tall. The SS trooper showed me his head and told me the head had to be like his. I went up into the mountains and I saw them. I have wondered what happened to the babies every since. They were pretty babies, blond hair and blue eyes.

When we first got captured, two of the guys said they were leaving and the SS guys said you better not and one of the Germans took out a carbine type weapon and shot one of the guys in the leg. Another one of our guys cursed the German who spoke better English than we did. So the German turned around and shot him. The other American who had been shot in the leg was moaning and making a lot of noise. The SS Trooper went to him, unzipped the American's pants and whacked off his penis right there and just threw it down on the ground. There was blood everywhere and he was screaming. The SS German then just shot him in the chest, dead. When he cut it off, the crazy SS guy just shot him and put him out of his misery. I thought he should have been buried with all of his body parts if possible, so I picked up the parts and put them back into his pants. We never heard what happened after that. I saw some atrocities during the war.

Horace with a downed B-24 at end of the war.

The Germans went from house to house and they kept us with them. They were keeping up with where the Americans were. They didn't want to be captured so they left us out there. That night the Germans had a radio and they just left us with the radio because they knew our troops were coming. Our trucks came for us. My friend and I got into the cab with the driver and he asked where the other boys were. I motioned to drive on around and he was aghast when he saw all of the bones that the rats had left. We got them all in the truck and returned to our unit.

I had not even thought of any of this until now. It doesn't matter now.

When the war ended I was in Iserlohn, Germany. We were out in the rural area and they told us the war was over. I did a lot of things but never got anything for any of it. I didn't even get a Purple Heart. I grew up so far south that nobody at home knew anything about frostbite. Sometimes we would have a little snow and it would get into the teens but nothing like the Bulge.

On June 28, 1946, I came back from Europe. I had to stay in Europe for over a year because I didn't have any points. I was a Corporal with a "T" under the rank when I got out. When I got discharged in Houston a guy told us if there was anything wrong with you, fill out this form and you would be discharged in the following hour. Nobody wanted to delay their discharge. When I got home I got a 10% disability and it was good for $12.20 per month.

I went to California looking for a teaching job and went to a school to talk to the principal. They told me if I didn't take the job they would not be able to have school because of no teachers. I know God set up the circumstances that took me to that school. I got their telephone number and said I would be calling them. I was offered a great job and we got a beautiful apartment in Santa Anna, California. You didn't use air conditioning there and it was never needed. It was a two bedroom fully furnished.

I have been here in the Veterans Center for seven years and really like it here. My hands were frozen, poor circulation, and they turn in. I make Kleenex box covers to keep them limber.

During the Battle of the Bulge, my feet were frozen, my legs half way up to my knees. We were pinned down for a couple of days and the Germans would shoot sporadic fire over us to keep us down. We had to lay there on frozen ice and I finally passed out from no sleep. I was afraid I would have to have my shoulder replaced because I went to sleep on frozen ice.

Horace on wrecked B-24.

A lot of veterans have returned to Europe. I am not sure I ever wanted to go back to Belgium.

After the war, I went to Holton, Kansas to visit Franklin Moore's family. I took them the pen that Franklin had accidentally traded with me the night he was killed. His father thought it was terrific.

What I remember most about the Battle of the Bulge was the weather. You cannot imagine how cold it was after I had just spent two years in sunny, warm, southern California. My blood was thin and it is no wonder that I had frozen feet. And I also remember stacking the dead men five this way and five that way until we got them high enough for cover.

Horace was awarded the ETO Ribbon with 3 Battle Stars, American Defense Service Good Conduct Medal Ribbon.

Benneau, BE, January 1945

Art Holtman

Place of Birth:
Leonardville, Kansas
Active Duty Date:
February 24, 1941
Unit:
Headquarters,
XVIII Corps
Division Location:
Aywaille-Spa,
Belgium
Arrival ETO:
August 27, 1944
Rank:
Tech Sergeant
Administrative NCO

ART HOLTMAN

In August, 1944 the Headquarters for the Corps Artillery, XVIII Corps arrived in England and learned that they had been redesignated as the U.S. XVIII Airborne Corps. The Corps was made up of The 82nd Airborne and the 101st Airborne Divisions, both taking part in Operation Market Garden and prominent roles in the Battle of the Bulge. The Corps planned and executed Operation Varsity, the largest airborne operation at a single location in a single day in history.

Here is his story. . . .

I was born on January 26, 1919 in Walsburg, Kansas, which is no longer in existence; it was a little village between Leoanardville and Randolph. My dad was a farmer outside of Leoanardville and I was born and raised on the farm until I was about 19. I went to a little one room schoolhouse a mile and a quarter south of home. There were 12 of us in the school and two of us in the eighth grade. We had two different teachers; the first was there for 4-1/2 years and got married. The second one had been the barber in Leonardville and they hired him to finish that term. I went to high school at Leonardville Rural High school, graduated in 1936 and then went back to school a year for postgraduate courses called a "normal" training course. I passed that and got a state teaching certificate in 1937 at age 18.

I taught school in the fall of 1937 in a little country school in Walsburg next to the one where I went to school. I taught a couple of my cousins there also. I think we had six grades and I made $50 a month. The second year I was offered a job at Cleburne. I was one of two teachers there and

I was also the principal. I had fifth, six, seven, and eighth grades and the janitor. I made $72 a month but I did not live at home. I stayed in a boarding room for $12.50 a month. My first teaching certificate was a two-year certificate and in order to keep it, I had to take some college courses in the summer.

I really wasn't that enthused about teaching so I got a job in Manhattan with the co-op elevator in 1939. I worked in the office.

The law for the draft was passed in September of 1940 and I registered drawing a low number, 270. I was drafted into the Army February 24, 1941. There were 57 of us when I was drafted and we were the first draft of Riley County. It was prior to Pearl Harbor and we were selectees for one year's training and then would be done with the Army at the end of one year. We almost had our year in when Pearl Harbor happened so that term was extended for the duration 6 months and for me it turned out to be almost 5 years.

I reported to Fort Leavenworth with the 57 other guys on the 24th which was a Monday. On Sunday, the 23rd, they had an event at the church in Manhattan. It was snowing that morning of the 24th when we got on the train in Manhattan and we headed for Kansas City to Union Station. There they put us onto buses heading for Fort Leavenworth. Before the day was over, we are in the Army.

We were not at Leavenworth very long, maybe two or three days, then eight of us were sent to Fort Knox, Kentucky. The rest of the group stayed intact and went to Camp Robinson in Arkansas and became part of the 35th infantry division. I got my basic training at Fort Knox and that's where I was on December 7, 1941, when we got the news of Pearl Harbor.

They needed pilots in the air Corps and I signed up for that. I was transferred out of the Army immediately into the Army Air Corps and was an aviation cadet. I got my flight training at Thunderbird field in Arizona outside of Phoenix but I was not very good. There were a lot of trainees washing out at that time so I went to an unassigned pool in the Air Corps. One day they came around and wanted to know who wanted to go back to the Army. I had been in the First Armored Corps at Fort Knox, so they said I could go back to armored if I wanted to. They sent me to the Second Armored Corps in San Jose, California, and it was one of the best assignments because it was right in town at the armory, and it was pretty good conditions. That was in the spring of 1943. I spent quite a bit of time here in the states before I went overseas.

The Second Armored Corps went to Fort du Pont, Delaware, in the spring of 1944, and from there to Camp Kilmer, New Jersey, sailing out of New York on the SS "Ile de France" on August 17, 1944. It was comparable to the Queen Mary and the Queen Elizabeth. It was a large converted luxury liner and they told us there 15,000 troops on the ship. Bing Crosby and his troupe were on the ship and they entertained us several times. The ship was fast enough that we did not need a convoy escort and we crossed over in about five days, landing in Clyde Scotland. The ship was so big that we had to park out from shore and they brought us to shore in small boats. From there we went by train to Swindon, England and we were in England for a couple of months.

When we got to England, we found out that we were no longer an armored outfit but an airborne outfit. We were called the 18th Airborne Corps and we received some airborne training there. The 82nd Airborne and the 101st airborne made up the 18th Corps Airborne. I was with the Headquarters Company of the Corps and we were not paratroopers, we were glider troops and we got about two months of glider training in England. I was promoted to Tech. Sgt.

In early September, we were transferred to France. We were at Corps headquarters, one size bigger than the division, so we were always behind the lines and a lot of times we were fairly close behind the lines but I never was in any real combat and our glider group never made any combat landings. During Operation Market Garden in September 1944, we were scheduled to go in on that battle but Montgomery insisted that the British Second Airborne take our place. The airborne group really hit trouble at Market Garden and had we gone in on our gliders we would've been blown into splinters.

We received three battle stars. First was the Battle of the Bulge, the second was the Rhineland and Ruhr Pocket and the drive into Central Europe. We ended up at Hagenau, Germany in the northern part of Germany, 50 miles from the Baltic Sea. We crossed the Rhine in March but the Remagen Bridge had fallen by the time we got there and we crossed on pontoon bridges.

Most of my company did not have as much service as I had because I'd been floating back and forth in the United States between service stations. They were using the point system to get out and I was high point man in the group. The war the Pacific was still going on and our outfit was scheduled to participate in the Pacific. In the middle of July 1945, we are sent back to United States and were given a 30 day furlough. We were supposed to report back to Camp Campbell, Kentucky, and from there on to the Pacific. The atomic bomb was dropped while I was home on leave.

When I reported back to Camp Campbell, we laid around doing nothing until I got my discharge on October 4th.

I was discharged at camp Atterberry Indiana and then came back to Manhattan. I went back to the farm over the winter to help my dad farm in the spring of 1946. After that I came back to Manhattan and got a job with the City Investment Company, a loan and finance office. They had a program for on-the-job training and that is where I was able to get my training in finance.

After five years with the investment company, I worked as the business manager for Nelson Clinic, a group of doctors. I was with them for about 12 years and after Dr. Nelson died they began to split up. I retired from the old Memorial Hospital in 1982. I was retired for about 10 years when I started delivering flowers for local florist and it was one of the best jobs I ever had.

I got married in July of 1948. My wife was from Clay Center. We had two children, a boy and a girl. Both of them are CPAs. My son lives close to Manhattan, my daughter lives in Phoenix. I have three grandchildren. My first wife died in 1977 and I remarried Jean a year and half later, and she has two daughters. Together we have nine grandchildren.

The winter of 1944-45 was a cold, long winter. The locals told us that it was the coldest winter in over 40 years. I was really happy when the Battle of the Bulge was over so that the Americans could get on the offensive again. Probably the thing I remember most was when the weather cleared and our planes were able to fly again, I was elated.

Art was awarded a Bronze Star, ETO Ribbon with 3 Battle Stars, American Defense Service Ribbon, and Good Conduct Medal.

This interview was taken from the Riley County Historical Society, Manhattan, KS, 2003.

Sgt. Lenhardt Homeier, WWII

Lenhardt Homeier, 2010

Place of Birth:
Ellsworth County, Kansas
Active Duty Date:
March 1944
Unit:
A Company, 2nd Engineers, 2nd Infantry Division
Division Location:
Malmedy-Elsenborn area, BE
Arrival ETO:
December 12, 1944
Rank: Sergeant

LENHARDT HOMEIER

The 2nd Infantry Division landed at Omaha Beach on June 7, 1944. They participated in the Operations around Saint Lo and fought their way to the Belgium-German border. During the Battle of the Bulge, they held positions around Elsenborn. They reached the Ludendorff Bridge on March 9 and provided security for the bridge crossing. On March 21, 1945, the Division crossed the Rhine and fought its way by War's end to Pilzen, Czechoslovakia.

Here is his story. . . .

I was born in Ellsworth County about 8 miles south of Wilson. My Dad immigrated in about 1896. He was born in Germany in 1877. My Mother's parents were Volga Germans from the Volga area in Russia so both of my parents were immigrants. I was born May 1, 1924, and at about four years of age we moved from Russell County to Lincoln County. We were tenant farmers. I grew up in Lincoln County, and I went to school there through the eighth grade. We were poor and we didn't have the money to go further in my education. I am the youngest of four boys and we had to grow up and work because things were tough.

My oldest brother and second oldest brother were both in the service but my next brother had health issues and he was rejected from service. I actually got one deferment because they allowed one person to stay at home but after my brother got rejected, they drafted me and he stayed at home and I came in later. It was 1944. I went in the service in March 1944 and went to Camp Roberts, California, to do my basic training. It was a 16 weeks program, and I completed it in July.

Then I was shipped to Ft. Benning, GA, for some advanced training and was there until the first part of November, 1944. I came home for leave and returned. From there we went to Ft. Meade, MD, and we were trained to get on and off the ships. We shipped out of New York Thanksgiving Day. Before I got on the ship, it was really cold, ugly weather at Ft. Meade and I got strep throat. I didn't want to leave my buddies so I told them that I was going to lie down in the bed and that they needed to get me up so I could go with them. They got me up and we loaded onto a ferry and on to the ship.

They immediately put me in the sick ward when I boarded the ship. We were out and ready to sail when an orderly came in and said if I would look out of the window I could see the Statue of Liberty through the porthole. And there it was.

Through the whole procedure I didn't have to stand guard duty nor did I have to do any duty. It was a 12 or 13 day trip in a convoy and it was a pretty rough trip because we hit some rough water. We landed at Southampton, England and that is where we disembarked. We went inland and they had a tent city set up. It had rained all the time before they set up the tents. The tents had no bottoms and everything was a sea of mud. We had to sit on our helmets and lean against the tent poles so we could sleep because we couldn't lie on the ground, it was pure mud.

Some guys got a great idea. There was an English farm a ways down from us so we went there and stole some of their thatching from their barn and lay it on the ground so we could sleep. The Army got charged for the roof and we caught hell for doing it.

We went across the Channel on a Liberty ship built by Kaiser. I was trained as a replacement in the infantry and we filled in where ever they needed people. Some of my buddies went into tanks and another went into the engineers, as did I. We got there just as the Battle of the Bulge started so we filled in for people who had just been hit. Then we loaded on box cars, a 40 and 8, and crossed France into Brussels, Belgium.

From Brussels we were taken out to our divisions. We didn't know which divisions we were assigned to until we got to the division. The night they picked us up they dropped me off at a farm house in the country they called "Bullingen" and it was the company headquarters was and where the higher ranked people stayed. As soon as I reported in they said here, take this weapon, you go out and guard. I didn't even know what directions things were and if someone would have said boo I would have croaked.

When we left the States we had no gear. Since I was in the engineers I hardly ever carried a weapon. The next day they assigned me to A Company and we were billeted in a town called Givet northwest of Bastogne. That is where I joined the unit and I replaced a fellow who had gotten wounded.

There was a guy from Pennsylvania who said, hey we already have a Homeier here, we don't need another one, but he spelled his name differently. Fred Reed told me I was going to be his buddy so he and I worked together. 2nd Engineers were part of the 2nd Infantry Division. We had the 9th, 38th and 23rd Infantry Regiments.

We never had any bazooka training so they took us out and trained us how to use them. They also trained us in pick and shovel but none of us needed any of that. That is where we operated out of until we had the Bulge stopped. We held the line there until the last part of January. We had so much snow, it snowed every day. We had Corporal Cotton Walker who had a snow plow on his truck and he would take one of us out with him for support to clear the snow and open roads. The snow was terrible.

We were just moving up to the front as replacements and we made a stop. We stayed in a glue factory. That is where we heard about the Battle going on from the troops who were coming back who were in the Battle. They were telling us what was going on, terrible stories they were telling us and very scary. They loaded us up on trucks and that is when we headed to the front. They had a very efficient system.

We acted as if we knew everything after we got there. After we got settled, my buddy and I went with the infantry for two days with mine detectors. I carried the bar and he carried the battery pack. We went with them in case they ran into a mine field; thank goodness they didn't. The first village we came to we stayed in a house over night. When we got up the next day there was a German prisoner tied up and sitting outside the house. I was the interpreter over there because I could speak German. I asked him about "minen: and he said, "ja, minen," meaning yes, there were mines. So we had to follow them for the next day but there never were any mines. We came to another village that was our objective for the day and our outfit came and picked us up and we went back to our unit.

When we were in the Ardennes Forest, there were 5 of us and we were called out to repair a dip in the road that went to the water supply depot. There was myself, Reed, the Sergeant, a Corporal and a few privates, 5 total in the jeep, 2 in the dump truck and the Sgt. In the other jeep. We went to the site and they dumped the gravel in the dip and we were in the dip with shovels filling in the hole and all of the sudden there was a whoosh, a mortar came in. There was a puddle of water and it landed in the water. We jumped on the other side of an embankment and another came in, they had us zeroed in. More came in. We had the dump truck on the top of the hill and someone yelled, "Get the dump truck out of here that is what they are after." And soon as the truck was moved the mortaring stopped. The Sgt. said it was clear what they were after, so we walked back to where we had to work and Reed was beside me. Another came in and a piece of shrapnel hit him right in the eye. He grabbed his eye and ran for the jeep. The Sergeant took off leaving his troops by themselves.

I lost track of that kid, all I knew was he was from Pennsylvania. Over the years I have tried to find him and have found all kinds of Fred Reeds with my computer. Finally I was looking at one in Nazareth, PA, and gave him a call.

I asked him if the 2nd Engineer Combat Battalion mean anything to you and he said it certainly does. I told him who I was and he was tickled silly. I told him we needed to get together. He said he and his wife were heading for Florida for vacation and when he got back we would get together. A week or so later I got a letter from his wife that he had died. That was terrible; at least I got to talk to him. He became a heavy equipment operator and he could do that with one eye. When he was wounded, I didn't know if it had gotten him in the head or what and never knew what happened to him. He was a wonderful young man and I sure wish I could have seen him again.

On our trip over, all at once something hit our ship and scraped along the side. We all hopped out of our bunks and lay on the deck. A guy named Shoemaker was on the top bunk and he beat me down to the floor. We never knew what it was but we could feel it jar the ship. The sea was awfully rough. We were in a convoy and sometimes a ship next to you would be above our ship and sometimes they would be below our ship. Guys were hanging over the sides puking their guts out but it never bothered me a bit.

We never had to lay wire outside our positions because we moved so much. Once we had to lay a mine field during the Battle. It was a barrier to stop tanks and advancing infantry.

When we came into the Ardennes Forrest, we were right behind the Germans. They had a big black kettle on a two wheel cart and the cart hung on the axle and was pulled by two horses. That kettle was still cooking when we got to the forest that is how close we were to them. Next to the kettle were two dead German soldiers laying there. They probably got wounded someplace and that is as far as they made it.

This buddy and I when the Bulge was on, there was a bridge that had not been blown yet and we were protecting it. Prior to us being there, someone had dug a pretty good size fox hole and they had covered it up with camouflage. We would put the detonator for the charges on the bridge in the fox hole with us. The bridge was quite a ways away and the detonator was connected to the charges by wires. If need be, we would blow the bridge. We had to stay out there for 72 hours watching the bridge.

When they came to pick us up there was a lot of snow on the ground still. The driver was an Italian who drove a pickup truck like vehicle with canvas top. I was sitting on the side and the driver was speeding down a muddy snowy road and he went around a curve and slid into a tank. The tank hit me in the back and pushed me out the back of the truck and into the road. Every since then I have had back trouble. That sucker pitched me right out of the truck.

The Ardennes Forest was thick. Roads were running through it. There was a headquarters in the Ardennes and we had to build a shelter for them. We had to call in the bigger engineers to bull doze it flat and then we cut down trees and covered it up so they could have bomb protections.

Once when we were in Brussels, they put us in a hotel. They told us around the corner was a shower. As we were walking down the street to the showers, a jeep came by and I looked over and thought, that is funny, an officer is driving that jeep and there are troops riding. I wonder to this day if they were Germans dressed in our uniforms. You just didn't see officers drive troops.

On the road that we were using for material transport, the Germans had it zeroed in and would shell the road when trucks would use it. So we cut evergreens about 4 to 5 feet high, took them in the snow bank to cover the truck traffic when it went by. It took us all day to do that.

There were a bunch of trees all along the road. The Germans would blow those trees down and lay them along the road to obstruct traffic. Then they would put antipersonnel mines in the trees and we would have to crawl in there and find them and remove them. They were detonated by trip wire. They also used a shoe mine that was planted in a hole and when you stepped on it, it would go right through your foot. We lost two officers to that. When we pushed into Germany, we pushed off in a town called Gemund. That town was obliterated from shelling and we were supposed to go down there and put in some bridges. This was south of the big dams and we were supposed to put in some Bailey bridges across this tributary from the big dams up north. There was a church over to the side.

We got the first bridge completed and a truck went across only to get blown up. They had the banks of the river mined.

We worked the whole day trying to get the bridge up. They brought back some German prisoners but had them walk through the river. Everything in the village was mined and booby trapped. I had to go in and find them and disarm them. We had a Regal mine, long, like a 2 by 4. We had a bunch of them stacked in the road to be destroyed. One officer got the idea of blowing them up right there and it blew a hole in the road.

After Gemund, they took us around the north end of the town. We moved at night. To me the sun came up in a different direction, I was so confused on direction, I just didn't know. We got in behind the infantry day by day and whatever we needed to do we did it. One night C Company had to go out, got surrounded by the Germans and we had to go down and bail them out.

I remember going along a road further along we came to a town where people were dressed in what looked like prison garb. And they were looking around trees watching us. We thought it was an insane asylum and found out it was a concentration camp. We didn't know. They were all over the place, skinny, ugly, horrible. The infantry had let them out. They were looking around the trees, never in their life had they seen anything like us.

I went across Germany on a tank. We rode the tank out to do road repair or other work. The poor guys in the tanks had no escape. It we sickening to see them hit. The hatches would be open, guys hanging out burned, dead. It was sickening.

The German people were amazed at the equipment we had. If we would go through a village, they would stand in awe, seeing the tanks, artillery and trucks. They would just shake their heads and some would say, we were supposed to beat that?

I don't remember crossing the Siegfried Line but I do remember crossing the Rhine River. We crossed right next to Remagen. The big engineers put a pontoon bridge across the river and we crossed it. We then went south to a town called Rhine Brohl and we held up there for a few days and they told us to find a house to stay in. Since I spoke German, I found a house for us and told the man what we wanted to do. The lady of the house told me when she was a little girl, troops in WWI were billeted right close by where we were. They were nice people. I had no animosity against them but the SS Troopers sure could be mean.

We went toward Leipzig, then south along the Czech border, entered Czechoslovakia at a town called Domalice. We stayed there over night, then they loaded us onto trucks the next day and they took up to Pilzen toward Prague. There was still fighting going on so I think we were in Czechoslovakia. I remember when we got the news that the war was over. When we got the news, a kid from another company got on a motorcycle and drove around like a crazy person and had a wreck and was killed. He went all through the war and then got killed on a motorcycle. We felt terrible about that.

We stayed in Czechoslovakia about a month. We had a real nice area where we stayed. We stayed there until we came home. It was sometime in May we were there.

I remember we were packing our gear to go to Japan. There was a saw mill there making crates to pack our equipment in. Then, when they were packed, they sent them to the port and we troops followed on another ship. We got a delay in route of 30 days and that is when they bombed Japan and the war was over. They sent me to Ft. Leavenworth, then to Camp Swift, Texas, for about a month.

The engineer group was a Texas group; I don't know where the Division was going to be sent. From there they sent me to Camp Chaffee, Arkansas, which is now Ft. Smith. I was there the remainder of my time and it was the best camp I was ever in. We didn't have anything to do the first week or so but later they assigned us a job and it was just like having a job as a civilian.

It was super coming home. When I came home, a friend of mine was getting married and I was able to get back for his wedding. I rode the train on the Union Pacific line back home, it was a main track. Then south of us through Hutchinson was a main line for Sante Fe and when we went to California we took the southern route. When we came back from California it was night I remember we ended up in Ft. Benning, GA. It was so hot, terrible trip, no air conditioning in those days.

We really didn't get welcomed home because people who were here were afraid that we were going to take their jobs. We were looking for the jobs that the other guys were holding and there were lots of ill feelings. I looked for work and couldn't find it. My brother ran a mechanic shop. He was the one who couldn't get into the service because of a heart murmur so they discharged him. So I went to live with him and helped him out some.

I got out in May, 1946 and we got married July, 1947. I had a job then in the flour mill in Wilson. It was owned by a company in Denver called

Old Trial Flour. It was owned by the Denver Colorado Milling Company. I started about May 1947. My wife's parents lived on a farm south of Dorrance. They had a son who was older who had never gone into the service. One day he came in and told me I was moving out to the farm. I said OK, quit the mill job, they loaded our stuff to the farm and suddenly we were farmers.

They left a few cows out there for us. We just about starved to death. That was our start. My wife's parents helped us get started. The first wheat I planted was in the fall of 1948 and was to be harvested in the spring of 1949. The wheat was looking great but we got a late frost and it killed the wheat. It was pitiful.

They started a GI school where they trained us in different things. It was a good thing for us. They paid us $50 a month and we had to go to class each week and we had to keep track of what we had done. It was good money for us because things were pretty tough. We had a big family, ten children. It was tough but we survived.

In 1950 the dairy plants started up in Russell so I bought a bunch of cows and I went into the dairy business and that lasted until 1985. We milked a lot of cows in 35 years. My family grew up in the dairy business and I was able to provide for them. I got my own education and in 1985 our son was married with a family and he moved out and farm. In 1989, we built a house on a vacant lot next to my wife's sister and I built the house we live in now.

I don't know what my parents thought when I had to go back to Europe and fight Germany. My Dad was in the States during WWI, he was 31 when WWI ended. He didn't have to go but his brother in law did. He didn't say much about it. He had a niece who stayed over there and his other brother came before my Dad did.

When I came home from overseas I came to the house one day and Mother wasn't there and I was visiting with Dad. He kept asking about different things about when I was in Germany. I would tell him about different things that had happened. I didn't realize it until after I had left. My Mom came home and asked me what in the world did you tell Dad? Just what he asked me about, I said. He was sitting there crying. It had to be hard on them. I wasn't too far from where he was born, Minden. I wrote in a letter home that I was close to where I was but that information didn't make it home.

I made Sgt. After I got back to the States. I got PFC very quickly because I was an interpreter. Guys would ask me how I got rank and I said it is just in what you know. Back then I could speak pretty good German.

My grandmother didn't speak any English so everything was German at home. My folks always talked German and in fact, my oldest brother couldn't speak a word of English when he started school.

We went back to Europe on tour in 1981. Our trip ended up in Holland and we had a guide who was fluent in all languages. He helped us get tickets to Minden, Germany. A woman asked us where we were going and I told her. Oh no, she said, you want to get off at Bunde and it looked like it was the wrong town. We had lunch and asked the proprietor what we were looking for. He brought out a map that showed where we needed to go. He gave us the map and we walked to a grocery store and they called a taxi.

We found the school where my Dad, his brothers and sister and her daughter went to school. We went back out and our taxi was waiting for us. We caught a train and headed back.

My son in law's parent's came from Germany and we went right through the town they were from. It was a town named Prim and just north of that is a town Schleiden. His uncle came from Feddesfate up north more. I didn't realize it but Dad knew these people when he lived in Germany. He was a young man when he came here. He told his mother he wanted to go to American. She told him here in Germany you walk behind the plow, in America you pull the plow.

The thing I remember most about the Battle of the Bulge was that they sent us over there without any winter clothes. We had light field jackets, no boots and it was just miserable. We lost so many men to frostbite but everyone thought the war was over when they made that big push from Normandy. We swept across France and everyone thought it was going to be a piece of cake. And it might have been except the weather changed. The Germans knew the weather would change. It got foggy and drizzly and we woke up one morning and there they were.

Lenhardt was awarded the ETO Ribbon with 3 Battle Stars, American Defense Service Ribbon, and Good Conduct Medal.

Troops from the 112th Infantry enjoy chow while moving toward La Roche, Belgium. *National Archives*

German leaflet urging American troops to desert during the Battle of the Bulge. *Author photo*

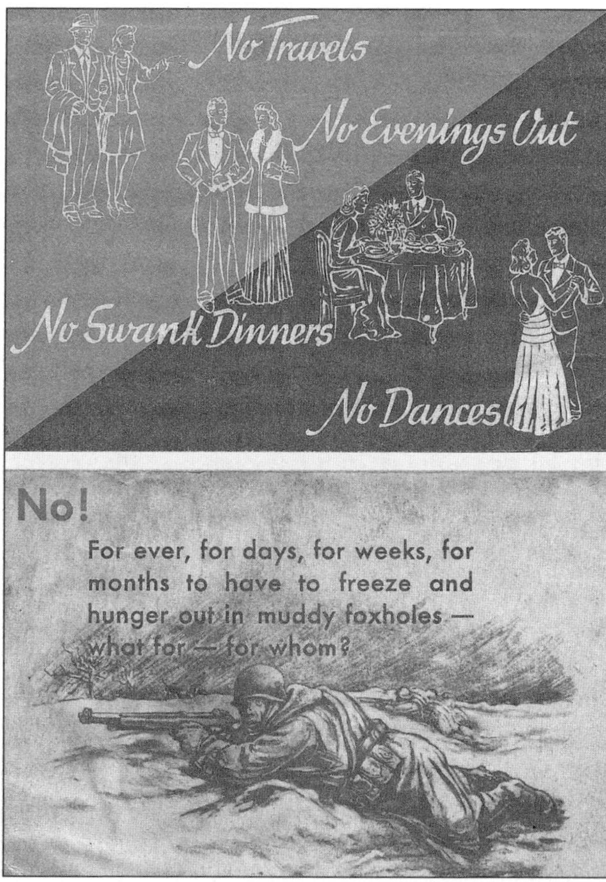

Dick in April 1945.

Dick in 2000.

Place of Birth:
Ellenwood, Kansas
Active Duty Date:
August 1943
Unit:
D Company,
120th Infantry,
30th Infantry Division
Division Location:
Malmedy, BE
Arrival ETO:
July 14, 1944
Rank:
PFC,
retired as Colonel

DICK JEPSEN

The 30th Infantry Division was federalized in October 1942 and was nick named "Old Hickory" because it had the flavor of the component units, the National Guard units from North Carolina, South Carolina and Tennessee. The Division landed on Omaha Beach on June 15, 1944, but some components landed behind the 29th Division when they were decimated in the initial assault. They were immediately committed to combat facing the formidable Panzer Lehr Division. The Division was designated the lead Division in the break through at St. Lo and the battle of Mortain and the Falaise Gap. They were known as the Workhorse of the Western Front and suffered more casualties than any other Army Division. The First Marine Division lost more in the Pacific. They held the area around Malmedy, Belgium during the Bulge. They crossed the Rhine on March 24, 1945, and proceeded to the Elbe River where they made contact with the Russians on May 4 at Grunewald.

Here is his story. . . .

My mother taught school at a country school south of Ellenwood. I was born on June 29, 1925, on the farm in Lincoln County, Kansas, and we moved from the farm to Hays, Kansas, where she went to college. My Dad had a mill, he actually had two, one on each side of the street. When I was seven or eight we moved to Nashville and I grew up in a beer hall. Seems like some of the people were always offering to give me a nickel to go out and pick a fight with somebody.

We moved a lot so I have to say that I grew up in central Kansas. I graduated from Raymond Rural High School in 1942 at 16 and could have volunteered for the service at 17 so I tried all the different recruiters in Hutchison. I tried to go into the air force but I had poor vision and they would not let me in. One guy told me to just wait until I was 18 so I could go into the army. So I worked on the farm and waited until I was 18. Two months after I turned 18 I went into the army. That was 1943.

My wife's name is Wanda Lee. She is from Ashgrove, Kansas. I met her when I was going to college and working on a ranch in the summertime. I was a cowboy on the ranch. I used to go to Lincoln County to rodeos and my friend Rex had a girlfriend. I told Rex that I didn't have a girlfriend and he told me I have got one for you. Her parents ran a grocery store so we went over to the grocery store and we started talking to her and acted like 18-year-old kids do, so I didn't say anything intelligent and then we went on back to the rodeo.

We met again at the rodeo in 1946 and one thing led to another. I did the rodeo as kind of a summer job and recreation. They would pick up a few kids like me I think more for entertainment than for competition.

We landed in Normandy in July 1945, the front lines were only five or six miles from the beaches. The 30th Division was up against the Panzer Lehr Division and we fought those bastards off and on all across Europe. The Panzer Lehr Division was still one of the elite German divisions at this point the war. It had highly trained troops and the latest in German war equipment. During the battle of Mortain and Falaise, we came very close to surrounding and pinching off the Germans and ending the war in France.

But the higher brass could get not together and they were afraid to pinch the gap closed because the Canadians or whoever might come in the other side and might start shooting at each other. I think that was the biggest reason why we were not able to complete that maneuver. We had the Germans surrounded. The men got out but the equipment didn't. We captured a lot of people but also a lot of equipment. Forty or fifty thousand Germans escaped the trap. We were sitting way back at Mortain for a week just to get the Germans off of us and then they took us another way to get further out east. I think we got in just at the end of this fight.

One of our generals wanted the French to get in on it and help pinch off the German escape, but the French wanted to go onto Paris and make a big splash. So the French army went on to Paris and left us trying to surround the Germans and pinch off their escape.

The Allied air support for this battle was extremely deadly to us. All the air support there could have been better. Our Air Corps was supposed to make their bombing runs parallel to the front but for some reason they decided to make the runs perpendicular across the front and this led to a lot of American casualties. We were looking for the planes to be coming from the north that they came in the other direction. They dropped a couple planeloads of bombs and it was enough to cover up the road. The next group of bombers came in and saw where the previous planes had bombed and they dropped their bombs where the first group dropped theirs. That was where we were. They blew the hell out of our division and one regiment in particular really got blown to pieces. We did not lose anybody in our platoon or our section because we were down in our foxholes watching. However, we did lose a lot of equipment. The bombs were shrapnel type and they dug very big holes, and some of the concussion bombs would dig holes.

There were things that I saw that mystified me. I saw some dead black troops at the side of the road. I don't know why I didn't see the uniforms at first and I did not realize they were Germans. They were covered with grass or something. I saw the black faces and I was amazed that the Germans would've had black troops. Later I learned that that's what happened to bodies when they lay out in the sun. It was really a contradiction when I thought to myself that Hitler was breeding a super race of blonde haired, blue-eyed Germans. Why in the world did he have black troops?

I'm not sure when I realized the magnitude of the Battle of the Bulge. We were ready and expecting a two week rest. On December 17 our guys were put on trucks and we thought we were going to the rest area. We got lost and we drove around. I heard somebody had a radio on one of the jeeps and they had civilian bands. The rumor came around that the Germans had come through our lines. I was able to get a map and I could see where Malmedy was, and I could see we were heading that way where we heard the breakthrough was and I thought there must be something to this. That's all I heard. We were riding in jeeps and trucks. The second day we drove all night and slept in the trucks and we got to Malmedy the next day. On December 18th we were in position on the front along the border of Germany. We spent the whole Battle in the Malmedy area. Our position was on a hill east to southeast of Malmedy.

The Germans had a number of vehicles that had been disguised as American. I saw one tank that was marked as ours. Two days straight we heard buzz bombs coming over the mountain. These were the first we had seen and now we knew what they looked like.

On December 29th, we moved to new positions southwest of Weismes and around 0330 we were awakened by lots of noise, tanks, machine guns and artillery. Bullets and tracers were flying over our heads and after a while we could see Germans coming over the top of the hill. We had set in on the reverse slope and were looking up. It turned out that it was an attack by Germans dressed in our uniforms. We moved our positions some because we could see the Germans over on the other hill and we were not dug in well. They attacked later up the road to our left and missed us. All of these attacks were made up of Germans in our uniforms using American vehicles.

The Germans had gone through Malmedy in 1939 and 1940 and we retook it in 1944 but there was not much in the way of war damage. On the December 23rd Malmedy was bombed by one of our bombers. I never understood why they would bomb an abandoned village that was well marked. It seemed senseless. Our 3rd Battalion and a lot of Belgian civilians were killed or wounded; the Battalion lost over 100 men.

We moved to set up a road block once and that night a German recon plane came over taking pictures. The next day we were blasted with heavy artillery.

The weather was terribly cold. Temperatures dropped at night to a minus 10 degrees and snow was over a foot deep. We spent a lot of time studying how to stay warm in such frigid weather. We were very tired all the time and fatigue was a constant problem because of trying to move in the snow. We finally got a rest after the Bulge was over and were issued winter clothes. Winter was almost over and I remember we only used them a few weeks.

The Germans were to the east of us and they were going south. They set up the command at Malmedy and we got there up along a hill line just enough to set up and I did not fire a shot. We were behind a slope. We waited for them to come up over the top of the hill for couple of hours. We finally saw them coming and in American uniforms. There was a whole band of them and Hitler had told them that if they ran into any serious resistance they should just bypass it and go on.

He said don't stop and we will send more troops to take care of them later. I'm sure that's what they did and they went on across through 4 miles of tough country. They had real good roads but some of the Germans took a straight shot right through the country. They weren't really hitting us; they were trying to bypass us.

After the Bulge, we made an early morning crossing of the Roer River. The night we got there we were making a night attack on another small town but the enemy was there and somebody didn't coordinate very well. We had a lot of air bursts and flares above us. They caught us out there in the open with tanks and artillery. We did not have a lot of casualties but boy you could see the rounds hitting up close. We did lose some people there however.

The Germans tried to cross the river and that's when they really made a serious attempt in getting through, but the guys blocked them off, and they went back across the river. They got pushed back across the river and at that point they were at the bridge. I fired at the tanks and everything else on the far side of the river and the tanks ran out of gasoline and they couldn't bring gasoline across the bridge because it was blown. We went over to help stop the Germans and some of them surrendered. Over three or four days it looked like they came back and forth across two or three times.

The German machine guns could shoot at a lot higher rate than ours could. I don't know what their rate was. Ours was supposed to shoot at about 400 to 450 rounds per minute but I think our guns really shot about 300 and 350. They would get dirty and slow down. They were not quite as fast when they needed cleaning. I think the German guns shot around 1000 rounds per minute and the sound of those rounds going by would make you pucker up. The more that you shoot the more dispersion you would get with the rounds because barrel would wear. We changed our barrels a couple of times but we should have changed them a whole lot more than we did but we just forgot about it.

When we crossed the Rhine River, we were positioned as a machine-gun post on the west side of the river and they told us the target was across the river in a certain church tower a mile or so out. We were supposed to aim on that. The artillery started about two o'clock in the morning and we were supposed to start shooting our gun about four in the morning. I had 25 boxes of machine-gun ammunition. The purpose was to identify our flank for the riflemen to go out into the woods. The tracers on one side or the other would let them know that was their boundary and they were not to cross. That was supposed to be their area and they could see where our artillery was hitting and that would help give them their direction. So in this case we shot 12,500 rounds and were supposed to change the barrel about every 3000 rounds.

We crossed the Rhine at Mehrum in Marine Alligators. When we were crossing the Rhine there was something going on all the time so we never thought of things like changing the barrel of the machine-gun. After we got across the Rhine we were about a half a mile up on the other side. We set up and could see a group of Germans in the distance. That night we took a lot of German artillery fire into our position, and we figured that group we saw running off was forward observers for the artillery.

When the Germans came through, they came through inexperienced divisions. An experienced division might have made a difference, I really don't know. I do know that our division, the 30th and the 2nd Divisions did stop them and didn't allow them to pass. I don't know what would've happened had more experienced divisions been there because the Germans slipped around behind the 106th Division.

During the battle of the bulge, when experienced divisions finally got in place the Germans were stopped. I think it was the upper staff, the planning staff that made some fairly large mistakes and panicked in a sense. They didn't seem to realize that we were fighting this war to win it. I don't know who made some of those decisions but they probably should've been sent back to officer school.

We were pulled back from the Duren area to rest and we never really thought of the end of the war or when it would be over never really crossed my mind. We didn't talk about when the war was ending at all at that point. I do remember when we broke out in Normandy we were all really happy and we were saying the worst can be over soon, we will be in Berlin in about two weeks. However, at that time somebody pointed out there was still another 500 miles to go.

It was all an adventure, every single day but it was a very unpleasant adventure. It was always uncomfortable. It was like living in the wild like an animal, always looking for protection against injury or death. Both were a constant threat. If a person could make it for a couple of weeks, they could learn enough battle savvy to protect themselves, learn what to look for. Rain and snow were miserable, no way to get warm or dry, just miserable.

I don't remember much about coming home on the Queen Mary. I do remember a couple of guys playing a poker game on the ship coming over but I never did play poker so I didn't participate. I was trying to learn poker after my basic training on a train from Oklahoma to California. Some guys on the train wanted to start playing some poker. This Chinese man sitting next to me said he didn't want to gamble in any event. He said that he would play some games with me and show me some card tricks. So he said I'll deal some cards. I didn't know much about poker but I drew a full house. He did it a couple or three more times and then went back to card tricks. He taught me a number of sleight of hand tricks. I tried to show my cousins when I came back but I couldn't remember all of them.

I think Monty was the best the British had. He'd gone through World War I and it seemed a lot of British boys were killed during the war and he was trying to keep that from happening again in this war. I think he may have been overly cautious on some things and based on some other people's writings. He didn't mind spending other people's troops but was stingy in spending his own. We went up through northern France about 125 miles with recon troops in front of us and took about a day, something like that. Then when we got up to the border the Brits were just to our side and we heard that Monty wanted to drop British paratroopers inside the German lines. They tried it two months later *(Operation Varsity.)*

I think General Bradley was a pretty sharp guy. I was not too crazy about him with the way that they bombed us at Normandy and maybe he needed to do more about that. He was a little overly cautious during the bombing because he agreed for the planes to go over our lines. Sometimes you don't know whether you should be cautious or aggressive. To me he was a really sharp guy. I know he made some minor mistakes but I don't know that he made any really bad mistakes. I didn't have very much on command structure. My participation in the war was a lot different than Bradley's but I think he did a good job. I had no personal contact with any of them.

When I was discharged, they had a whole room full of recruiters and before you got out you walked by each one of them and they would try to recruit you. They told me that they would make me a sergeant if I stayed on active duty and I said I don't think so. I wanted to go home. I decided to talk to one of the recruiters because he was with a reserve division and signed up for the reserves.

While going to Kansas State, I was in ROTC so that when I graduated I would get a commission. Plus I also needed the $20 a week to live on. I joined the Guard in 1956. When I joined, I told the sergeant what had happened and they looked into it but could not find any record of it. I know I was in that division because I signed the papers and have them. They just couldn't find them. When I graduated college I was a Second Lieutenant and I joined the company in Manhattan as one of the company officers. In 1963, I joined a reserve unit here and I finally retired as a colonel.

What did I learn? War is terrible but our way of life must be defended. Being enslaved and losing your freedom is much worse; the French, Belgians, the Dutch and I think even the Germans learned that.

I have four children, two boys and two girls. The two boys are engineers. One of the girls is a nurse the other a CPA. I have nine grandchildren and several great-grandchildren. I graduated from Kansas State College in 1950, got a Masters Degree from Kansas State University in 1962 and a Doctorate from North Carolina State University in 1974.

Dick was awarded three Bronze Stars, the Combat Infantryman badge, a Presidential Unit Citation, Meritorious Medal, two Belgian Fourrageres, French Croix de Guerre and the European Theater Ribbon with 5 battle stars. He was also awarded the Good Conduct Medal, American Theater Ribbon, Army of Occupation Ribbon, Victory Medal, National Defense Service Medal, and Berlin Crisis Call up Ribbon.

(Editor's note: I knew Dick Jepsen for only six short months and felt an immediate bond that seemed to transcend the differences in years. I remember sitting across the room at the American Legion in Manhattan for the 65th anniversary of VE Day. When the speaker mentioned the bond between veterans Dick turned toward me and smiled and nodded his head, seeming to salute another veteran from another time. He embodied what was great and daring in his generation, tough but understanding and ready to fight for what they believed was right. He passed away on August 26, 2010, a soldier to the very end.)

Bob Knight in Nancy, France, waiting to come home.

Bob Knight, 2010

Place of Birth:
Anthony, Kansas
Active Duty Date:
1943
Unit:
2 Platoon, I Company,
334th Infantry,
84th Division
Division Location:
Marche, BE
Arrival ETO:
November 16, 1944
Rank:
Staff Sergeant

ROBERT KNIGHT

The 84th Division landed in France in September 1944, and the "Rail-splitters" were sent immediately to the Geilenkirchen, Germany sector of the Siegfried Line where they knocked out 112 pillboxes and bunkers in Hitler's West Wall. During the Bulge, the Division was relocated to Marche, Belgium to help stem the German assault. While fighting across Germany, they liberated the concentration camps at Hannover-Ahlem and Salzwedel. By the end of the War, the Division made contact with the Russians at the Elbe River.

Here is his story. . . .

I made it a point to try to forget a lot of the War and it works pretty well. When I first got home, any loud noise and I was ready to get under the table. Talking to other GI's, you shared a common bond and we would talk. But it wasn't until our granddaughter, Allyson was in Junior High in about 1996 and she had to interview a WWII vet for a school project that I thought about it in any great depth. We were living in Colorado and she was in Wichita. She asked very good questions and I had to write the answers. It got me to thinking about it and it got to be very emotional answering the questions. Then she asked me to come to her class and show some of the things I brought home and share some stories.

> *(My son, Allyson's father, got interested in my Army experiences by reading the interview. He never had heard much about it. On our 50th wedding anniversary in 2003 he wanted to take*

> *Marion and me to Europe so he asked me where would you like to go. I remembered the areas I fought in so I got out my 84th Division history book and printed off some of the maps and we outlined some of the areas I would like to see. He arranged for us to stay in a house in Marche, Belgium which is where I started my Battle of the Bulge experience.*
>
> *That house was within half a mile from where one of my memorable foxholes was. We were able to travel to all of the areas where I was engaged. The people to this day are so appreciative; we liberated them twice. You cannot imagine how appreciative they are.)*

I was born on November 4, 1924. I grew up in Anthony, Kansas, southwest of Wichita about 9 miles from Oklahoma. The town had about 3,000 people. The area was made up of wheat farmers. We were town folks and I went to high school there.

I graduated from high school in 1942 and went to K-State for one year. I went to school the first semester and into the second semester and got my draft notice that I was supposed to report. It was so close to the end of the school year I applied for a deferment, and when they finally finished processing the deferment the school year was over. I got drafted. I was studying Civil Engineering so I was eligible for a program called Army Specialized Training Program, ASTP, for college guys.

I went to Camp Fannin, Texas, near San Antonio where I took basic training. Then they sent me to Drexel Institute of Technology to study, but after one semester there they decided they needed infantrymen more than they needed college students. We knew that going into the service was inevitable and we felt very fortunate that we were going to college. But deep down we didn't understand how it could keep going because they needed troops on the ground.

I was sent to the 84th Infantry Division at Camp Claiborne, LA, along with a whole lot more of ASTP guys. Here we were a bunch of college guys in with this old regular Army outfit. It was pretty hard to get accepted. We were not the regular guy that they were used to and that was a difficult situation at first.

The 84th trained as a unit at Claiborne, and went overseas as a unit all at once. We landed in England in September 1944. By that time our forces had gone across France and were stopped at the Siegfried Line.

We were in England for a while, and then took a small landing craft across the Channel and we still had to wade to shore. We got on a train and went across France to Holland. November 16 is when our company hit the front at the Siegfried Line near Aachen. We were fresh green troops and even though the old timers had been stopped we didn't know the difference and fought our way through it. We didn't know what to expect but we had practiced these things in Louisiana. We started to advance just like we were taught and all of the sudden the guy next to me fell over dead and I thought, boy that could have been me, that really could have been me. This is it.

You just keep going until finally you have to take some kind of cover and wait until dark. I remember on Thanksgiving Day, there was a hill code named Mahogany Hill that was overlooking a town we wanted to take, Prummern. We were supposed to take this commanding hill and there were pillboxes on the top. We started up the hill and I got pinned down in a beet field, they were sugar beets and I was pinned down in one of the furrows between the beets. I couldn't even raise my elbow without hearing a shot come by. I was there until dark and there were people all over the field. Some were wounded. Some were dead.

It was mostly machine gun fire and single shot rifle fire. I didn't know the whole situation about the town or what other outfits were doing I just knew we were supposed to take Mahogany Hill. I read later that another company was attacking the town from another direction.

> *(On our family trip I found the field and it still had sugar beets growing. When I was pinned down in the field I was trying to eat some of the sugar beets, didn't have anything else to eat. I don't think we took the little town but I think the other company took it.)*

We fought our way on to the Roer River but the Germans had blown up a dam somewhere and had flooded the river and it was too wide to cross. We were in a waiting mode. Then one day orders came that we were moving out, going back. When we got out of our holes some British took over our line. A British soldier took my foxhole that I had dug. We had been on the firing line for a while and thought we were going back for a rest.

They loaded us on trucks and we headed for Belgium. We were so happy that we were out of that for a while. Usually when we were on trucks they stopped every so often for a nature break but they just kept barreling along. I had the GI's probably from drinking water from ditches so I couldn't wait. I took off my helmet and I asked the guy next to me to take the liner out of it. I had on a field jacket, an overcoat, two pairs of pants and two pairs of long johns. I started getting those off. After I got it all off, I sat on my helmet going through this town about six inches above everyone else. I was sitting toward the front of the truck and if I dumped it over the side it would splash on everyone so I passed it to the end of the truck and they dumped it for me, and then passed my helmet back to me. I used some of my precious water to wash is out and passed it on again to be dumped. I was not too popular but I had no choice. We just were not stopping.

I think we were going through Brussels at the time and if we were, it was out of the way. But nobody knew where the Germans were so maybe we were playing it safe. Nobody knew how far the Germans had come into Belgium. We got to this town of Marche, Belgium. However, I heard later our kitchen truck didn't make it. The German's had taken the road.

They unloaded us at a school and took our whole company up to a classroom while our commanding officer went to a meeting. We thought we were coming back for a break and here we were all in one classroom thinking we are not going to get much of a break here. He came back and told us that the Germans had broken through; we don't know where they are or where they are coming from. We were told that some were dressed like GI's and they have some of our equipment so be careful. If an MP directs you somewhere make sure it is really an American MP. I never saw any Germans dressed up like us.

Our mission is to defend the town of Marche, so much for going back for a break. We will form a perimeter around the town and defend it. We don't know which direction they are coming from or anything. So we went out and dug into the frozen ground to form defenses. I was a Staff Sgt. I started at the first battle as a PFC but after a matter of days I was a Sgt and after a few more days I was a Staff Sgt. Squad Leader. Attrition was very high. We had all quickly learned the importance and value of every single man.

(I remembered that all very well so on our 2003 trip we went to Marche and I was telling this story to the people whose house we were staying in and the man said he knew where that school was.

House with 84th Division Patch, Marche, Belgium.

He said, "First I want you to meet someone who lives next door." The woman next door was 80 and during the War she was 17 or so and lived there. He took us to her house and she was so appreciative. It turned out that right before Christmas in 1944, her parents had some GI's in for dinner. She remembered that they had duck.

They must have been artillerymen. They told her that after the dinner they had to get back to their post because the Germans were coming.

Then our host took us to a guy's house that had the 84th Division patch painted on the side of his beautiful two story house. It was 15 to 20 feet in diameter, another indication that they had not forgotten. He was a 17 year old kid during the War and lived in the house. He knew the countryside well and made contact with the Americans, showing them routes they could go and places they could hide. When he found out I was an 84th guy he brought out the wine and we celebrated at 9:00 in the morning. He took us out to his back yard and we talked about the War and looked at maps. I told him I wanted to find the school house and also my foxhole around Verdenne. I remember our company headquarters was in that little village so after finishing the wine, we all went to the school yard.

I remembered the classroom quite well. The school didn't look like I remembered. The Principal came out and found out who I was and why I was there. He invited us in and he showed me what the school looked like in 1944. They had added on to it several times.

He called a local priest who was a history buff and we all went to the classroom. We stayed in the classroom for a while, looking at things the priest had brought and then we all headed out to find my foxhole. We drove from Marche to Verdenne and started to walk down the main street, which today is named "Noel 1944," People along the way would ask who I was and when they found out they invited us into their homes for wine. Some of them joined our group and walked with us. It was really something.)

I remember I had been chased out of this foxhole by German tanks that crept out of the woods in front of us, all painted white. The tanks were coming toward our lines. Our line was pretty thin, about 100 yards apart;

we had a lot of ground to cover. We had just gotten there but the whole unit had not gotten there yet.

When we got there we dug in. During the night, the Germans captured my platoon leader, a Second Lieutenant, and the runner with the radio who was supposed to keep in touch with people. They captured those two guys and the ones in the hole next to them. We followed the footprints in the snow and we could see that they went back into the woods behind us, so we knew that some of the Germans had gotten behind us.

That morning at daylight, the tanks started coming toward us. I was a squad leader and I was trying to figure out what to do. We weren't much resistance against the tanks. We had this guy in our platoon who was the BAR (Browning Automatic Rifle) man and a Native American about 30 years old and here we were 20 years old. We called him chief. He was from the old group and we looked up to him. All of the sudden he yelled, let's get out of here! Simultaneously, everyone got up and ran into the woods in hopes we could scatter out and get away. We had no anti tank weapons, only rifles.

When the tanks were attacking they were shooting at us with machine guns and I tried to change my pace and path to keep from getting hit. We got into the woods and found a little barn and we got in to hide. So we knew then for sure that the Germans were behind us and in front of us, and we also knew that sooner or later the Americans would counter attack and here we were in between them. We didn't want to get shot by our own people.

Several guys volunteered to go out and scout the situation. We never saw them again and don't know what ever happened to them.

The German tanks ran into the anti tank weapons behind us and were destroyed. Our forces, which were in reserve, regained the ground back to the original line. I went back to my foxhole and someone had been in it because some of the things I left were gone. All this happened on Christmas Eve.

We didn't have close air support, but suddenly the skies cleared and huge groups of B-17's came over. We all cheered their arrival, you know. The Germans never did get into Marche. The counter attack came from the rest of our battalion which was in reserve. They were waiting to see when and where the Germans penetrated because we didn't know where the Germans were. They had the armor support that we didn't have.

We didn't know the big picture at all and didn't know what was going on. We did not know about Bastogne. We thought we were going to the rear for a break. You just didn't know anything about the big picture.

You only knew you had to take that little town or take that hill or road junction.

Whoever was coordinating the whole thing had a difficult problem because communication was so bad. Everywhere you would go there would be all these wires nailed to trees, just reams of wires from our guys trying to maintain contact. And there were little signs that would say 3rd Battalion, 334th or I company. Cowhide was the 334th, blue 3rd Battalion, Item was I company so there would be little signs "Cowhide-Blue-Item" with a little arrow pointing toward our route. They were trying to keep people where they were supposed to be.

I remember the P-47's and P-51's and occasionally a P-38. We really liked the P-47's because those guys could dive in and loosen them up to the point where we could take them. We felt like the tide of the battle changed when the weather cleared and air support came. We knew that they were stopped right there where we were but we didn't know what was going on in other places.

After a few days we went on the counterattack and moved to a different area within 5 miles or so of where we were. We were attached to the First Army and Patton was the Third Army and we knew he was coming from the south and we were coming from the north to pinch off the Bulge. We knew that much, so there was a group of towns we were supposed to advance through and they were coming from the other way. We were starting out at the perimeter and trying to knife in to cut them off. We were taking all those little towns in the snow and cold, oh man!

We took the town of LaRoche. *(On our 2003 trip we went to a Battle of the Bulge Museum at LaRoche. They had examples of rations, bazookas and all manner of Army gear from WWII. It was interesting to me.)* We took the town of Beffe and advanced through that area. It was very hard. We were outrunning our supplies and it was so cold. When the kitchen truck did catch up with us they would always have clean socks. We usually had two pair of socks, one on and one in your jacket drying. And your long johns, I wore two pairs and that is four services, one on each side and then turned inside out. It was a forested area with pine trees and at night you would try to get under the low hanging branches to get out of the snow and there was always a bed of pine needles.

I also remember we were able to stay in a lot of Belgium farm houses. Most would be a house with a barn and the yard for the animals, a little compound. Nobody was there because all of the civilians had evacuated. I don't know what they did about feeding their animals. I remember once

I went into a barn and slept under a cow because it was warm. I was hoping she didn't want to lie down.

Once we met up with Patton, our duty in the Bulge was over. We had the 2nd Armor attached to us. Patton also had infantry and armor and our outfit met up with them. Once that happened we got sent to where we were back at the Roer River in Germany.

There we started training to cross the river. We crossed it on February 14. I will never forget that date. While we were training we were in a holding pattern along the same line we had established along the Roer River. They made sand table models of where we were supposed to go. The boats would hold a full combat ready squad with all of our packs and equipment. On the table they showed each squad's boat and where it was going to be. There were to be string lines to our boats because they were going to cover the area with smoke so we would be invisible.

We went back to the Meuse River and practiced rowing across to get used to the current and you learned that you would end up downstream a ways.

We practiced all of that and on the morning of the attack we got up before light and I got my squad and we found our boat and we were carrying it down to the river. A shell either hit our boat or right next to it, it was so confusing, and I don't know what happened. Our squad got scattered so a few other guys and I got together, found another boat and we got across. We landed on the other side. Our objective was to take this little town of Baal. We got up on the road that was headed toward Baal and there were trees down on the road and general confusion. Someone discovered that these trees had been mined so we had to make sure we didn't go across the trees and trip the mines, we had to figure out other ways.

We got to Baal and there was mass confusion. I had already lost contact with my group and there was nothing left to do but go ahead. Here we were in this little town and not knowing who was there. We were in a little two story house and we heard voices and we didn't know if it was Germans or our men upstairs. By daylight our tanks started to come and we just started following them. This is my own personal experience; I don't know what other people saw.

We took a few more towns and formed a little beach head. We started the drive for the Rhine and our battalion ended up being the lead for the whole thing from the crossing to the Rhine for which we got the Presidential Unit citation.

(When we went in 2003, I remembered distinctly where I crossed the river. There was a bridge across the main street of this little town of Linnich. Over to the side was a school with an oval track and I knew my boat was between that bridge and that track. There was a brand new bridge but it was still obvious since it was main street so I said let's see if that track is still there and it was. So I had my two guide posts for where my boat was. It was all together different because we had to walk down the bank to the river but when I went across in the war, the river was higher and there was no bank.)

When we got to the Rhine I got wounded and got sent to the field hospital in Liege to remove some shrapnel, then on to a hospital in Paris. I was wounded on March 2nd. We had crossed the Roer at a little town of Linnich. Our drive to the Rhine started after the Roer crossing, and we took Krefeld on the way. We were mostly riding on a truck because they were on the run. Some of the time we would have to return fire but mostly we rode. We got to this little town of Moers near Duisburg on the Rhine in the middle of the night. I was walking behind an American tank and the Germans were shooting at the tank. I saw a flash and felt the concussion simultaneously. I was thrown through the air and into a ditch. My helmet was gone, my rifle was gone. I lay there for a minute and started to move everything and felt like I was alright. I picked up a rifle and found my helmet and put it on. I started to catch up with my outfit and I felt my leg get warm and I saw blood. My buddy said you better stop and wait for the medics which I did.

Before you had to go back to your unit from the hospital, you got a three day pass and I was on a three day pass in Paris when VE Day happened. If you ever saw a celebration, we had a celebration. I checked into a hotel and never saw that hotel again because I was out all the time. They turned the lights on in Paris. They turned on the fountains. It was really a celebration.

I rejoined my outfit from the hospital. By that time they had met the Russians at the Elbe River and I rejoined them there. The Russians took over that area and we were sent back to Mannheim, Germany. We were in the Army of Occupation. In the middle of the night we would knock on German doors to search for contraband to make sure that there wasn't any insurrection brewing. We weren't supposed to fraternize or socialize with the Germans.

My squad was in a German house we took over, we told folks that they would have to move out and we were moving in. The lady of the house wanted to know if it would be Okay to come back and feed the chickens every night and I said yes. She sent her daughter back every night and we got to know her. We probably weren't supposed to but you're bound to talk a little. We took care of her place. We were using their dishes and we were very careful. Some people were not.

From Mannheim our battalion was sent to Karlsruhe in the same capacity and somehow I got invited back to the people's house for dinner after they moved back in. They were on very scarce rations but they wanted to share it with me. They had ersatz (German for substitute) coffee made of some kind of grain, maybe rice. The daughter actually corresponded with me when I got back to the States. She was an engineering student also. They were grateful that we didn't trash their house.

All of the people I came in contact with would say, Ich nicht Nazi. Nobody was a Nazi. And that is probably true of most of the people.

We were in Eberbach just outside of Heidelberg along the Neckar River and we were training to go to Japan. The news came that they had dropped the atomic bomb, the Japanese had surrendered and the War was over. I remember jumping up and down on the mattress in the German house we were staying in. Here I am a 20 year old guy acting like a 9 year old, the war was over and I was still alive. We came back to the States on the point system. The days in battle, the number of battles you were in, whether you got wounded, all counted as points. They sent you back according to how many points you had and the most points went home first. I was in the middle of the pack.

I got sent back to Nancy, France to await my turn. Three of us squad leaders were selected to run what was called a Sports Palace in Nancy, France. There were basketball courts, boxing rings, ping pong, a dance floor, everything. We were to do the daily running of the Palace. So we three Staff Sergeants ran the place. The motor pool had an ambulance that we could run around with. We were with an Air Force unit so we had an Air Force band and had music all the time. When it was my turn to come home I was supposed to go to Frankfurt to get a train to Le Havre. I didn't have any way to get there so they gave me a jeep and a driver to take me there.

We sailed out of Le Havre to Camp Kilmer. We went by the Statue of Liberty when we came in. It was really something to have been in such a rugged place and then to see all of the civilization here. I then went to Camp Chaffee, Arkansas by train and discharged by the Air Force because

Bob at Rue Noel 1944 or "Christmas 1944 Street" in Marche. *Photo by Bob Knight*

I was attached to an Air Force for all that time in France. It was interesting because I was an infantryman.

But, that did give me the right to sign up for the Air Force Inactive Reserve. If there ever was another war, I did not want to be in the infantry again so I signed up in for the Air Force Inactive Reserve. I went to K-State, graduated in Architecture, came to Topeka and got a job with an architect here and was working there in 1950.

I went home from work one night and found an envelope in my mail box addressed to Robert Knight. It said by direction of the President you are to report to Camp Francis E. Warren, Wyoming for active duty on a certain date about a week out. I thought, that can't be but that's what is says. It listed an Air Force MOS. I asked my buddy who had been in the Air Force what it was and he said, oh my God it is a tail gunner in a B-29. He was joking.

I went to the local recruiting office and they confirmed what the letter said. They got me a train ticket to Cheyenne. I was already an architect. They sent me to Lackland Air Force Base in San Antonio as part of air installation, the group who takes care of the physical plant on the base. I was assigned to the job of remodeling the Commanding General's office. The project seemed to be going along well then one day I received orders to go to Korea to be a map maker.

I went to the Warrant Officer I had been working with on the remodel job and showed him my orders. He was the base personnel officer. I said "I've been overseas; it looks like there would be someone who hasn't had that opportunity." He said "I agree with you. We are going to build an air base in New York and we could use you there." So I was on new orders to go to Sampson Air Force Base near Geneva, New York.

During that year, I met Marion who was a kindergarten teacher. She and I got acquainted and eventually we got married. We have two boys and one girl.

I was a partner in our own architectural firm in Topeka. I retired in 1987 and moved within a month to Colorado. We lived there for 20 years. We moved back here three years ago.

There is a lot of detail that I can't remember but you made it a point to forget a lot of things and it works pretty well.

Bob was awarded the Purple Heart, Presidential Unit Citation, ETO Ribbon with 3 Stars, and the Combat Infantry Badge.

Virgil in WWII.

Virgil in 2010.

Place of Birth:
St. Joseph, Missouri
Active Duty Date:
May 7, 1944
Unit:
4st Plt., G Company,
2nd Battalion,
317th Infantry,
80th Division
Location:
Neiderfeulen,
Heiderscheid and
Borscheid,
Luxembourg
Arrival ETO:
September 28, 1944
Rank:
Sergeant, Mortars

VIRGIL MYERS

The 80th Infantry Division was activated in July 1942, and landed in France on August 5, 1944. They fought their way to the Luxembourg area and fought from there into Belgium. They liberated the Buchenwald Concentration Camp. During the war they experienced 277 straight days of combat and ended the war in Czechoslovakia. It has been confirmed that the 80th fired the last shot of the War.

Here is his story. . . .

A trip back to Luxembourg is really a trip to take because they treat you like royalty when you're over there. And they make you feel that way too.

I started with my unit on the 28th of September 1944. The 80th Division originated with the Virginia, West Virginia, Pennsylvania and New York reserve units. That's where most of the reservists came from. And then after they landed in France, the 15 July 1944, they went through the battles and began to bring in recruits or replacements. Then they began to have people from all over the United States.

After I went into the Army, I only had a 12-week training period. We landed on Omaha Beach and that was four months after D-Day. When we got to Omaha Beach, they would post your names on the bulletin board. Now when we got there, they told us, you watch the bulletin board and your name will appear sometime in the next two or three days. So they said if you miss it, you could be court-martialed. So you know in World War II, we were so naive we believed everything they said. I was in a replacement depot there at Omaha Beach for a day and a half and we looked at the bulletin board five or six times a day. At noon on the second day, my name came up along with 11 other guys. We were to get together and move out at 1:00 o'clock.

They loaded twelve of us on a truck, and took us to Pont-A-Mousson, France. That was northeast and it was about a six hour ride. When we unloaded out of the truck we were right below Pont-A-Mousson, France, Dieulaurd, France, and that's where we crossed the Moselle River at Trier. Trier is on the border of Germany and Luxembourg and Pont-A-Mousson, France, are about 50 miles west of the German border and it's on one of the navigable rivers in France.

When I got to my unit, Sergeant Smith asked if anyone knew anything about firing a 60mm mortar and Ken Mauer, who had taken basic training with me, said mortars are a lot better than being on the line because we can set up 150 to 200 yards back of the line. Well we volunteered and when he did, Sergeant Smith said Sergeant Fidenas is going to be real happy to see you because he lost all five of the one squad day before yesterday from an artillery burst. And so we weren't real sure that we'd made the right decision at the time.

But the day after that, our first objective was to take a work camp just on the east side of the Moselle River, and as we set up and got close to the work camp, they set up a defensive line and Ken and I started digging a foxhole about 150 yards back from the line. The lieutenant who was in charge of the rifle platoon came back and he said what are you fellas doing back here? We said, we're setting up our mortar and he said what are you setting it up back here for? He said how in the hell do you think that I'm going to be able to direct you onto the target with you clear back here? He said get that thing up there by me and we were then only about 20 yards back of the riflemen, so that's where we stayed all the rest of the time during combat close to the rifle platoon lieutenant that was directing us.

The only time that we ever used the mortar men in the defense and to man the line without the mortars was when we crossed the Rhine River. After we crossed the Rhine River and got into Casteel, a little village east side of the Rhine River, they said we had to clean up the houses man to man.

They told us to put our mortar in the Jeep trailer and go with the riflemen. That was the first time that we had ever done that.

We moved so much we didn't have that much of an opportunity to fit in as a defense. We would usually set up and then they would direct us to a target that they wanted us to hit, such as a house or maybe a clump of trees or something that they thought a machine gun mist was in. That is what they would try to get us to fire in. We got pretty darn good at selecting the distances too. No aiming stakes much, mostly by guess and by correction.

We had the correction wheels on that would move the tube, with a 60mm mortar, if you just moved it a little bit, you would adjust its fire 10 yards. We used the correction wheels that were on the legs of the 60mm mortar. You see in movies where they held the barrel and would fire, but that never happened much. We did that a couple times in the Bulge when we were firing so close to us that we had to get a high elevation and down where we were only firing 150 to maybe 200 yards in front of us. Then we would hold the barrel but other than that we didn't.

When we reached Luxembourg on the 18th of December, we were just northeast of Luxembourg City at Gonderage and they split our division and the different battalions went different directions. They sent the second battalion over to Stencil, which is about nine miles directly north of Luxembourg City. It was sleeting and the roads were so slick that the trucks couldn't maneuver on those crooked roads so they stopped them.

We all got out and we started walking north through the woods. We stayed all night about three or four miles north of Stencil in a barn and a house. We had190 guys got in that barn that night and we were laying like stove wood because we hadn't got into combat yet. Early the next morning early we started out and as we approached Ettelbruch, there was a line of German soldiers that was going down a road towards Bastogne or Arlon. We didn't know where they were headed but they were marching with their guns slung over their shoulder just as if we were not even in the country.

They didn't know that we were near there and we had four TDD tanks that were supporting our outfit. The 318th Regiment was on our right and they fired first knocking out a weapons carrier at the end of the column and one of our tanks knocked out a big weapons carrier at the front of the column. The German troops were stuck on this hill that goes up west of Ettelbruch, and the bank behind the road was real steep and up high. As the guys in the snow would try to climb, we were picking them off like ducks.

The firefight lasted about an hour and a quarter and they finally surrendered. We took 132 prisoners and that was our first introduction into combat. From that day, we set up a defensive line there because we thought maybe there was going to be more Germans coming down the road but there weren't any more.

The next morning they moved us over between Mertzig and Neiderfelden and that's where we set up a defensive line. Then on the morning of December 23rd and 24th, right north of Niederfelden, and this ridge today is known in Luxembourg as Bloody Nose Ridge because for two days we were trying to get up to Borscheid and Welschied. We got stuck out in the open country and the Germans pinned us down.

Every time anybody would move out of their foxhole, they would get creamed because the Germans had artillery behind Borscheid on a mountain and then on the north side of Welschied and Borscheid. They were across the valley and were firing over the valley onto an open area with mortars and with artillery pieces.

We suffered really heavy casualties on that and finally on the second afternoon, we pulled back down towards Neiderfelden and that was our Christmas introduction. We didn't take Borscheid until we came back and went around and took Heiderscheid, Tadler, and Ringel and then we came up the mountainside to the east from Ringel towards Borscheid. That's where Captain Mankowitz told a machine gun sergeant Adam Hiser, and me, the mortar Sergeant, to go up the fire lane and said he didn't need us until we got to the top of the hill. A fire lane is about 150 feet wide cut in the forest as a fire break in case the forest catches on fire.

We didn't know that there was an 88 dug in at the top of the hill. That morning Adam and I started out with 26 men and as we got about halfway or maybe two-thirds the way up the mountain, they fired on us. They had an 88 at the top of the hill. We had a lot of new replacements in our group and they started to run. You don't do that when you're fired on. You drop to the ground.

They ran to the edge of the woods and the Germans had booby trapped the trees with Bouncing Betties mines. These Bouncing Betties would jump in the air about 40 inches when they were set off and they would explode and just rain the area with shrapnel. Out of the 26 men, four of us got out that afternoon after about an hour and a half of firefighting and they finally knocked the 88 out with artillery and tank fire.

Only about an hour and a half, that was all, and so many of them had been riddled with this shrapnel that they couldn't walk. It was about eight below zero that morning and we had 15 inches of snow on the ground. We always felt that some of them actually froze to death from shock after they got hit.

We kept hollering at the new guys, don't run, don't run, don't run! But they did. It's just natural. I remember the first of October when I first got in combat that that was my first feeling was to run to safety. You think trees are safe but they weren't because they had booby trapped them.

They had no experience and who knows how they were trained but in actual combat, you would find that some of the things that you had in basic training were not true.

It was the same way with setting up our mortars. In basic training we always set them up 150, 200 yards back of the riflemen. In actual combat you had to be up because of communications. You had to be up where the lieutenant was and he had to be up with his men so you were 25 yards or maybe 50 feet back of where the riflemen were in order to hear what he wanted you to fire at.

They always told you when you went into combat, always remember that a rifle bullet will go straight and if you can get in a defilade place or behind a rock or behind a tree it's safe. I'm sure that that's what those young fellas thought, if I can get behind a tree, I'll be safe.

But they didn't take into consideration the booby traps and we had been in combat long enough to know that if they started firing at you in a certain spot that usually they had a booby-trap set up for you to run into and that's what they did that morning.

We were hoping that it would get dark so that maybe they would let us pull back but the first night they didn't let us pull back. The second night we got orders from division to pull back over the hill towards Neiderfelden. We were about two miles north of Niederfelden.

We were up about two miles so they pulled us back about a mile to get out of the artillery range because they were about three miles on the other side of the river. By three miles and then a mile we could get out of their direct artillery range because they would fire most accurately about four miles.

The 88's were feared because it fired in a direct line like a rifle. Their muzzle velocity was equal to an M-1 and our artillery shells fired in an arc. An 88 could fire direct four miles and hit a target like firing on a level lot. Until the Bulge, we didn't have an artillery piece that would come within a mile of what the 88 would do, like our 37s, 57s, 75s; they couldn't compete with an 88. During the Bulge we got some TDs (tank destroyers) with Long Tom 90s that would fire direct and knock the turret off a tiger tank.

You were basically hit before you even knew it was even fired. You could hear our artillery coming. You could hear it, but an 88, by the time the shell hit, you would hear them fire. Oh, they were mean. And they just backed up all across France; they just backed up a mile at a time to keep out of range of our artillery. That's why the Air Corp came in and were so effective was because they could get to the tank or an artillery piece where ours couldn't reach it.

How were the divisions aligned? The 80th Division was between Heiderscheid and Ettlebruck. That was our field of advance. The 26th Division was on our left going north the 4th Armored and 35th were right next to the 26th the 35th was advancing towards Bastogne and the 4th Armored was in support of the 35th at that time.

The 80th, the 26th and the 35th were going forward on the south salient. The Germans were coming down the valley, coming down the Wiltz River Valley. So that's why the First Army was pushing down south but Montgomery would never move unless he was sure he had more combat troops than the enemy. Patton was south and we moved up to where we actually met them. The American forces met west of where the 80th was at Hofalsese.

Pintsch was the last village that we went to in central Luxembourg and the 35th then came around us on the north. Most of the Battle actually took place east of Bastogne. They came up this highway to Bastogne, and it was a battle all right from the German border to Bastogne.

The 80th was continually facing new recruits who were always coming in and hitting us and we could not understand why we kept getting new German soldiers coming in all the time. Well there was nobody over here; they had all pulled back by that time.

At the Battle of the Bulge we caught 32 kids that were 15, 16 years old. We had a rifle firefight for an hour. We asked them, why didn't you give up? You knew you were overmanned. They said, we had four SS Troopers with us and they told us if we gave up they were going to kills us. The SS were the policemen of the German replacements right at the end of the Bulge fighting. They were there to keep the replacements fighting.

When the Germans started The Bulge, the Americans only had one company covering about five miles of the river on our side. They had been preparing for this since the middle of November and they didn't want anybody to know that they were over there or active. Because they were bringing up — they brought up 500,000 troops and hid them for two weeks before the Battle of the Bulge started.

We had two separate incidents of running into Germans who were dressed up like Americans just as we got to Neiderfelden. They were ahead of the Germans who were marching down the road and were trying to get us to turn south to come down this road rather than to come up the other way. They wanted us to get away from that, because they knew if we got up the road and blocked their road junction, it would stop their supplies from coming over to Bastogne and around. So they were trying to show us the wrong way down here and Captain Mankowitz, pulled one on them. The captain had him pull his pants down and he had on German underwear.

We realized the magnitude of this whole offensive when we were on Bloody Nose Ridge, we realized just how vulnerable that we really were because we were losing men and we weren't getting replacements. They wouldn't bring new replacements up while we were in a battle.

After this battle, our company was declared ineffective so they attached us to another company to make us a workable company until we could get replacements a week later. It happened to us twice, once here at Neiderfelden and north of Neiderfelden and then it happened again in January when we were trying to go into Borscheid.

The morning when our fighters began to show up, there wasn't a cloud in the sky, and the airplanes started coming over, coming over, hundreds, hundreds of them. They were flying and man, we were yelling. By 10:00 o'clock in the morning, the sky was covered with a white cloud so thick the sun wasn't shining through it. That went on for two days.

Crossing the Rhine in a boat. *National Archives, ww2-117*

The first time they came in was when we were at on a hill Borscheid. We had moved back and they were firing over the hill into a valley right along the river. I can still see a great big barn. A bomb went into that barn and the whole roof just came up like a mushroom. I can still see it.

When we crossed the Rhine River and we were going north of Wiesbaden off of the Rhine River was the first time we saw a jet plane. This jet came up the road that went up this valley between the hills that came off of the Rhine River. That jet came down and all of us hit the ground because we didn't know what it was. It just scared the hell out of us because it made so much noise. It was probably a thousand feet high but it seemed like it was just right over our heads. And that was the first time that we'd seen the jet plane.

The second time I saw a jet plane was when we went on north from Wiesbaden to Kassel. Our objective was to take the east part of the city so our battalion was clear out in the countryside and as we were walking up the road, all of the sudden vroom, a jet took off of the top of the hill. When we got to the top of the hill here was a big hill that had been leveled off and it had metal planking down with holes in it that made the runway. The jets were coming up out of that hill. They had the inside of that mountain dug out and they had an elevator that came up like on an aircraft carrier and when it hit the top it took off and it took off right across us. Again, we all hit the ground because we didn't hear it coming until it passed us.

> *(When my son was in the Service, he was stationed at that airport. Bobbi and I went over there to visit him and that was the first time I got to go down in. They had it all off limits to everybody but the company commander took me down and that elevator went down seven floors in the mountain. It was around Kassel.)*

We came from Kassel and came back south to Erfurt and then started east. We went through Eisenach, Gotha and Weimar. The 319th Regiment took the City of Weimar. They took the surrender of it because the Colonel in charge of the 319th sent a message into the mayor that if they would surrender the town we wouldn't bombard it. He gave him an hour and a half and an hour later the mayor and a lady came out on a bicycle and they surrendered the city of Weimar to the 80th Infantry Division. Our job was to set up a police station in the town just to keep control and that's where we ran onto Buchenwald Concentration Camp.

We saw these guys with striped suits and we went into town to investigate. Captain Mankowitz told Percy Smith and me take a Jeep driver and see if we could find out where in the hell these guys were coming from. There were not many of them but they were all over town and they were hunting things to eat. They weren't hurting anybody or anything. They were 6' tall and didn't weigh 80 pounds.

So we asked four different civilians if they knew where these people were from and no, they didn't know anything about them. We asked two kids that were about eight to 10 years old up by the railroad station, who these men were. They said way out in the country, way up on a hill.

We started out in that direction with the Jeep and it was just dirt roads at that time. We got out there and we drove about four miles and we thought those kids have conned us. But we kept driving up around this hill. We got up to the top and all of the sudden here is a whole 350 acres of barracks and wire fences.

We didn't know it at the time but we came up to the back gate of Buchenwald. It was the supply gate rather than the front gate. That was the part of the camp that was called the death camp. If people got sick and they thought they were going to die, they put them in that part of the camp called the death camp. It was just one story barracks that had canvas sides, they weren't even wooden sided barracks. So we saw these fellas and they were just lying against the fence. Percy could speak a little German and he asked what this camp was. One fellow said "Ine minnet" and he ran into the barracks and came out with a fellow that said "I'm from Lithuania. I've been here four years." He said, "I can speak a little English," so Percy asked what is this camp? He said it is a work camp. Percy asked if it is all Jewish and he said a few, but not all. Some are from Poland, Bulgaria, Romania, France, Netherlands, Belgium, Luxembourg, they came from everywhere.

Whenever the Germans went into a village, they took everybody between 18 years old to 45 years old and they sent them to these concentration camps.

We went back to Captain Dankowitz and told him what we found and said those people don't have anything to eat. He called Regiment and Regiment called Division and the next morning they had five truckloads of food and medicine. They set up a first-aid station and there was an evac hospital that was attached to the 80th. I don't remember what it was but I think it was 357th. Anyway they came in and set up an aid station also.

But my job with my squad was to gather up these guys, take them back out to the camp but they didn't want to go. They didn't want to go but it was for their own good. As we came back down to town in the Jeep, we would pick these guys up and take them back out to the camp. We picked up one guy coming out of a house eating a huge potato.

He was eating that potato and he got about halfway to the Jeep and he just fell over. We picked him up and took him out to the camp and told them what happened. Captain Bob said probably he was so hungry that he ate too much and this shock on his body might kill him but we never did know whether it did or not.

(I was invited back this year, April the 12th to the Weimar and the Buchenwald 65th anniversary and they asked me at the banquet if I would say a few words about how I came upon the camp and what my feelings were. I told the story about the people we saw when we came into town, and tried to find out from the town's people where they were from but nobody knew, but they did know.

The kids finally told us where they were. When I finished, a guy at the second table held his hand up and I said, "Yes, sir?" He stood up and said, "My name is David Kane." I was a 16-year old Jewish boy that was sent to the death camp because I had Typhus Fever and they didn't expect me to live for four days. I was standing within 50 feet of you three fellas when you drove up in the Jeep at the back gate. You saved my life."

I said, "No, the SSers had already left." He said, "No, they knew you were putting the pressure on the German army at that time and they didn't want to be caught here, for you to find them. They left and I figure you saved my life because the next day they sent me to a hospital in Gera. Four days later they sent me to England and I have been an American citizen since that time, I've been a Jewish rabbi for 45 years in California."

That's just unbelievable. And he came up and hugged me and he was crying. Well, I did too because the emotion was just so that it was and I have a picture of he and his wife and son with me.)

I got the Silver Star when we crossed the Rhine River. The Rhine River, 12 of us fellas and an engineer picked up a wooden boat with a five horse motor and went across the river. We got about halfway across and they started using 20mm antiaircraft guns as an antipersonnel weapon and they hit the front of the boat killing the guy in the front seat. Then they killed a guy in the back seat as they were racking the boat.

When those shells would hit the water, it sounded like a 20 foot 1X6 hitting the water to me. It was midnight when we crossed and when those tracer bullets hit the water, that green, you could see that bullet just zigzag down through the water at the side of the boat. The German tracers were a kind of a yellowish green.

When we finally landed on the east side of the Rhine there at Castell, there was a rock wall that goes for blocks and blocks along a railroad track. All of us except Lieutenant Lyle got into half a bomb crater that came down to the water line. He had just been made a 2nd Lieutenant and was one of the first that got to stay with his outfit. Before that, if you got a battlefield commission you were moved to another company. So they let him stay with the 80th Division.

As we got to the other side, we were held up in that damn bomb crater. There were only 10 of us left at that time and every time we'd start to get out they would fire at us and we couldn't get out.

Finally a German came up with a potato masher and he was drunker than a skunk. He staggered up and started to throw that potato masher in our hole and one of the guys shot him. He threw the potato masher over into the water and it splashed water back on us when it exploded.

By the next morning, the tanks across the river began firing across at the warehouses. They finally either knocked them out or they moved on.

When we started to get out of our hole, not quite a block away in the corner of a street was a machine gun nest. It was right in the corner of that place and every time we would start out, they would fire at us. Sergeant Ragsdale was a rifleman, and he was meaner than hell. He finally said if somebody will go with me around to the left, and somebody else will go around the right, and you fellas in the hole just hold your rifles up and fire at those bastards to hold them down, we can get out of their line of fire and get them.

And that's what we did. We had grenades and as we got up close, he and I both threw the grenades into the machine gun nest knocking it out. We all got out. We were to meet at the end of the river bridge and the Jeeps would be there. We got there and they were there like they were supposed to be. The Captain said you fellas in heavy weapons, just throw your weapons in the Jeep and you go house to house with the riflemen. That was the first time we did.

We got into town about a block or so and we came upon five barracks in a courtyard. They told us that the Germans had probably pulled out by that time. Fedinas told me to take the bottom floor and he took the next floor because these barracks were about 200-250 feet long, three stories tall.

The first story was half above ground and the rest of it was below ground. I went down the steps and I started back in the hallway and I saw some guy way back at the backend, stick his head out around the door. I hollered for him to halt and come out in German. I said American soldiers were all around the house. He didn't come out for a minute but finally he stuck his head out again. I said come out. I said nobody's going to hurt you. I said, come out. So he stuck a handkerchief around the corner and came out. Another guy came out, then another and another. They just kept coming out and they were German soldiers. I thought at first they were just civilians, you know, pilfering.

So I started backing down the hallway. I wasn't scared at that time. I got a whole line of guys coming out of the doorway and I said American soldiers were all around the house. Don't try anything. Nobody will get hurt. I could speak a little German but very damn little. By that time I was beginning to get a little nervous and a guy said where are you from, Sergeant? I said, who said that? He said, I did. He held his hand up. I said come here. He came over to me and he was a nice looking young man about 30 years old.

He said, I'm an American citizen. He said I worked at Ford Motor Company for nine years in Detroit and I came back home to see my parents. The goddamn SS got me and they put me in the army. If I tried to leave they would shoot my parents and grandparents.

I said, line these people up and march them out front. There are GIs outside and if they don't try anything no one gets hurt. He said they're ready to give up. As we were marching them down the hallway, he said, my company commander; my no good son of a bitch company commander is in there with some other officers in that office there. Fedinas came down the steps and he says what the hell do you have here Myers and I said I don't know. His company commander is in that office and by that time, some other GIs came down the steps. They took them outside and lined them up.

Fedinas and I went into the office and knocked the door open and said you're our prisoners. There was a Colonel, a Major, a Captain and a First Sergeant in there. The Colonel said, we will give up but you have an officer of my rank come in and we will give up. I said we don't have an officer of your rank around here and we can't and won't get one.

He said well then you go find one and then we'll give up. Fedinas had had enough and hit him in the gut with the end of his M-1 knocking him in the chair. He says you no good Nazi bastard, if you aren't up and out that door, you're another Nazi casualty. He was a Russian that came to the United States when he was 14 years old by his uncle. The Germans had killed his parents and his grandparents in Russia.

He hated the Germans and that Colonel knew that he meant what he said because he spoke in a Russian accent. He got up out of that chair and marched outside. I said there was a map on the wall, just about the size of that wall. What is that? And this fella said, that's the artillery placements around Mainz, Wiesbaden, Castell and Frankfurt. There are 22 of them up there.

As he went out, the colonel wanted to know, he said can I keep my weapon? Fedinas says yes, but if you touch it you're a dead son of a bitch. He knew it too but he wanted it to protect himself from his own men.

We went outside and two Captains came down. I told them what had happened and we went in the room with the two captains. When they saw the maps, he got the radio and said they would give the coordinates to the dive bombers and they can get all of them. He asked me what my name was and I told him.

That was March 28th and never heard a word until May 8th, my dad's birthday. At that time, I was First Sergeant of G Company. I opened the mail and here was a citation for the Silver Star and they presented it to me at the next Saturday's formation.

(We crossed the Rhine at Mainz. We walked down the steps on the west side of the river, right where the Hilton Hotel is today. There was a school house that was right up on the bank and we were standing in front of that school house. My wife and I took a river cruise from Vienna to Amsterdam and we stopped there overnight. When we got out, we were down where the wall was. I said to the tour guide, that's funny, I walked down some steps in World War II to get into a boat to cross the river and she said come with me. We walked up the river about two blocks on this walk and here were the steps where I walked down. I looked across the river and there was a white spot on that wall where they repaired the bomb crater where we crossed the river and got in.)

When I came home I was married and had a three year old daughter. In August I called my wife from Geneva, Switzerland, from a USO. It took me five hours to get a telephone call through from Switzerland to the United States. I had to reverse the charges on that. I could only talk three minutes. I told my wife, I think that we're going to get to come home around the first of the year. I said I don't know when it is but when I get to the States, I'll call you. When we landed in Camp Kilmer, we stayed there two days, and I called her from Camp Kilmer and told her I was on my way to St. Louis to Jefferson Barracks. I was familiar with a hotel in St. Louis so I told her what hotel to go to.

When I got there, I called the hotel and she wasn't there. But they said wait a minute. She was living in St. Joseph, Missouri and she tried to make a reservation but they had to put her into a hotel over in Manchester. That's where she was and that's where she thought I was going to be. That damn telephone call cost $27.

We lived in Topeka for 10 years two different times. I lived on 29th Street Terrace which is just a block west of the Topeka Country Club and Burlingame. Later we lived on College Street just north of Huntoon.

My wife's name was Emma but everybody called her Bobbie. We have a boy and a girl and my son served four years in Army intelligence and deciphered codes.

What do I remember most about the Battle of the Bulge? The thing I remember most about the Battle of the Bulge is the snow and the cold and the damn artillery shells hitting the ground and the shrapnel flying. The ground was frozen so hard that it would make a dent maybe eight inches deep. When the artillery shells would hit on the hillside, the shrapnel at night looked like firefights because the shrapnel would hit rocks and would spark. I can still see those sparks at night from those artillery and mortar shells going off.

Virgil was awarded the Silver Star, Bronze Star, Purple Heart, Combat Infantry Badge, Marie medal of France, French Legion of Honor, Medal of Honor Luxembourg, American Service Medal, and the European African Middle Eastern Service Medal.

Berlin, August 28, 1945.

February 2010.

Place of Birth:
Colorado
Active Duty Date:
April 1943
Unit:
Band, 17th
Airborne Division
Location:
Givet, Belgium
Arrival ETO:
December 23, 1944
Rank: PFC

CHARLES NEALE

The 17th Airborne Division was activated in April 1943, and arrived at the Battle of the Bulge on December 25, 1944, defending the area from Givet to Verdun. After participating in Operation Varsity, the Division pushed on into northern Germany at the end of the War.

Here is his story. . . .

I spent a year in Vietnam as a civilian employee. I was in charge of all of the entertainment in Vietnam. If you ever went to a USO show, I was in charge of those in 1966 and 1967. I lived and worked out of Saigon for a year. We flew in helicopters at treetop level to some of the outposts where we had shows.

Vietnam was totally different than our War. I used to entertain all of the dignitaries who came over to watch the War. We would watch bombs in the distance and anything else they wanted to see. I lived in a hotel in Saigon, the Saigon Hotel. I lived and worked in all of the places that later made headlines as the war came to an end. I worked with the Australians quite a bit along with other USO entertainers.

I remember the 1967 Bob Hope Show. It was 98 degrees but the show never missed a beat. He had Ann Margret, Jerry Colona, and others. I worked for the Army and was a liaison with the USO organization as a civilian employee.

I was born February 24, 1925. I grew up in Colorado and graduated from the University of Colorado after the War. I was born in Sterling and grew up in the Brush area. My Dad was a school principal and I went to the junior high when he was Principal, and I was known as "PK," Principal's Kid.

I entered the Army in April 1943, at Fort Logan, Colorado. It was what was called Voluntary Induction. I knew I was getting drafted and if you volunteered, you could get your pick of units. It turned out that it wasn't true, I ended up in the 82nd Airborne and the 17th Airborne, neither was my pick. I was in the band and played the Sousaphone.

I learned to play the Sousaphone in high school. First, I wanted to get into the Marine Corps Band but because of my poor eyesight I ended up in the Army Band.

My basic training was at Camp McKall, NC, where we had 13 weeks of basic. After basic, we went to Camp Forrest, Tennessee, and then overseas. I was assigned to the 17th Airborne Division at Camp McKall. It had just been activated in April 1943.

We went to England on the Queen Mary in the fall of 1944. Part of my training was to be jump qualified. It wasn't a requirement but we did it anyway. The division had its own jump school so it was easier to go to the qualification training. Our jump training was in England, and I did qualification jumps there.

While we were in England, I got into London quite a bit. It was 90 miles away and we would hop on a train and spend the weekend. During the air raids we had to go into the air raid shelters when they were bombed.

I remember we were stationed next to a German POW camp and there was fear that they would revolt when they heard about the D-Day invasion. We were all armed just in case but nothing happened.

We were flown in at night to an airport close to Reims, France, just before Christmas 1944, to fight in the Battle of the Bulge. We were just outside of Compiegne, and then moved later.

During the Battle of the Bulge I was in the 17th Airborne in the band. We were a separate unit. Our non-band job was to guard the General's headquarters at Epinal in the Champaign area. Our headquarters was in Sanvic during the battle, about 20 miles from Bastogne. The HQ moved around but was generally in a city and we guarded the HQ.

We traveled with the command post. We went to all of these little towns in France to urge them to contribute to the war effort. We would play a concert in each town. It lasted for about six months and we stayed in the town's buildings overnight.

We flew into Berlin and occupied it for six months. I remember we had a VE parade in Berlin. They had an English Band, a French Band, an American Band and a Russian Band all playing together. We played our own numbers then we played some together. We got into East Berlin several times surreptitiously by slipping in there at night and then going to night clubs. I remember flying over Berlin and seeing how destroyed it was. It was in shambles.

A lot of the clean up was done by the time we got to Berlin. The war ended in Europe in April, and we got there in August of 1945. We stayed there until the summer of 1946 and then we flew back and were discharged. During that time, we played a lot of concerts in German and French towns to help raise morale. We had big crowds at our concerts. My favorite town to play in was Reims. It was in the Champagne area and a very lovely city. We drank Champagne out of a GI cup, not very classy.

We all came back together as a Division and we had a victory parade down 5th Avenue. Lots of people came out to greet us home. I have never seen so many people in my life, there must have been 3 million people. Our parade went all the way down 5th Avenue in New York City and back all the way to 105th Street. I had my Sousaphone going all the way. I really like to play the Sousa marches.

We were all discharged then and I got on a train and went home. I had called my family and they knew I was on my way home to Colorado. I went to school at Colorado University and got my Masters at K-State. I taught high school music in Colorado then came here to Junction City to work for the Army. I worked for the Army for 30 years as a civilian.

I met my wife, Mary Anne playing together in the Symphony Orchestra. We both played violin in Colorado in 1946. It made playing in the Orchestra much more exciting. I played in local orchestras and jazz bars for a while but don't play anymore. We were married in 1950. We have one daughter who lives in Guatemala and they have one child.

It was strange to get home. I was so glad to get back and start living a normal life again. I lay out one semester before I returned to college. I really enjoyed my time in the service. I never felt threatened, it was a great adventure. It was an interesting time in our country's history.

I first realized the magnitude of the Ardennes Offensive when they moved us over to France. We knew things were rough. I remember how cold it was. It was a deep, penetrating cold that never went away.

Charles was awarded the ETO Ribbon with 3 Battle Stars, American Defense Service Ribbon, and Good Conduct Medal.

Don Nixon, 82nd.

Don on his 90th birthday 2008.

Place of Birth:
Manhattan, Kansas
Active Duty Date:
February 21, 1941
Unit:
Service Company, 505 PIR, 82nd Airborne Division
Division Location:
Viellsalm, BE
Arrival ETO:
January 3, 1945
Rank:
Staff Sergeant

DON NIXON

After participating in the invasion of Italy, the 82nd Airborne Division jumped into France on June 6, 1944. It participated in Operation Market Garden, the Battle of the Bulge and Operation Varsity. At the end of the War it was pushing toward the Elbe River. Interestingly the shoulder sleeve patch for the 82nd Airborne is AA, America's Army. It earned its name during WWI when it was staffed entirely by draftees from all of the then 48 states. They became nicknamed "All American Division.

Here is his story. . . .

I was born on a small farm in 1918 at Zeandale, Kansas. When I was ten months old, my father moved the family to a farm five miles from Stockdale. I attended school at Stockdale first grade through Junior in high school. Like most other farm kids growing up in the Depression, my brother and sister milked the cows, fed livestock and did farm chores. When I was 15, my father moved the family to a farm a few short miles south of Fostoria, Kansas, where I, finished my senior year.

My first paying job was in a box factory in Manhattan. In addition to giving me a few dollars of spending money, I was able to save enough to buy a pickup truck.

I was not surprised that I got drafted and I was in the first bunch that got drafted out of Geary County. We were sent to Little Rock almost a year before Pearl Harbor. I had an old 1935 Ford pickup and the old thing usually would kick right up and take off. The morning I was to report, I had to catch a train in Junction City at 4 AM. It was 2AM when Dad and I started for Junction City. That morning it started and just died, it wasn't going to start. So I took the fuel line off the carburetor, no gas, the line was frozen; it was really cold that day.

We didn't have a blow torch so I wrapped a gas-soaked rag around the gas line and lit it. I moved the rag up and down the gas line until I got gas and finally got rid of the fire. It is a wonder I didn't get blown up. I was able to pull a tube of ice from the pipe. The roads were nothing but ice but we made it on time. My Dad drove the truck back home and I guess he made it back OK

I was drafted in the first Geary County Draft, February 21,1941, and was shipped to Ft. Leavenworth, KS. They gave me a physical exam and aptitude tests for four days. From Leavenworth we were transferred to Camp Robinson, Arkansas for basic training.

When we got there they had just set up the buildings and fences but all the streets were mud. They dumped gravel in the streets from dump trucks and we had to rake it into the mud to try and make the streets passable. I guess it kept us out of trouble, I was also jumped qualified in addition to my basic infantry training. I was assigned to the 60th FAB.

Right after Pearl Harbor, the Army changed from the square divisions to the triangular divisions so we were assigned to the 35th Infantry Division. The 35th Division was made up of National Guard units from Kansas, Missouri, and Nebraska and they landed at Omaha Beach on July 6, 1944. I was not with the 35th when it landed in Europe because of an accident. I was hospitalized when they left and had to go to Europe later as a replacement.

I sailed from Camp Shanks, NY, to Glasgow, Scotland, on the "White Star," took a train to South Hampton, England, and then crossed to Le Havre, France. From there, they put us on trucks at Le Havre and it seemed like we rode forever. I went right out to the boonies when I first got there. I was assigned as the Motor Sergeant for the 505 PIR, 82 Airborne Division.

We had tents but we didn't use them much, we moved all the time and by the time they were put up we had to move. We moved in trucks, I was usually the last one in the convoy. There was lots of snow and it was very cold. It was a rough winter; it seemed like 2-3 feet of snow with drifts higher. Lots of snow but we never had any special winter gear. We had our regular overcoat and boots, very cold. I remember we ate C-Rations and K-Rations but we were lucky when we got C-Rations because they were better. K-Rations were just a bunch of junk in a cardboard box, all dry stuff; C-Rations were canned stuff. We laid the cans on the truck exhaust manifolds and when you arrived at your destination, they were hot. That's where we carried the C-Rations. They came in flat boxes with baling wire.

I carried a .45 automatic pistol and an M-1. I never got my thumb or finger caught in the M-1. Pretty lucky, a lot of people did. I had to stand guard at night and had to check to make sure that the guards were awake.

I had 65 trucks, from jeeps on up to weapons carriers that I had to keep running. Maintenance, we didn't do much unless they broke down, when we got to where we were going we got them patched up. We didn't have much trouble with them really. They just kept on going. We would just fix them unless they got hit. We lost two drivers and two 2 and a half ton trucks loaded with ammunition about a quarter mile apart. They both hit landmines.

By the time we got into Germany, our vehicles had almost all been replaced. Autobahns were really good highways. They were not as rough as I-70; they built some good roads over there. Hitler was responsible for getting autobahns built; he wanted to get troops moved fast.

The severe winter was rough on the vehicles. I remember when I got to the Bulge it was January 3, 1945, and the 82nd had consolidated their positions outside of Spa. I was very cold but we were able to keep the transportation running in spite of dead batteries, frozen gas lines and battle damaged vehicles.

We crossed the river on February 17th. The Roer was a pretty good sized river, it is one where we built a pontoon bridge, and just as we were ready to cross, here came one lonesome airplane and dropped an egg right on it that blew the bridge up. It was a jet. We had to start all over.

I remember General (James) Gavin, 82nd Division Commander. He was looking around at a whole bunch of us standing around and it was the first jet I ever saw, they had them before we did. And he said, 'all that and not a shot fired.' I actually saw the jet; it went right in front of us. I did not know they had jets, it was an odd airplane, no propellers.

We spent most of March performing security duty in and around Cologne. The Rhine is a big River; we crossed it around April 15 on a bridge and stayed in Cologne until the end of April. I went through Cologne, went through the city and could only see what you could see from the cab of a truck, we were not there very long but it was really devastated. Buildings were in shambles.

When we got into Germany, there were not any large towns between Cologne and Berlin but I do remember Bleckede. We saw a few USO shows on the way and I remember seeing a couple that had some of the popular women of the time. I thought Mickey Rooney was a dud and didn't think much of him.

In April we moved on over the Elbe and into Berlin. We were stationed in Berlin for several months as occupation troops. I saw quite a few Russians but didn't think too much of them. They didn't think too much of us either. They were a little bit better than the Germans because they weren't shooting at us but they would like to have been. Berlin was quite a place, we had a section, the English had a section, the French a section and the Russians had a big section. Every time you tried to go someplace a Russian was there trying to stop you. I wasn't surprised at what happened later, I thought that the Russians would become the big enemy, I think they wanted everything.

How did I find out the war had ended? On April 30th the 505th crossed the Elbe River at Bleckede I knew when we went across the Elbe the German Army was ready to give up. The Germans stopped shooting and we knew it was over. The German 21st Army surrendered to the 82nd Division on May 2, 1945.

Berlin looked like a war zone; most of the buildings downtown were flattened. But I did not to spend a lot of time in Berlin. When we were in Berlin I was picked to go on a plane to Nice, France but we didn't know what we were going to do or where we were going. We found out when we arrived that we were going to be hotel operators. There were 154 hotels there. It was where people went for recreation and I got to be an assistant hotel manager. I knew about as much about that as a hog knows about a wheel barrow. I was there for 6 weeks. My work was not hard and I didn't have a lot of it. We had to keep a lot of records, lots of them; it took me about 10 minutes every day to get my work done. A First Lieutenant was the hotel manager and he had four hotels. I had to go to all four to get records of how many people they had and vacancies. It didn't take me long. The beach was within walking distance and I went to the beach everyday when my work was done. There were lots of joints that we hit on the way.

What was it like coming home? I flew back to Berlin, got on a cotton picking train and went back into France where I started. I was there for about a month. I was loaded on a liberty ship at Marseilles and sailed right to U.S. When we got out a ways, the ship ahead of us, an old Liberty ship, had a large gap in the side and it kept getting bigger, so at the Azores we got it patched. It was like being in a banana crate. Once they said we made it one mile in 24 hours. It seemed to take forever. It was a large ship with over 2,000 GI's aboard. Coming back, made you want to stop boat riding.

We came into Camp Caffey, Arkansas, by train, 22 December 1945. There were guys who had cars, private vehicles and there were five of us heading for Topeka. We saw a driver and wanted to know what he would charge to take us to Topeka: $100 per car, $20 each. We told him to keep on driving. That is how we got to Topeka. Then I put my thumb up and the first guy coming along was my neighbor two doors down. He loaded me up and dropped me right at the door. I didn't call them, my sister picked me up. I was discharged and out. They gave us $300 to spend so we had money.

It wasn't hard finding a job when I got home. I got a job in a body shop for a while, and then went to California, then back to Kansas where I worked for the John Deere Dealer in Frankfort, Kansas. Then I lived in Topeka and returned back to Manhattan. I worked at Ft. Riley in civil service in aircraft maintenance and retired March 2, 1979.

I met my wife, Opal, at a dance at the American Legion in Manhattan in 1993. We were married in Vegas in 1997. I spend a lot of my time gardening now and spending time with our children and grandchildren and great grandchildren.

Don was awarded two Distinguished Unit Citations (Presidential Unit Citations) and three foreign decorations: the French Forragere, Netherlands Military Order of William, and Belgium Forragere.

Paris, France, 1945

Robert as a civilian.

Place of Birth:
St. Louis, Missouri
Active Duty Date:
March 9, 1943
Unit:
HQ Company,
328th Infantry,
26th Division
Location:
Eschdorf,
Luxembourg
Arrival ETO:
September 30, 1944
Rank:
PFC

ROBERT PEARSON

By Pamela (Pearson) Westmeyer

The 26th Infantry Division was called the Yankee Division because it was made up of units from New England National Guard units. The Division was shipped directly from the U.S. to France and landed on Utah Beach on September 7, 1944. In October, they were in defensive positions in the Slonnes-Moncourt sector. They entered Sarreguemines on December 3. In December, they were resting in Metz when they were called into the Battle of the Bulge moving north to Luxembourg. They captured Arsdorf on Christmas Day, regrouped and continued to the assault taking Grumelscheid crossing the Clerf River on January 24, 1945. On January 29, they were shifted to defensive positions in the Saarlautern area. Advancing through Germany, they moved into Austria and Czechoslovakia liberating the Gusen concentration camp as the war ended.

Here is his story. . . .

I was born September 14, 1923, in St. Louis, Missouri. I lost both my mother and my father by the time I was eleven years old. I lived with my brother and then with my sister. I considered them my legal guardians and did what they told me to do in most cases. I went to grade school at St. Theresa's grade school, a parochial school. I moved into my own place when I was about 16 years old. I remember my first job was delivering groceries for my brother's store. I also worked at the Curtis Wright airplane factory in St. Louis during high school and was working there when the War started.

A friend of mine tried to talk me into joining the Navy Air Corp as a pilot. Thank God I didn't or I probably never would have come home. I took the Navy pilots physical exam and failed the depth perception test when they checked my eyes. He said "Do you want to be in the Navy, Army or what?", and I said, "I don't care, put me in anything." So I was then drafted into the Army in March of 1943 out of St. Louis. They sent me for my Basic Training at Jefferson Barracks, Missouri and after I was finished with that, they put me in the infantry.

I was in Basic Training when I was nineteen and they assigned me to the Army Specialized Training Program (ASTP). I was at ASTP at Kansas State University in Manhattan, Kansas, for nine months as an officer cadet. I completed this program on March 4, 1944. Then they found out they needed cannon fodder so they put us in the infantry and they cancelled the ASTP program. After ASTP we trained by taking 30-mile hikes, walking over sand in California and shooting guns. Then they took us out of that division and set us up in this special recon. I went to Ft. Leonardwood, Missouri, to San Luis Obispo, California, then to Jefferson Barracks and ultimately to the east coast at Fort Meade, Maryland. During the fall of 1944 I was transferred on a ship to South Hampton, England, as a replacement draft.

I was broke when I left California to come to the East Coast for transport overseas. We boarded trains and rode across the country from San Louis Obispo to Ft. Meade, Maryland. Most of the way across country we played poker on the train.

We were only at Ft. Mead three or four days and right before we left, they got us out of bed and said, "Okay we want you to know, that as of 12:00 o'clock tonight, anybody missing is no longer absent you're AWOL and you could get court martialed for missing movement." I didn't hear that and I was out somewhere in Washington, DC. As I came back the girl that I was with got mad at me or something and she left me there on the highway to Ft. Meade, Maryland, at 3:00 or 4:00 o'clock in the morning with a chance of being AWOL the next day. I'll never forget that I walked into the camp, and just to be a joker, I laid down in my cot and took my blanket and pulled it over my head and then threw it up and said, "Well that's nice, I had a good night's sleep, let's get going."

I think we may have embarked at Boston on a boat with a name like Rosa Lee or something like that. It was one of the biggest ships but it didn't have arms for fighting on the water.

We started out alone and as I remember when we got about 3-400 miles from England, they had a Navy boat that came out to meet us to help keep the U-Boats at bay. We landed at South Hampton, stayed there a day or two, and then were moved to a camp somewhere in England. We could hardly keep track of where we were because you just went where they said.

In England, we were placed into a replacement depo. I was a replacement for the infantry. So when we moved from England to mainland Europe in France I was assigned to 328th Infantry, 26th Infantry Field Division. It was called the Yankee Division because it was from the Northeast and started out as the Massachusetts National Guard. In France, they split us up. There was Wilson and O'Bryan and Pearson. Wilson was a little older than the rest of us and I remember when they put out that movie Saving Private Ryan it reminded me of Wilson. This Wilson had three brothers that were killed. When they come up to get him, they just took them right out of combat.

I recall when I first entered combat. The Battle of the Bulge was during a pretty severe winter. I remember dirty, muddy fields, and ungodly weather. People were freezing in the muddy, wet, soggy ground. All I had were the socks that I put on in England. I had to wrap my feet because they were freezing. I know a guy who used to play a professional instrument of some kind and he couldn't play again because he had his fingers frozen.

As time went by before Wilson was transferred, he, O'Bryan, myself and another guy formed a team and we would scavenge any booze or trinkets we could find when we seized a town.

When we got there, we were supposed to have a seven-day or a 10-day leave. My brother-in-law was in France as well and was looking for me and asked every person in the 328th if they knew where I was. They said, "Well I don't know what to tell you but if you're going any further, you better get some different equipment because from here on its Germany." When we got there the next day, the Germans attacked and the Battle of the Bulge began. So they canceled our leave and pulled us right back to Luxembourg.

After the Battle of the Bulge, on one occasion we were shooting across and into some fields around Saarlautern and the Germans were shooting back at us with machine guns and rifles and everybody was looking for a place to hide. I went to a stone building and crawled beneath the lip of the building because I didn't want to get hit. Some kind of bullet or projectile came through the open area and buried itself into the wall tearing out some of the wall and it hit part of my eyeball. It didn't hurt that bad and I didn't mind it, so I let it go for awhile. Finally I went to an aid station and they put me in a hospital in France. It was a small town and it had German soldiers right in the town with us at the same time. I've had some work on my eye over the years but not that serious.

When the War ended, we were in Vlacova Brese, Czechoslovakia. We were transferred to troop camps in Linz, Austria. I remember being in Austria and we were staying in the barracks and just running around enjoying ourselves. About that time they left Germany and were pulling out. Some of the troops were going home and you were sent home based on the number of points you had earned (time in theatre, medals). When I received my Purple Heart I got 5 extra points and that allowed me to come home earlier.

When we left we went to the town of Marseilles on the southern coast of France. There they put us on a ship and took us to the United States, put us on a train and then took us home. They treated us real good. Boy, it was something else. We got home December of 1945.

I feel funny telling a story like this with the idea that so many things have happened and after all, 50 years is a long ways back.

I was worried, actually none of us were crazy about leaving Europe because the war was over and we liked it. We did not like the idea of getting on a boat and going to Japan. Most of us felt like we had done our share; let somebody else do the rest. But naturally in those days you were in for the duration. And of course that's when the atom bomb hit and it ended the War

When we hit the States, we each had our barracks bag and we disembarked in Boston, Massachusetts, and then to Camp Kilmer by New York. I went home on a train by myself to Jefferson Barracks, Missouri, because St. Louis was my residence.

The thing I liked most about the Army was getting out. I don't feel like I was a great soldier and I didn't feel I did enough but still I didn't want to volunteer and do something stupid. The thing I really did not like about the Army was that it seemed that the wrong people got credit for things and it was really hard for anyone else to get credit for things they did. In addition to that, the Army was keeping me from doing the things I wanted to do. But at the same time, I thought it was a valuable experience and I would never have gone to college if it wasn't for the GI Bill. I had a lot of friends while I was in the Service but I think the fact that several of them were from the New England area made it hard to stay in touch.

In St. Louis, I stayed with my brother and sister-in-law and I didn't know what I was going to do. I had some college credits from when the

Army sent me to college at Kansas State University as part of the ASTP program. I ran into some old fraternity brothers and I decided this was the place I wanted to stay.

I graduated with a Bachelor of Science in Business Administration and Accounting and then about a year or 18 months later I got a Masters in Business and Economics.

After college, I went to St. Louis and I wasn't successful and I wasn't happy. So I went back to Kansas State and worked at the student union. Soon, I took over the student union as manager.

After the Army, I owned my own business which brought me a lot of joy and satisfaction. My businesses evolved, but usually centered on being a butcher and a salesman.

I met my wife at a fashion show at Kansas State University. We have been married over 50 years.

I am the luckiest guy in the world. My three kids have done great. One is in real estate and her daughter has already graduated from college and my son has a boy who is graduating from college and my other daughter has her Doctorate from Washburn University.

I didn't do anything special in the War, really, I just did my job. I have three battle stars, one for the Battle of the Bulge, the Rhineland, which are the Rhine River and Remagen Bridge and Central Europe which took you from the Rhine almost to the Elbe River. I also have a Purple Heart. The medal that means the most to me is the Combat Infantry Badge. People have asked me about my medals and as far as I'm concerned, I'm not real crazy about opening medals and pinning them on my chest. It is so hard to say after all these years what I felt in combat, but I think what it all amounts to is the fact that it was a job you had to do and you went in and did it.

Robert was awarded the Purple Heart, Combat Infantry Badge, ETO Ribbon with 3 Battle Stars, American Defense Service Ribbon, Good Conduct Medal, and the WWII Victory Medal.

Footnote.com

Paratroopers invade Southern France at H-4, D-Day, August 15, 1944 to start Operation Dragoon.

Frank, WWII

Frank at the WII Memorial in Washington, D.C.

Place of Birth:
Abilene, Kansas
Active Duty Date:
June 23, 1942
Unit:
Demolition Platoon, HOQ, 509th Parachute Battalion
Division Location:
St. Vith, BE
Arrival ETO:
September 14, 1942
Midnight jump into Avellino, Italy
Rank: Sergeant

FRANK RHODES

The elements of the 509th Parachute Battalion were activated on October 5, 1941. On November 8, 1942, the unit spearheaded Operation Torch, making two combat jumps the invasion of North Africa. While attached to the 82nd Airborne, the 509th jumped behind enemy lines in Operation Avalanche, the invasion of Italy and in December they were in the Mountains with Darby's Rangers.

In January 1944, they were and an amphibious landing at Anzio, Italy and in August, 1944 jumped into southern France during Operation Dragon. During the Battle of the Bulge, the Battalion, attached to the 82nd Airborne, held off two Panzer Grenadier Battalions at Sadzot, Belgium for nine days.

Then, after moving on the offensive to St. Vith, the battalion was depleted, leaving only fifty-five members of the original 700 members. By March 1945, the 509th Battalion was disbanded and its members moved to other units.

Here is his story. . . .

I was born in Abilene, Kansas, and we lived there for a couple of years. My Dad was in banking and we moved to Tampa, Kansas and lived there until I went on to junior college. I grew up in a little town. I think people who didn't live in a small town missed a lot.

I went on to work for Bank of America and moved to California. I had just been to work for a few months when the war broke out so I volunteered. I was 21 years old and the year was 1942. I volunteered for the paratroopers and I didn't know what I was getting into. I worked with a few guys from Nebraska who were from little towns just like me. We went to a movie one night, something about the paratroopers. We got out of the movie and I looked at them and I said, "I am going to join the paratroopers."

One said," No, don't do that!"

One said he was going to get into the Air Force and the other one said he was going to go into the Navy. So I got into the paratroopers, the other guy got into the Air Force but the last one flunked his exam. So that's how I heard about the paratroopers and how I ended up there and I was lucky to go into a great outfit.

I was in the 509th Paratroop Battalion, you probably have never heard of it. It was the first parachute battalion to go into combat for the United States. When we started we were the 2nd Battalion of the 503rd Parachute Infantry Regiment but they changed it to the 509th Battalion. It was formed on February 14, 1942, Valentine's Day. We were not part of a division. We were under General Mark Clark, the Allied Commander in Italy.

Nobody really knew we existed. They have heard of the 101st but not us. The 503rd was part of the invasion of North Africa but I was not in that invasion because I joined them right after that. We were in the invasion of Italy, and then fought on up to Casino.

I had my thirteen weeks of basic training at Camp Roberts, California. They sent me to Ft. Benning, Georgia, where I went to jump school, demolition school and communications school right there at the fort. When we were done, I couldn't get a furlough but they put together two, three day passes so I got to come home for six days. After I was home for five days I went back and they were getting the guys ready to go to Camp Kilmer, NJ, to ship out. There were 120 of us and we were not in a company yet. We were just new paratroopers.

It took us two weeks to get to North Africa. When we got there we were assigned to the 2nd BN 503rd. I was assigned to the demolitions platoon until after the second battle and then I was assigned to the machine gun platoon because they had lost so many machine gunners. There were so many hit so I was a machine gunner for the next battle. After that, I went back to the demolitions platoon and stayed there for the rest of the war.

After Africa, I made the combat jump into Italy. We jumped into Italy in the night. It was kind of hurry up deal. The Americans were about to get kicked off the beach at Salerno and part of the 82nd Airborne jumped behind the American lines to help them. I jumped into Avellino, Italy, only to find the DZ (Drop Zone) was occupied the night before by the 6th German Armored Panzer Division. We were about 20 miles behind the German lines where there was a little road junction. They thought we could stop the Germans from coming down the road junction. There were 52 of us left out of about 200 and we were behind the lines for ten days.

Actually, there were only five of us who got there alright and when we got to the town of Avellino there were Germans everywhere so we backed out. After that about 50 of us got together but we could never get into Avellino again. We would do some fighting at night and then go back into the mountains during the day. They were afraid to come after us. Seven of us got together and headed back to our lines together.

When we got back we were in the British lines and we didn't know the password because it changed every day and we had been back there for ten days. All I could remember was that the first day was "California grapefruit" and the second one was" long johns" and the third was "Notre Dame Quarterbacks" and that is all they gave us.

We hit the British lines late at night and they had been expecting us. Some other paratroops had come in earlier. They hollered for the password and it was "Buckingham Police" so we were lucky that we knew it. We really lucked out, they picked out something that everyone would have known and was current went we jumped in.

We got back from our first jump into Italy and they took us to a little town outside of Avellino. They staged us for an attack on Mt. Croco on November 11. We jumped off at 11 AM and we reached the top of the ridge. We took the Germans off the ridge by evening and we lost quit a few guys in that attack. After we got to the top I was talking to Captain Siegel. We found a cave where the Germans had started to fix supper on a charcoal grill in the cave and we ate their supper.

We stayed there for 30 days. They took me down because I had frozen feet. They hauled me to the hospital in Naples and after I got out I went back to Pichcanolli, a little town south of Naples. Then they took us to prepare for the Anzio landing. I was in the first boat that landed, could have been the second boat but we both landed at the same time. We got off of the boats at 4 AM in the morning on January 20 and headed for Rome, Italy. It was colder than the dickens's and we hit a sand bar going in and had to jump out into the water. After that I kept wet for the next three months.

Anzio was bad and we lost a lot of guys there. We had so many guys wounded. We heard one guy call for his mother. For two or three days we listened to him and finally we said we're going out there to get that guy.

So we got the ambulance with the Red Cross flag on the front and another guy and I were riding on the running board. A Medic with a Red Cross on his helmet was with us. There was a road that ran across the front lines between the Germans and us. My friend was waving the Red Cross flag so hard, back and forth, that he broke the staff off of it. When we got up there the Germans saw us. They all got out of their foxholes and started picking up their dead too. I don't know how many we picked up, 15 or so, we had the ambulance filled with dead bodies. We finally came upon this guy who had been calling for his mother. We put him on top of them and I don't know if he lived through the night or not.

I got yellow jaundice and was in the hospital for 30 days, everyone had to stay for 30 days. I was sick for two weeks, everyone was. When I got out I went to Rome the day that Rome fell to us and we stayed there for a few weeks.

From there we jumped into Southern France. We were there for 90 days and I got my first Purple Heart there. We lost quite a few guys there in Southern France. They called it the Champagne Campaign. People said there wasn't much fighting but that depended upon where you were. If you were the one not getting shot at it wasn't much of a problem but if you were the one getting shot at it was a problem. It was bad. There were a lot of guys killed in Southern France.

I made two combat jumps, one in Italy and one in Southern France. I think I am the only person in the whole US Army who made two combat jumps and landed exactly where I was supposed to land. I was just lucky.

When we got over the beachhead at Salerno the Germans started firing at us. In those days they had no radar, except on the lead plane and no lights. So when the Germans started firing on us the planes started to scatter so they told us to jump at midnight. We jumped at midnight; we landed right where we were supposed to land. I was very lucky. It was very uncommon.

Operation Dragon started on August 15, 1944. It was the invasion of southern France. On the jump into southern France, we jumped closer to the beach in France. I was seven miles inland, it was 4AM and it was foggy. It was so foggy, and when you are up in the air you can hear everything. You cannot believe how much noise we heard from the ground. I could hear the guys before me going through what sounded like evergreen trees, I thought it was water and it scared me but there wasn't much I could do about it. I landed right where I was supposed to land. I landed about 50 miles west of Nice.

When I got the first Purple Heart, it was on a patrol that wasn't even in my area. But they wanted a good patrol leader so the Colonel, who was a good friend and good friend of everybody, called me and said, "I want you to take this patrol up there." And just before we left, we had a new Lieutenant come into the unit so he said, "I am going to put him in charge of this instead of you. He needs the experience."

We went up to the France-Italian border and we met three French soldiers. With them there were sixteen of us. We met a First Lieutenant and a guy from the Aberdeen Proving Ground. That was the place where the Germans had first run into the Maginot Line. They took the new Lieutenant and a photographer to take some pictures because they had run into this fortification that was built into the side of a mountain, 4 or 5 stories high and wide. We got there and started up the road, but sixteen people are too many for a patrol. They had some new shells that they wanted to try out on the fortifications to see what they would do. They wanted the photographer and the other guy to go up with us.

They assured us that the Germans were gone and had been gone for a while. We got up about 100 yards from the fortifications when all hell broke loose. The Germans shot everything at us they had. Three of us got back out of the 16 that started up. I was wounded, not seriously. I was shielded from the fire by boulders. The shells were flying everywhere. You see in movies how a machine gun will kick up a line when it fires. I saw it go by the first time and it looked like a picture. I watched and it went by the second time and I watched and on the third time and I thought I had better get out of there. I jumped and dropped 20 or 30 feet into this mountain stream. I am not a swimmer but it wasn't very deep anyway.

I didn't know I was wounded. I dove down into the stream and pulled out my .45 because my rifle had been blown out of my hand and when I came up I started firing at the Germans. I knew they were coming after me. I was lucky because I was going downstream and as I went around a curve there were two friends of mine. One wasn't hurt at all but the other had his hand blown off. The uninjured friend was helping the other so I tried to help too and we took him back to the compound. The village was Eisola. My friend with his hand blown off said, "You are bleeding a little."

I didn't know it so I looked. The slug had disintegrated and I was hit with lead and brass from the slug. My eyes were OK but I was wounded in the head. That was my first Purple Heart.

A day or two later the Colonel told me that they had left the camera from the guy at the Aberdeen Proving Grounds. He wanted me to go up and find it so I picked Poindexter to go with me. We went back up there but I was a little shaky and I thought I saw the camera in the middle of the road. I thought it was a trap, maybe not but I didn't' want to go close to it. We didn't find the camera so we made it back to our lines.

From there they sent us in a train up through the Rhone River valley and a road ran right along next to the railroad tracks up near Reims. We passed a place where the American artillery caught the Germans in the open and there were hundreds and hundreds of vehicles, dead people and dead horses. You could walk on them, everything everywhere. We were supposed to jump across the Rhine but we were only there about 10 days when the Bulge broke out.

We wore the jump uniforms of the paratroopers all the time. I carried an M-1 all the time except at Anzio. It was so flat you did most of your walking and fighting at late at night. You could barely walk or see so I carried a Tommy gun, a Thompson submachine gun. They were heavy and they had two big 50 round magazines and those things were heavy. They had .45 slugs and they were pretty good scatter guns. I don't know what they weighed but we were so loaded down that you couldn't even get into the plane by yourself. The paratroopers today are so much smarter than we were.

As I said, I was so loaded I couldn't get into the plane by myself partly because I was in demolitions and I had 50 pounds of TNT in my pack. I had two bandoleers of ammunition, plus an ammunition belt loaded, and M-1, a pistol, a canteen, and two 17 pound anti-tank mines on my legs. Today they drop it to them by rope. I thought we were smart but we weren't. We carried the boxes of B-Rations, cardboard boxes of rations, usually a couple of those with me.

We were in France when The Battle of the Bulge started. It was so darn cold and I froze my feet the second time. I grew up in a little town and we had a dentist who had been in WWI and had been gassed. He was not in very good health so my Dad made it a point to tell me that if you ever go to the Army and anything happens to you, be sure and tell them about it. So when I got discharged I told them I had frozen my feet twice.

In France, we did see some Germans dressed up as Americans. I remember the day the Battle started, December 16th, we were in a little town and they woke us up, it was very early morning, about 4 AM and put us on weapon trucks. It was so cold; the wind was blowing the snow. They trucked us to a small town and we couldn't figure out where we were and then discovered that we were in the wrong place. The Germans had changed the signs around and we had gone to the wrong position. We stood around in the blowing snow for a while then they put us back on the trucks and they took us to the town of Aywaille.

We got to Aywaille and they dropped us off at a big stone bridge. If you ever saw a big stone bridge in any of the Battle of the Bulge movies, that is where they left us. We weren't supposed to let any Germans through along the road and over the bridge. We didn't know where they were coming from, there were roads coming from different directions. There were about 20 of us that they put there and the Germans had not yet arrived. We waited for two or three days but they never showed up. The Germans got stopped before they got that far.

They moved us on closer to more places and on to St. Vith. I remember the day the weather cleared and our planes began to fly. It started in the morning and planes flew constantly all day long in celebration. From St. Vith, we ran patrols every day. St. Vith was cold and there was a lot of fighting.

I guess I was too good at patrols or something, so they kept getting me to go on patrols. I remember one patrol that may have been on the last day I was in combat and in the last days of the Battle of the Bulge. We came up to a field, probably an eighty acre field, and the snow was a probably eighteen inches to two feet deep all over. We were in some woods, then there was this field, and if you looked east for probably a quarter of a mile, more woods. Then up to other side of the field, probably a half of a mile away, was more woods. We had, I think, five tanks with us, and they wouldn't go out in this field because they thought there were some German tanks across the field. They were probably right and it was probably a pretty good idea not to go out into the field.

The colonel called us and wanted a patrol go around and see what was over there. I was supposed to be running the patrol, and I think there were probably four or five us. We stayed in the woods on our side of the field, down to the corner, and went across the woods and then up the other side. We had to change point man in the patrol about every hundred yards because the snow was so deep that it just wore you out breaking a path. We were in pretty good shape in those days, too. But I happened to be in the point and we were not to engage the enemy or to fight. All we were to do was to see what was there and get the hell out of there.

Honest to God, I came up ten yards from this German tank. They had two tanks there together, but they were camouflaged and painted. The Germans were a lot smarter than we were. They camouflaged their tanks, and we didn't and I didn't see them. Of course, they were right in the middle of all kinds of woods, big trees, little trees, sunflowers, you know, just trash. And there were two guys sitting up in the turret of this tank, and they saw me just when I saw them.

They were scared and I was scared to death. I hollered to the other guys to get out of there. The guys in the tank slipped down in the tank. On the tanks, their machine guns are mounted in the front of the tank, so they couldn't fire at us with their machine guns. But they turned the turret around towards us and started firing at us.

They must have fired twenty shots at us out of that eighty-eight. After this was all over I decided the only reason we were alive is because, in the tank they lay out either their high explosive shells or their armor piercing shells depending upon what they think is coming toward them. They knew there were our tanks across the field and they had laid out armor piercing shells, waiting for our tanks to come up.

They must have had the armor piercing shells all laid out, so that's what they were firing at us, because that's what they had available. They would go clear through a tree but they wouldn't set the shell off. I mean, it wouldn't explode; it would just rip through things but would not explode. They had to hit one of us directly with the big shell, and it would have blown a hole six inches across if they'd have hit one of us.

I was probably more scared that day than any other. I knew the war was coming to an end anyway, and I didn't want to get killed then toward the end. That was probably my last day of combat.

This was when I got my second Purple Heart. A medic named Maddox, a good friend of mine, and I was being shot at and he said, "Hey, you've been hit." It was nothing really, so he saw blood and he said, "you got a Purple Heart," so he wrote me up. So that was the end of the war for me.

Our Battalion was still the 509th. I was on a patrol, three of them that day, and we got back to our battalion and they said that this was the last battle that the 509th would fight. The Army is transferring us to another outfit. The next morning I was assigned with three new guys, I didn't even know their names, to the 508th in the 82nd.

We got on the truck to go and one of my buddies, Lt. Shaw, gave me a Major's duffle bag and there were two quarts of scotch in it. So I pulled it out of his bag, put in my bag and went to the hospital with it. The day I went to the new outfit I went on sick call. They put me in an ambulance and took me to a hospital right outside of Paris.

I put the two bottles of booze in my pillow case in the hospital but the nurse found it and gave me the dickens. She took it from me. The day I was discharged after about two weeks and here came the nurse. She had a box with the two bottles of booze. The war was over for me this time.

I was assigned to a Machine Records Unit. On my records it said that I was a machine worker because I worked for the Bank of America. I was in a little town outside of Paris, St. Germaine. I had about three months there and reported to a major. He told me, "Rhodes, I don't have a thing for you to do. We have an inspection once each Monday morning. That is the only time I want to see you."

I could walk out the front door of the barracks to a subway and I could be in Paris in 10 minutes. I spent the last three months in Paris, almost every day. When I got ready to come home, I was the only paratrooper in our group when they came out with the point system. I was in this Machine Records Unit and I had more points than anyone. I had six battle stars, two Purple Hearts and a bunch of other junk so I was the first one to come home. However, I didn't have my service record so another guy got to go home, then another before I finally got my service record and I finally got to go home. They put me on a train in Paris for La Havre. When I got there I was all by myself, I reported in, there were hundreds of tents and I finally found a Sergeant.

"What are you doing here?"

"I am supposed to report here to go back to the States," I said.

"We just cleaned out a group to go to the States, they are all gone," the Sergeant told me. "You can put your bag in that tent and you will be the first one on the next ship. That might be a couple of weeks before we get enough guys to go."

I threw my bag into the tent and he came running back yelling, "Hey, get your bag, you are going."

It was typical Army. When one guy gets into trouble they kick the whole group out. When a guy got on the ship he dropped his bag and it blew up, had a concussion grenade in it I guess. When they kicked him off they kicked the whole group off. So they called me to take his place.

I went back to New York to Camp Kilmer where I had left. I landed on July 1, my Mother's birthday was July 4th. It was about 10 miles out of New York and I could see the sky line there from the Camp. I called my parents from Camp Kilmer. At the time we had a central operator and I told the operator that I wanted number 6, our number. The phone rang and rang and finally Delia, a gal I had known my whole life said, "Is this Junior?"

I said, "Yes."

"I can't get your folks but if you will hang on a minute I will get a hold of them."

I waited for a couple of minutes. I found out that Delia lived in the telephone office. She had slipped on her housecoat, got in her car, drove to my house and rang the door bell until she got my folks awake and said, "Junior is on the phone." You don't get that kind of service today do you?

I got in after midnight and the next day they woke us up early and fed us so much you cannot believe. They had a bunch of trains there and about 10 AM I got on a train that was headed to California and went through Kansas City.

The troop train was packed. We got into Kansas City on the 2nd, they really moved. They sent me to Leavenworth. The next morning I got up and they were set up to hand out hundreds of discharges but there were only four of us so we got our papers in no time.

Someone asked if anyone wanted to go to Kansas City and I said I would like a ride. He dropped me off at Union Station and I called my parents to tell them that I was in Kansas City and would be home that evening. I caught a train to Abilene on July 3. That evening, Mom, Dad, my brother, sister and families met me at the train station. It was the best birthday present my Mom ever had.

From Mrs. Rhodes: "We were high school sweethearts. Frank was a year ahead of me. I went to a country school and my Dad was a farmer. When I got to high school Frank and I started dating. He left for college and I had one more year left. I went to Wichita for nurses training. I went to Mercy Hospital in Denver and worked for a short time. There I enlisted in the Army Nurse Corps. I was only in fourteen months and the war was over. We stayed in touch and got married in 1946. We have three daughters, two in Topeka and one in Troy, Kansas. All three are married and doing great."

So many things happened to us at the Battle of the Bulge none stick out in my mind more than the others. I was on patrol numerous times and that wasn't the place to be.

Frank was awarded the Silver Star, 5 Bronze Stars, 2 Purple Hearts, and 2 Presidential Unit Citations.

Don in WWII.

Don in 2010.

Place of Birth:
St. Joseph, Missouri
Active Duty Date:
June 1942
Unit:
F Company,
317th Infantry,
80th Division
Location:
Luxembourg
Arrival ETO:
August 1944
Rank:
Lieutenant

DON RICHARDS

The 80th Infantry Division was activated in July, 1942 and landed in France on August 5, 1944. They fought their way to the Luxembourg area and fought from there into Belgium. They liberated the Buchenwald Concentration Camp. During the war they experienced 277 straight days of combat and ended the war in Czechoslovakia. It has been confirmed that the 80th fired the last shot of the War.

Here is his story. . . .

I was invited to Germany for the 65th Anniversary of the Relief of the Buchenwald Concentration Camp scheduled for mid-April 2010 but was unable to make the trip due to a total knee replacement in my right knee.

I was born November 25, 1922, in St. Joseph, Missouri. My parents moved to Manhattan when I was in the third grade. I went through high school there, then enrolled and graduated from K-State in September 1943. My father was a flour and feed broker and traveled northeast Kansas, representing several flour mills in Salina and Abilene.

In June, 1942, at the end of my sophomore year at Kansas State, I enlisted in the Army as part of the advanced ROTC program. With double summer schools in 1942 and 1943 I was able to take enough courses to graduate a year ahead of schedule in September, 1943. In the summer of 1943 while in school I was called to active duty and was billeted in the women's dormitory just a half block from my home.

Immediately after graduation I, along with 10 others who had enrolled in summer school, were ordered to Fort Benning, Georgia, for officers training. After completing OCS and getting my commission as a second lieutenant in April 1944, I was ordered to the infantry replacement training center at Camp Robinson, Arkansas. Three months later I had my orders for the European Theater of Operations.

I reported in at Fort Meade, Maryland, and shipped overseas from there as a replacement officer. We landed in Scotland and took a train down to Southern England. After a week here we were shipped over to France, landing on Omaha Beach since none of the French ports were open. This was in August of 1944.

I didn't know what unit I would be assigned to until I arrived at the last replacement depots. I went from one depot to another, moving ever eastward. In early September 1944, I had my orders to report to the 317th Regiment, 80th Infantry Division.

The division was at Pont a Mousson on the Moselle River, between Mets and Nancy in Eastern France. This was my first taste of combat crossing the Moselle under heavy fire, and holding the beachhead for several days and nights until the Fourth Armor Division could get across and through our front lines on the hills above the Moselle. The 80th had tried crossing the river several times before but could not hold the ground on the other side of the river as it was very well defended by the Germans. Our battalion objective was a hill near a little town of St. Genevieve. I can remember that very vividly. We were counter attacked every night for about a week. Several nights the attacks came from our rear. I led a daylight patrol into enemy territory one day and was able to spot an ambush before we were fired on. One of my most vivid memories was the afternoon that I led the battalion in attacking a hill and thinking well, this is going to be my last day on earth. It was a rainy day and we advanced up the hill without receiving one single shot. When we arrived at the top, we found Germans with machine guns and rifles under their rain gear in a shallow trench trying to stay dry. Because of the rain they had not heard us coming up the hill.

We didn't carry packs. We had them staged and they were always brought up later. In combat we tried to get a few winks of sleep on the ground in a shallow slit trench. Several days after the crossing we were flushing snipers out of some woods when I was shot through the upper right thigh. The infantry has a philosophy that the odds are that you are going to get hit; you just pray that it is a million dollar wound that doesn't kill you but lets you go home. Unable to walk, I was carried to a field hospital not far from the front lines. I was operated on there and then I was flown back to an army hospital in western England for major surgery.

In December 1944 just before Christmas, I was released from the hospital and ordered back to France to rejoin the 80th. On Christmas Eve I was on a troop train that was stopped in the freight yards of Paris. It was a sad evening for everyone on the train until someone spotted a wine car on the next track and a couple of the GI's climbed on top of the wine car and broke into it. Soon everyone was celebrating Christmas with a helmet full of green wine. After Christmas Eve, we were transferred to 40 and 8 box cars as we headed to eastern France. At Thionville we were trucked up to Luxembourg City at another replacement depot. It was there that I was reassigned back to my unit during the Battle of the Bulge. In checking in at the regiment, I was examined by a doctor and it was apparent that my leg injury would keep me from going back to the front lines I was assigned as a liaison officer to division, to carry orders each night from the division to regimental headquarters. Every night I would get our orders around 8:00 or 9:00 or sometimes 10:00 o'clock and my driver and I would just head out in the blackout and travel 10 to 30 miles to the regiment in our jeep.

After the Bulge, the 80th headed east with the Third Army, crossed the Rhine River at Mainz and then headed up north to take Kassel. The division then came back south and headed east along the Autobahn working with the Fourth Armored to take Erfurt, Gotha, Weimar and Jena. We were in on the liberation of Ohrdruf and Buchenwald concentration camps near Weimar and were almost due south of Berlin when orders came for the Third Army to swing around and go south into Austria. Our division crossed the Ingling River into Austria at Braunau, birthplace of Hitler. A few days later, May 7, the war ended in Europe. Our division was stopped at the Inn River in the middle of Austria to avoid clashing

Photo by Don Richards

with the Russians who were coming from Vienna. The day after the surrender, our division was approached by Germans from the east side of the river. They wanted to surrender to us, rather than the Russians. Our division gave them three days to get as many troops as they could across the river. In those three days the entire SS Sixth Panzer Army of 100,000 men surrendered. At the end of the war, the Third Army had 12 Infantry Divisions and six Armored Divisions.

One of my unusual experiences was late in the War. My driver and I were lost and were traveling on 40 miles of German roads in enemy territory carrying orders down to the regiment. We pulled into this little village about 10 or 11 at night. As we stopped in the middle of town I could hear some carousing over in a building across the street. I was looking at a map with a flashlight when suddenly a figure appears next to me in the jeep, it was a German soldier. What does he do? He salutes me. I quickly returned the salute k and I yelled at John, my driver, "Let's get the hell out of here". We had gone right through the German lines there.

At the end of the War, our division was in Austria for about a month and then we were relieved by the 11th Airborne. We came back to Fussen in Bavaria. While there I saw a notice on the regiment bulletin board asking for officers with photo experience to volunteer for the 166th Signal Photo Company to replace high point photo officers who were going back to the States. The signal photo company was scheduled to ship out from Europe to the Far East to cover the invasion of Japan.

I applied for the assignment and was accepted. I left the Division in late June and joined this Signal Corp in Camp Orleans south of Reims, France. Two months later I was in charge of two box cars of photographic equipment headed for Marseilles, France, to load up for the Pacific. As the train pulled into Marseilles, the atom bomb was dropped and the War was over. It took me six weeks to sign off on all the paperwork on this photographic equipment. By the time I got that done, I found out the photo company had been demobilized. I was assigned an administrative job at Signal Headquarters in Reims. A month later, I was assigned to Western Base Section in Paris. I came home in June of 1946.

I went back to Manhattan for a couple of months. A job opening came up in Parson as the sports editor for the Parsons Sun and since I was going to be married in Parsons in August I took the job. My goal was to study professional photography and specialize in photo journalism. All the photo schools were full when I applied in July but an opening came up at the Fred Archer School in Los Angeles. I could start in January 1947. So in January of '47 we moved out there for two and a half years.

My first job after completing photography school was at Stevens College. I taught photography for two years and then accepted a job Topeka as the assistant editor of the state magazine, "To the Stars". We moved to Topeka in June of '51, two weeks before the great flood. The next year I became editor of the magazine. I edited the magazine for 12 years and changed the name in 1957 to "KANSAS" which is now published by the Kansas Department of Commerce. .

In 1964 I had the opportunity to join the staff of The Menninger Foundation in public relations, and served as Director of Information until I retired in 1978. After retiring from Menninger's, I entered the travel industry, organizing and leading tours to international and domestic destinations. I and my wife have conducted more than 26 tours to destinations such as China, Russia, Africa, Italy, Greece and most European countries.

For the past 60 years I have operated a business as a freelance magazine and advertising photographer as well as a publisher of color postcards and books for four presidential libraries and historical societies.

Don was awarded Purple Heart, Bronze Star, ETO Ribbon with 4 Stars, Combat Infantry Badge, and Army of Occupation Ribbon.

Russian soliders in Austria. *Photo by Don Richards*

Jim Sharp, Sgt. of the Guard, Nuremburg Trials.

Jim Sharp, 2010

Place of Birth:
White City, Kansas
Active Duty Date:
June 1944
Unit:
B Company,
1st Battalion,
18th Infantry,
1st Division
Location:
Honsfeld, BE
Arrival ETO:
January 10, 1945
Rank:
Staff Sergeant

JIM SHARP

The 1st Infantry Division entered World War II in North Africa participating in Operation Torch in November 1942. The Division then was part of Operation Husky and Gela in Sicily. On D-Day, June 6, 1944, the "Big Red One" assaulted Omaha Beach. The Division moved on through France to Belgium where it was in the Battle of the Bulge in December 1944 in the Ardennes. On January 15, 1945, the Division again attacked the Siegfried line and occupied the Remagen bridgehead. On April 8, 1945, the Division crossed the Weser River and on to Cheb, Czechoslovakia, where it remained to the end of the War.

Here is his story. . . .

I was born April 25, 1924 on a farm near White City, Kansas. I entered the Army June 17, 1944. I first went to Fort Leavenworth then to Fort McClellan, Alabama, for my basic training. When I went to Fort McClellan, I didn't know what was going on. I had an agricultural deferment from the Draft because I had one brother in the Air Force and one in the Navy.

I played quarterback on the high school football team and the guys I played football and basketball with were going to the War. My Dad had an auctioneering business in Herington, Kansas, and he could not run the farm without me. So while I was gone he rented the farm out and moved to Herington.

Before I left, I met my wife to be, Marilyn, everyone called her Mickey. She was from Parkerville and I lived near White City and White

City was a shopping center for Parkerville. They had a farewell party for a guy we both knew and she was at the party. I had my Model A Ford parked on the street with the window rolled down and Mickey and some other girls walked by. They didn't see us in the car and she said something about a can of oil in the gutter and I said I didn't think so. She picked up one and showed me. I tried to get a date with her but her Mom wouldn't let her date until she was sixteen.

After basic training at Fort McClellan, I had no more training. When I was done they said I was ready for combat. A couple of Paratroopers came over from Fort Benning with shinny boots and wings and it was too much to resist. Two friends of mine and I decided we were going to be paratroopers. I think it was the thing that saved my life when I volunteered for the paratroopers. They shipped us over to Fort Benning. We didn't think there was anything worse than the infantry.

So those two guys went in and I flunked out. They found out I had a bad arch and the doctor said he wasn't going to redline me unless I said so. I asked what the risks were. The doctor told me that I would be coming down with 100 pounds of equipment on my back and my arch probably would not stand the strain. I figured that there was nothing more important than my health so I decided not to go with the paratroopers, but I felt bad about it.

But it probably saved my life, because it delayed me an extra month. If I had not volunteered for the paratroops I would have been in Europe three weeks earlier and been at the beginning of the worst of the Battle of the Bulge. I went to Europe as a replacement. On December 18, 1944, I arrived at Fort Meade, Maryland, for processing to go overseas and we went by train to Camp Miles Standish near Boston.

On December 31, 1944, I sailed out of Boston Harbor at dark in a forty ship convoy sailing on the *Thomas H. Berry*, an old Cuban mail liner. We had a forty ship convoy because of submarines out lurking around. It would only hold two or three thousand troops and was very slow, probably the slowest boat in the convoy. They were dropping depth charges to keep the subs away and they sounded like they were coming through the side of our boat.

We sailed directly to Le Havre, France. The Germans had captured and killed so many of our men and there were big gaps in the bulge, so they were pushing us hard to get to the front lines. We loaded onto 40 and 8 boxcars and headed for Belgium. That is when I first heard about the Battle of the Bulge. I knew there was a big battle going on because the French engineer asked me where I was going and I told him to the front lines. He said, "You had better get there in a hurry because the Germans have broken through the lines and they are winning the war." The magnitude of the battle was still beyond my comprehension.

We went through Givet, France. We got off the train there and mounted trucks moving to a monastery compound that was full of nuns. It was the first time I saw buzz bombs. They went over and they sounded like my Mom's Maytag washing machine. They were very low and I couldn't understand why nobody shot it down.

I arrived with my unit around January 10, 1945, right in the middle of the battle. I didn't know which division or unit I would be with until I met my platoon sergeant. I was assigned to 1st Squad, 1st Platoon, Company B, First Battalion, 18th Infantry, 1st Infantry Division, the Big Red One. We were in the Ardennes most of the time.

The first time we came back for a rest was at Combalinlatour, Belgium. We took showers for the first time in weeks and we got cigarettes and candy bars.

I don't remember that I was ever a PFC. I jumped from Private to Sergeant, then to Staff Sergeant because so many people were getting killed or wounded. Once you were there at the front lines, there wasn't anywhere to go. Someone asked me about my diary I kept and asked if it wasn't illegal to keep a diary. I told him that there wasn't anyone coming up there arresting us, we were the one's enforcing things.

I had a Bronze Star and about three Purple Hearts but I didn't turn any of them in. They were pretty minor wounds and I didn't go to the hospital. Our First Sergeant was the one who turned in the Purple Hearts and he probably should have turned in more, but I was more worried about life and death.

It took a long time to feel like I could look out for myself in combat, I may not ever have felt like I could take care of myself. Every day was different and every day was life and death. One day you might be riding on a tank, the next on patrol. As you got more experienced you got called on more. All I knew was what they taught me and I didn't know if it was right or wrong. A lot of people said they were not ready to go out and kill people and go on patrol but who is? You had to do it and you might be wrong. There were a lot of people who were killed because they were wrong. We had one Company Commander, Lieutenant Yarborough, who sent us to attack the wrong position. But I didn't know, I just did what they told me to do and was fortunate to survive.

I got called on patrol the very first day I was there and put on point. That first patrol had about seven people and I guess I did OK because I was on patrol five to seven times over the next month. I asked Sergeant Zilish why I was on patrol all the time and some of these guys have not been at all. He told me that when you go on patrol, you want qualified people who were not going to get anyone killed. That didn't help me any but that is exactly what I wanted when I was on patrol. I didn't want any goofballs who didn't pay attention.

One or two of us would carry submachine guns. If you had seven men you were out on a reconnaissance patrol. We were only gathering information and not out to engage a larger force in combat. The officers, sergeants or patrol leaders had maps and compasses but I never had one. Intelligence would tell us what was happening but most of the time they didn't know either.

Riding on a tank was hell. It was frightening. They gave us notice at 4 AM and told us we were going on attack at 5 AM. They gave us snow capes and we put those on. Sergeant Zilish told us we were going on attack at 5 AM and he told me that I would be on the first tank with him. Those tanks started up at 4:30 and as soon as they got started the artillery and screaming Mimies started coming in. I thought surely they wouldn't attack with all of this going on.

It didn't stop the attack and all of this artillery was coming in, and we climbed on the lead tank. I climbed on and noticed all of these Jerry cans tied on the tank and they were full of gas. I was hanging onto gas cans and I saw all of these tracer bullets going by and I thought we were on a death trap. It was bitter cold, at times close to 0 degrees and we started out.

It was a tough job hanging onto the tank; we were going 20 miles per hour. We captured the town and there I saw my first dead German. The tank machine gun was spraying bullets as we moved ahead. The German was trying to climb over a barbed wire fence and he was hanging grotesquely over the fence. I think we were with the Third Armored Division. The 3rd and the 9th were the two that we worked with the most but I really don't remember much else about what was going on there.

One time we were sent into a town called Wilnsdof to bring back prisoners for interrogation and we only had 12 people on the patrol. We took off about 10 or 11 at night and went through a reforested area where the trees were about 6 to 8 feet tall. It was scary.

We captured some but we got into town and got surrounded, The Germans came in from all sides. Poor Sergeant Angel, a Blackfoot Soldier from Washington, our platoon leader and leader of this patrol. He got into a battle right in the middle of town when they got into a German command post. He got killed along with four other guys.

Sergeant Angel and the others got into the command post and a German half track was hidden in the shadows between the CP and a house and they couldn't see it. The half track had a machine gun on it and got several of our guys before they knew what was happening. Two of our guys were captured but we later captured them back.

Another guy and I were out at the edge of town setting up a roadblock, protecting the patrol from the incoming traffic. There was traffic coming in and we had a machine gun set up in the window of a building right on the road. There were two of us and we had two prisoners in the basement. In the middle of the night I could hear these German trucks and didn't know if we were supposed to fire on them or not. We had guys we were supposed to be protecting and it was a rough situation.

I could not really tell when the battle turned in our favor. I only knew what was going on in my foxhole and maybe the guy's next to me and what I could see out front. Most of the big changes took place south of where the First Division was and we were fighting to hold the Germans back. We helped relieve the other outfits. The Germans were ahead of us and we were trying to pinch them and we did.

Most of what I saw of air support was mostly around the Ruhr River. When we got a bridge across the river I saw a German jet plane. They were trying to bomb it and did. When the jet started to return back to Germany, two P-47's were circling to get him but as they turned the German pilot saw them and made it past them before they could get turned around. It was fast and I thought their planes were lot superior to ours.

Once, I saw a P-47 crash what appeared to be about a half mile away but it turned out it was three or four miles out. We ran over to help out but nobody was alive in the plane. Apparently it had been in a dog fight. They told us the patrols were out and the Germans might attack but as far as any action there was only artillery.

In the Ardennes we found several small huts about 50 yards into the forest. When we checked them out we found them full of dead American soldiers in bunks, apparently killed while they slept. They appeared to be bedded down for the winter and were not going to fight for the winter. Someone told us that might have been why the Germans were able to surprise us when they attacked because the front line was asleep in these huts. I don't know if that is true or not. They were little Hansel and Gretel houses like a story book house.

About a mile from there, we found a German farm house where a terrific battle had occurred and there were dead Germans and Americans lying all around the farm house. It was very strange to see. Most of the time neither side left their dead on the battlefield, both sides removed them. Our Sergeant told us that this was where the Germans broke through at the beginning of the battle. I just don't understand to this day because the house was out away from other battle areas.

I did not have to fight any of the fortifications and attack any pillboxes along the Siegfried Line and I am glad of it. I remember the Rhine and Wessel Rivers. I do remember going into Bonn. It wasn't nearly as damaged as St. Vith or Aachen. The thing I remember most about Bonn was when we thought we had it cleared, the snipers appeared all over the place. We were sent out on three man patrols and it was the first time a German woman approached me. You were not supposed to fraternize with the German women as it was a courts martial offense.

Two nineteen year old German girls approached me with a bottle of wine. We were on patrol and I could speak German better than the others. They stopped us and they handed me the bottle of wine. We had been warned about the possibility of it being poisoned. They wanted to give me the wine as though we were going to be friends. One guy said we should take it. We told them that they needed to take a drink before we would take one. Then one of the guys pointed out that we were not supposed to be talking to the girls. These girls were trying to be friends with us and we told them no.

About that time we heard rifle shots and it discouraged the girls from flirting with us Americans. We moved ahead and found a huge bomb shelter down the street. It was a massive concrete building and we believed that the German sniper had gone in there so we opened it up and looked in.

The stench was terrible and it was loaded with people. They were still afraid to come out. I was the patrol leader so I stepped inside. It stank inside and the people didn't look too happy anyway. We asked them where the Nazi soldier was and they said they didn't see one. There must have been a thousand people in the shelter, no lights so you couldn't see very well. We decided we weren't going to find a German sniper in there so we moved on.

When we were done, we took up defensive positions on the hill. It was where we had surrounded the Germans but they broke through our lines. I was out on outpost with a guy named Queener. Our job was to fire three warning shots if we heard them coming. In the middle of the night we heard them coming but we couldn't tell which way they were going. Before we ever fired three warning shots all hell broke loose. They broke into our perimeter defense and there were people running all over. You couldn't tell who was who in the dark. We didn't know if they survived or got killed or what, we couldn't see anything.

Early the next morning we could see Sergeant Zilish coming out from the perimeter to look for us. He didn't know if we had survived and on the way he captured a German who was hiding in the bushes. He was one of the Germans who survived.

The D's (*displaced persons*) came through the lines and gave our intelligence people the idea that the Germans wanted to surrender. Our Battalion Commander, Lt. Colonel Leonard and five jeeps of us went into Siegen under white flags made out of sheets we got from the beds of the houses there. We put a white flag on the front of each jeep and went through the German lines. When we got there they told us they were going to take us prisoners and they made us lay all of our weapons down into a ditch. The German soldiers were walking around us wondering what was going on.

They blindfolded the Colonel and took him further into the town. I could see all the way into town and I could see the civilians out in the street and little kids walking around the jeep. It looked like a little party going on, celebrating the capture of an American Colonel. There wasn't anyone there who was ready to surrender so we had to get out as quickly as possible. They didn't surrender then but we captured quite a few later. The weather was miserable. It was raining and we returned to our lines and went out into the forest to dig foxholes.

We crossed the Rhine on pontoons and they had been badly shot up collapsing several. The river was high because the Germans had broken the dams up river. I saw the boats bobbing up and down and I decided to take my field pack off because I didn't want to be dumped into the river with it on. The river was swift and the debris was catching on the pontoon. It was swaying around and I wanted to be able to swim out and the pack would have been like an anchor.

After we crossed the Rhine River at Remagen we moved into Erpel. There were a lot of air battles taking place. We could see the German planes but they didn't get into a position to bomb us. We got to the other side and they took us into a railroad tunnel and it was night. We could hear all kinds of action outside the tunnel and we wondered if they had counter attacked.

There was a Sergeant who started shaking, falling to pieces, battle fatigue. I told the Sergeant that he needed some help. He told me that he told them he wasn't ready to come back to the front but they didn't listen to him. He eventually had to go back.

We moved east toward Wellesberg where the second and third platoons got into serious trouble. The Germans had them pinned down and ordered them to surrender. My squad only had eight men left and we were ordered to attack the town and get them out. We rushed toward their location which was about three blocks away and we were under continuous fire. One of our guys took a tracer bullet in his pack and set it afire. For a moment he was a walking torch, but we got it out and were able to move ahead. We set up in covered positions around the house and were able to bring so much firepower on the Germans that they thought we were a lot bigger force than eight men.

They put up heavy resistance but we ran them out and took 14 prisoners. The survivors of the other platoons were very grateful. For this action our squad received the Bronze Star.

We came upon an area that was surrounded by high wire and guard towers. I was more concerned with looking for snipers and enemy positions and I had not really noticed the encampment until we got right to the front gates. I noticed people inside the wire and they waved at us and seemed to be friendly. As we got closer to what seemed to be the main gate, people began streaming out of the gate and greeting us. Most of them looked thin and haggard. We later learned that this was the Nordhausen Concentration Camp. They hailed us as their saviors and hugged us and slapped us on the back. But we did not have time to visit as the battle was ongoing.

We took a long truck ride from Nordhausen and eventually reached Sangerberg. There was a heavy rain falling and our squad occupied a house near the edge of town on a black top road. While on guard near a black top road, we saw three Jerries coming down the road with their hands up. We captured them and questioned them about where they came from. They told me there was a German Army camp just up the road. I got a couple other soldiers to go up to the base with me to investigate. We found the small base mostly abandoned, but with office equipment, supply room, barracks and beds intact. We captured four more Jerries and brought them back to the POW cage. We went back up again to investigate and off in a distance in the forest a Jerry machine gun started firing into the barracks and kicking up dust all around us. We had heard rumors that the war was about over, so we got out of there in a hurry. The next day we heard on a tank radio that the war was over. We were in Czechoslovakia close to Cheb.

Immediately following the war, I assisted the Police and went out in pairs in the middle of the night and arrested German leaders who caused the war. In October, 1945, I was assigned to Special Guard Duty to the Four Power International Military Tribunal at the Nuremburg War Crimes Trials for Nazi leaders of the Third Reich.

I was supposed to be there three months but was extended another three months. During that time I was the Sergeant of the Guard in both the Courtroom and the Prisoner Cell Block. I was able to talk with a number of the famous war criminals and collect the autographs of more than a dozen of the Nazi's on trial.

When I left Europe it was April, 1946. I wasn't able to call anyone to tell them that I was heading home but I did write them a letter. We were down around Berchesgarten about a week or so before we were sent to Bremen to ship out. I came back on the ship "Fayetteville Victory" an we came into New York Harbor passing the Statue of Liberty. One of the interesting things on the trip involved a little German boy who was smuggled aboard the ship by a soldier in a duffle bag. When we arrived in New York he made quite a hit with the newspapers.

When we came into New York Harbor, there were USO girls on a boat singing patriotic songs to welcome us home. They told us we would be able to call home from Camp Kilmer, NJ, to tell them we were on the way.

I arrived home, alive and free on May 9th, 1946. On May 9th, I went directly to Parkerville to see Mickey.

That summer I went to work with my Dad who set me up with a combine and tractor and I did custom combining of wheat. Then in the fall I went to Kansas State University. Thanks to my Dad I had a job so I didn't have a hard time.

I was able to get on the ground floor of information system development and served as Information Systems Manager for Kansas Farm Bureau.

I have two boys and a girl with seven grand children and two great grand children. I am keeping busy writing another book.

The thing that sticks in my mind the most about the Battle of the Bulge was the bitter cold weather and freezing while huddling in my foxhole. You could not get out of your foxhole without getting shot by artillery. You had to sit in your foxhole and suffer. Our hands and feet were frozen. We tried everything. I took a little canister of oil and heated it up. I took my boots off and rubbed my feet in the oil but it seemed like everything we did to make it better made it worse. The boots we had were not made for winter. We had one blanket and one shelter half. Later we got a snow cape and I really thought I was going to freeze to death before I got shot.

I also vividly remember jumping off a tank and almost instantly it was hit and blown up by a German 88. I am glad I went to war and lived to tell about it, but I wouldn't take a million dollars to go there again.

Jim was awarded the Bronze Star, American Service Medal, and the European African Middle Eastern Service Medal with 3 Bronze Stars. He has written a book based upon his diary, ***Diary of a Combat Infantryman****, available from Amazon.com.*

Deetra Driver at the 1st Infantry Division Monument in Cheb, Czech Republic. It is dedicated to those who fell in the Rhineland Campaign. *Author photo*

Chapter 6

The Tankers

Few EXPERIENCES IN WAR are more terrifying than facing an enemy tank. German tanks presented formidable obstacles to the infantry as did the American Sherman tanks. There were nine American armored divisions consisting of an integration of tanks, infantry and artillery engaged in the Battle of the Bulge. A reconnaissance cavalry squadron became the eyes and ears of the armored division, scouting out in platoon and company sized patrols gathering information and capturing prisoners for interrogation.

Each Infantry Division was supported by a tank battalion. They supported the infantry in coordinated tank-infantry attacks, perimeter watch and provided direct fire in support of operations, a more personalized form of artillery support.

Carl Shell grew up in the coal mines of eastern Kentucky and joined the Army to avoid the life of a coal miner. He was assigned to the 14th Tank Battalion, 9th Armored Division. On the night of December 15, 1944, Carl was wondering what was happening on the German border close to St. Vith, Belgium. For days there had been strange sounds from the forested Ardennes and he had seen German reconnaissance planes slowly checking out his tank's position.

National Archives photo

Kenny Luigs was a Sergeant in the 89th Cavalry Squadron, the reconnaissance unit for the 9th Armored Division. Things were pretty quiet in Luxembourg and there was no inkling that anything was about to happen. He and his unit were enjoying a warm relationship with the local population and were anticipating a Christmas season that would be marked by a quiet rest.

On the morning of December 16, all of that would change for both Kenny and Carl. Carl suddenly saw German troops attacking his position, suddenly coming out of the woods unabated. Over the next several weeks, both Kenny and Carl would engage the enemy at different locations and under unanticipated circumstances as their units maneuvered to gain an advantage on the attacking enemy.

Herman Westmeyer was still in England with the 11th Armored Division. His 490th Field Artillery Battalion along with the rest of the Division was rushed into Belgium arriving in time to help turn the tide of the German onslaught. They arrived on Christmas Day but Herman was only to see the remaining days of 1944. On January 1, 1945, he was evacuated severely wounded.

Harlan Henry arrived as a replacement for the infantry and was assigned to the 83rd Infantry Division. "It was colder than hell, I mean it was miserable. I was in the infantry for about a month." Harlan had a chance to become a tanker and jumped at the opportunity and was assigned to the 774th Tank Battalion in Support of the 83rd.

Carl White arrived in December, 1944 and was assigned to the 749th Tank Battalion in support of the 44th Infantry Division. The main German attack occurred north of the 749th position but some German units attempted to drive toward Strasbourg and were stopped. His unit along with the 71st Regiment of the 44th Infantry Division was awarded a Presidential Citation for a battle near Rimling, France, for action between December 31 and January 2.

Upon arrival in France in early January, 1945, Jack Gragert, a tanker with the 8th Armored Division headed for the German-French border. They were needed there to help block the German drive to take Strasbourg.

Knowing that the German's supply lines were stretched thin and the 7th Armored Division had taken a debilitating shot, as the Battle of the Bulge progressed, the Americans allowed the attackers to create a longer bulge in their positions. Instead of facing the German offensive head on, the Germans continued west and failing to capture the Americans gas supply for their tanks at Stavelot, they ran out of gas, literally. The Allies began to counterattack and closed off the bulge trapping the attacking German units.

Captain Orval Abel

Place of Birth:
Emmett, Kansas
Active Duty Date:
1940
Unit:
Column Commander
A, CCA,
67th Regiment,
2nd Armored Division
Location:
Billers le Bouillet, BE
Arrival ETO:
June 28, 1944
Rank:
Lt. Colonel
Train Column
Commander

ORVAL ABEL

The 2nd Armored Division was activated in July 15, 1940, at Ft. Benning, GA, as a heavy armored division. It was known as "Hell on Wheels," a nickname that came from its first commander, General George Patton. The division participated in the war in Africa in 1942 and the invasion of Sicily in 1943. In November 1943, it was shipped to England to train for the invasion of France.

"Hell on Wheels" landed on Omaha Beach on June 9, 1944, and headed toward Belgium with Patton's Third Army. Crossing over the border around Schimmert, Germany, the Division began to attack the Siegfried Line around Marienberg, Germany, in early October. With the launch of the Ardennes offensive, the 2nd Armored Division pulled out of their positions along the Roer River and moved to new defensive positions around Houffalize.

With the conclusion of the Battle of the Bulge and after a refitting and rest, the Division drove to the Rhine River, crossing it on March 27, 1945. They ended the war along the Elbe River at Schonebeck, Germany, on April 11, 1945. The 2nd Armored was the first American division to enter Berlin by July.

Here is his story. . . .

I was born on April 21, 1909, in Emmett, Pottawatomie County, Kansas. We lived in Silver Lake, Kansas, where my Dad worked as a banker and farmer.

In 1935, I was commissioned a 2nd Lieutenant in the Infantry Reserves when I graduated from Kansas State College. After I graduated I took a job teaching high school in Waverly, Kansas.

I met my wife, Lucille Byarlay, in Green, Kansas, in 1929. My mother's sister lived across the street from Lucille's family and that is how we met. We were married Sept.14, 1935.

In July 1940, I received orders to report to Fort Benning, Georgia for active duty on September 1, 1940. The orders stated that the length of service would be one year. It sure didn't work out that way. At Fort Benning I was assigned to the 2nd Armored Division. My wife and I purchased a house trailer as living quarters near Ft. Benning because housing was very scarce. The location of the area where the house trailers were set up had many other Army families as residents. Our oldest child fit right in with the other youngsters; the youngest had been born in August 1940 and it would be a few months before she was able to show much interest in the other children.

One of the experiences during the assignment at Fort Benning that I remember was learning about the attack on Pearl Harbor. It was during the afternoon of December 7, 1941, and we were watching a soccer game — I don't remember who was playing — and the loud speaker interrupted the game. The announcer at the stadium told us of the Japanese attack on Pearl Harbor. Our oldest daughter who was four could recall that announcement years after it occurred. The commotion must have made a great impression on her.

We trained at Fort Benning, at Ft. Knox, Kentucky, and Fort Dix, New Jersey. Our division took part in the maneuvers in Tennessee, Louisiana, Texas, and the Carolinas. During this time I completed the Armored Wheeled Vehicle Maintenance Course and the U.S. Army Armor School for Preventive Maintenance. I was trained as a Column Commander in the Division Trains of the 2nd Armored Division and was assigned to the Division Trains in the 67th Armored Regiment in CCA.

Orval leaving for Africa 1942

The Division Trains was the unit responsible for keeping units supplied as well as protecting the supplies from enemy destruction. Division Trains often had to run supplies hundreds of miles, sometimes through enemy-held territory, to provide gasoline and oil to keep the tanks running, to supply the ammunition for the weapons, and to have provisions for feeding the soldiers. The men who procured the supplies, received training also to prevent the enemy from destroying or absconding with the supplies.

In December 1942, the 2nd Armored Division received orders to go to French Morocco. My wife bought a house in Clay Center, Kansas, after I left in the fall of 1942. We owned that home for 50 years. Our ship landed in Casablanca on Christmas Eve, December 24, and we ate our Christmas dinner on board ship before we disembarked. In the spring of 1943 we moved to Arzew in Algeria, then on to Tunis.

On July 15 a small group from Company C of the 67th Armored Regiment left the port Aux Poules on a ship bound for Sicily. Other members of the Division Trains were serving as guards at the POW camp in Port Aux Poules or continuing to train at the small arms ranges.

On November 12, 1943, preparations were finalized for boarding the British transport "Orontes" and to make way to Liverpool, England. The quarters on the ship were very good as the men had individual double-decked bunks and mattresses. We had no enemy activity during the trip. My assignment during the time spent in England and on the European continent was as Division Trains Commander. This assignment continued until January 1946.

The 2nd Armored Division landed on Omaha Beach on June 9, 1944, and I landed on an LST on June 26. I was the Train Commander for A Train, assigned to CCA and my job was to keep the supply train moving and keeping the tanks loaded with gas and ammo. As we moved east, we had to shuttle gas and supplies from Omaha Beach and as we got farther east, the supplies were difficult to get.

Orval on right in Europe, Fall 1944

During the day the Germans would send scout planes over our lines and mark where the tanks and fuel tankers were positioned. They would return at night with waves of aircraft to strafe and bomb our positions so at night I changed our positions. When they came each night they bombed the old positions and missed us.

We moved toward Belgium and we were in Germany ready to

cross the Roer River when the Germans attacked in the Ardennes. It was the Battle of the Bulge. I first heard about it when we got word that there was a big tank attack by the Germans but initially we didn't know what was happening. Soon we saw German tanks coming directly toward our tanks and us. Many had been destroyed. We withdrew back into Belgium and on Christmas Day we were set up around VII Corps Headquarters northwest of Marche, Belgium.

The Division Trains were at Merkstein, Germany, when the breakthrough of Germans occurred on December 15, 1944 and we moved to the area south of Huy, Belgium on 21-23 December. I was billeted in Villers le Bouillet. As the Germans were pushed back, we saw lots of prisoners.

When I think of the Battle of the Bulge, I think of digging trenches in the frozen ground, the cold and snow, difficulty sleeping. Then there was the movement of all forces and trying to keeping warm and eating C Rations and the tragic loss of life; lost good friends and great troops. Our biggest problem during the Battle of the Bulge from a medical standpoint was frostbite. This is one time when Division Trains failed to provide adequate clothing. We had no overshoes. We were fighting in snow, sometimes it was up to 3 feet deep. The temperature went down to well below zero every night, and fog would roll in from the sea and coat everything with ice. No planes flew and we lost our air coverage. We dug trenches in the frozen ground and into the snow. It was so cold and it was nearly impossible to sleep. The only food we could get was C rations. It seemed that everything was moving all the time, infantry, tanks and other vehicles, very confusing, you couldn't tell which side was winning or what was going on anywhere but where you were. You had to move all the time to keep warm and the tragic loss of life was staggering, good friends and great troops.

I remember when the weather cleared and our planes began to fly again. Everyone was excited and we hoped that this would get us back on the offensive. I didn't like having to pull back from ground we had already taken.

After the Battle of the Ardennes, Train Headquarters was in Eaneux, Belgium. We took time to train and re-equip our Division because we had lost so many men and our equipment needed a lot of repairs. We also were able to deliver new tanks to replace the ones we lost during the battle.

We were preparing for attacking across the Roer River. All of the Division insignias and vehicle markings were removed by February 3, 1945 while we were in Banheit, Netherlands. The Division kept its identity secret until February 22 when the attack by the Allies began in order to be able to cross the Roer River into the Rhineland. The Division started crossing The Rhine River at the end of March, 1945 and I crossed on April 1, 1945.

On April 3, 1945, I was wounded in a skirmish when we were attacking toward Hamm, Germany. I spent nearly three weeks in the 23rd General Hospital in France and returned to my unit April 27, 1945.

Our Division was the first American Division to reach the Elbe. It was on April 11, 1945. We were ordered to halt at the river because the Allies agreed that the Russians would be the first to attack Berlin. The Russians came up to the other side of the Elbe and the war was over. In July we entered Berlin, the first American division to do that. The Train Headquarters was given an area of occupation at the Hermann Goering Werke at Immendorf.

I arrived back in the States on Sept. 22, 1945. It was wonderful to have my wife meet me at the train station in Manhattan. Getting reacquainted after almost three years was great. I did have several months of leave so I spent the time being a home owner and going over the house my wife bought in Clay Center.

I returned to the U.S. Occupation of Germany in March of 1946. My family could not accompany me as Lucille was pregnant and the physician said that it would not be good for her to travel. My new son was almost a year old when I returned in 1947. Later I attended Command & General Staff College, Regular Course Equivalent, a Course on Military Justice and stayed in the Army. I retired a Colonel in September, 1960.

Orval was awarded the Bronze Star, Purple Heart; American Defense Medal; European-African-Middle Eastern Campaign Medal; American Campaign Metal; WWII Victory Medal; Entitled to wear the Belgium Fourrageres; WWII Occupation Medal (Germany); Service stars for Normandy Service, Northern France Service; Sicilian Campaign. Rhineland Campaign, Ardennes Campaign, Central Europe Campaign.

The major crossroads in Bastogne today. *Author photo*

Jack Gragert, 2010

Place of Birth:
Topeka, Kansas
Active Duty Date:
1944
Unit:
A Company,
18th Tank Battalion,
8th Armor Division
Location:
French-German border
Arrival ETO:
January 1945
Rank:
Staff Sergeant
Tank Commander
T/5

JACK GRAGERT

The 8th Armored Division was activated in April 1942 and landed in France on January 5, 1945. They immediately headed for the Battle of the Bulge, arriving at Pont-aMousson, France, to help stop the German drive for Strasbourg. On February 27, they crossed the Roer River and the Rhine River on March 24, 1945. The 8th liberated Halberstadt-Zwieberge, a sub camp of the Buchenwald concentration camp, on April 12 and continued on through Germany ending its campaign in Pilsen, Czechoslovakia.

Here is his story. . . .

I was born on January 15, 1924, and I grew up here in Topeka. I was drafted in 1944 and trained as a tank gunner. I went through about half of my time over in battle as a gunner and the rest of it was a tank commander and platoon sergeant and was promoted to Staff Sergeant. I was assigned to Company A, the 18th Tank Battalion. We were part of the 8th Armored Division. We were always 8th Armored and never part of any other organization. I went into the unit the day I was drafted, went right to 8th Armored Division.

I'll tell you kind of a funny story. I got my orders I'd been drafted and I was to report to Ft. Leavenworth. I would have seven days after I went to Leavenworth to be inducted so I would have seven days to come back home, get my affairs in order, so I could go back into the Army seven days later.

First, I was going to join the Marines. So I went in this room and they swore us all in as Marines. Then they told us we would leave tonight.

I said what the hell are you talking about? I got seven days. No you don't, not in the Marines you don't. You go tonight. I told them, hey man, I'm a married man. I got a wife and kid at home, I haven't quit my job. I said I can't go tonight; I got to go home and get things done. So they told me, get the hell out of here, you're out of the Marines, you either leave tonight or you don't go.

So I said, okay. I went right across the hall, joined the Air Force. I sat there and I pulled all of the rods and the strings and everything in a test, and passed everything. They said okay, report back here in seven days. So seven days later I go back into this room again and these guys stood up and told us how glad they were that we belonged to the Air Force. And then we mounted a troop train and I was in the Air Force, sworn in and all. From there, we got on the troop train. We got in the high hills of Louisiana on this troop train, the chaplain came by, giving us a long speech about how everything was nice and this was beautiful country.

He said, I want to congratulate you tankers. I looked at the guy next to me and I said, what the hell's a tanker? I don't know. I said, what are we doing in Louisiana? He says I don't know.

I said in Shreveport there is an air base and that's probably where we're going. But I said I don't know what they mean by "tanker." They pulled us out of the hills of Louisiana and down into Camp Polk and that' where we got out of that troop train. We all lined up in straight lines and trucks came by and picked us up, took us to the companies, and dumped us out. Hell, we were in a tank corps and didn't know it. So I was in the Marines, I was in the Air Force and I finally ended up in the Army Tank Corp before I even got a uniform. It was unbelievable.

I walked into the orderly office and Capt. Brendegan, stood up and welcomed me there and he said, you're from Topeka, Kansas? Yeah, I am, I said. How far are you from Kansas City? I said, oh, not very far, 60 miles or so maybe. He said my mother's in the hospital in Kansas City. So I got to talking to him about that and right away he made me an acting corporal. Just because I knew the Kansas City area where his mother was, I was made an acting corporal. So I was acting corporal for about a month and then I got derated out of that and about three months later I was a buck sergeant. Later, I went on up to staff Sergeant.

So I went to Louisiana and went through the tank basic training. That is when I got assigned to the 8th Armor.

Leesville was a hell hole! They had all the stores there. There was Air Corp, Navy and Army, all of us in that one area. You couldn't walk down the street without bumping into somebody. I mean it was so crowded you couldn't even move. And going to stores, they did everything in the world except put a gun on you to buy something. I had two good buddies from Topeka there and I noticed them one day at the camp. They had really good looking outfits on instead of the Army issued.

I said, how in the hell did you guys get those? Oh, they said, we bought them in Leesville. You bought them in Leesville? Yes, we did, they told me. Come to find out they stole them in Leesville. They went in, tried them on and then put their ODs over the top of them and walked out. I should have known that they would have done that because this one guy's name was Boots Hennessey. He was ornerier than hell. I went to school with him and he and this other guy he was always with, I can't' think of his name right now. Years ago when we were all in high school, it come out in the paper that Boots and his boys were arrested for stealing chickens.

We finished our training there in Camp Polk, Louisiana, and went from there to Camp Kilmer, New Jersey, and shipped out on November 7, 1944, on four different ships. We landed on November 19 in Portsmouth and Southampton and moved to Tidworth Barracks in England where we drew our equipment. We shipped out on LST's on January 3 to Rouen, France, and were part of CCA, 8th Armored Division. We were sent to the vicinity of Pont-a-Mousson.

I remember we traveled 100 miles to free these guys that were bottled up in that little town of Bastogne but we were diverted toward Strasbourg. The Germans were trying to take it. From there we went to Ninneg, Germany, about thirty miles southeast of Luxembourg. It was bad, just a damn war. I mean it was a war.

And I think we lost at least two tanks there. But I don't know, you know it seems like you would think going through this kind of thing I would never forget. I tried to forget all of it. Well, you know it's kind of like that saying, why should I remember something I've been trying to forget for 65 years? But now so much I really have forgotten, then so much is like it happened yesterday. But I know there are things that are so vivid, it's like it just happened when I got out of the car. And then there are some things that I just have forgotten. That sometimes come back to me but don't. I just don't remember very often.

You know we used to have those big thick Navy ropes that we draped around our turret and then come around to the back of our tank. And we'd go into battle, infantry then would jump onto the tank, grab those ropes, swing up on the back of our tanks and ride into the battle with us. And boy, I'd be looking through the periscope, see those guys shot off that tank like flies you know?

An infantryman would say he wouldn't be caught dead in one of those damn iron coffins, no way. I said, hey, if you'd seen as many guys as I have shot off my tank, you wouldn't be saying that. I'd say if I was an infantryman, I think I'd desert and run off and hide.

My buddy who I went with to the Marines went to the Pacific and he went through a lot of battles. The only wound he ever got, he told me, was when the Japs started bombing them they jumped in a foxhole and got bit by a snake. The back of his leg from his heel up about halfway to his knee was just black where the snake bit him. I know I would have rather been shot. I am scared to death of snakes anyway.

I was buttoned up in that tank and I know we went right into some towns killing everything we saw there and that's all I can remember. I don't know how anybody else got in there or anything else. But I know we eliminated a lot of the Germans in there and I know there were other units coming in after we went in. But I don't know who they were or what.

The thing I remember the most about the Bulge, was that German tank blowing the whole right side of my tank off and putting me out. I saw him before he shot us. We fired at him twice but he made one shot and just took the right side boogie wheels and sprocket all right off, just sheared it right off the tank. We bailed out of it and took off running and next thing I knew that tank of mine was blowing up. Because he not only took the right side of it off, he hit it and destroyed it.

What I remember most about that was when we got hit I was looking out the periscope. I was the tank commander then. I was looking out of the periscope and there was an infantry boy crouched down with his rifle behind a tree right off the highway where we were. When it was all over, the tank was gone and everyone came out of that tank. I don't know why but the first thing I looked for was that infantry guy. All that was there was a rifle with nothing but the stock laying there.

I believe to this day although I am not sure, it was him across the road sitting down in the ditch. I couldn't tell anything about him but I think that was him. I just don't know. I couldn't tell whether he was German, American, or who he was. I still think to this day that he was the one that was by that tree. I believe that the blast on the side of the tank got him and threw him over the tank. I still think it was him but I don't know for sure. I mean he was blown all to hell.

Three different times I had three different tanks shot out from under me. That last one, the last tank, got hit in the motor compartment and in the gas tanks by a bazooka. A gunner got hit and it kind of dazed him a little bit. I opened a turret hatch, started out and he was still sitting there. I grabbed him by the hair, started pulling him by the hair out of that tank and finally he came around and came out. My driver, before he got out, the bottom of his field jacket was starting to burn.

I came up on top of the turret, everybody got out and I jumped off the turret, hit the ground and started running to a ditch, that's when I got hit in the back with shrapnel, a tree burst. It knocked me down but I made it to the ditch and I crawled up the ditch to a bridge, a little culvert bridge. A guy started running across the bridge and I didn't know it but there was a machine gun, a German machine gun in an old building. My driver was on the other side of the bridge, he said later behind me dirt was flying up from machine gun bullets and I was running my tail off but they couldn't catch me.

After the Germans were pushed back, after the Battle of the Bulge, we went into Luxembourg, and then we went on toward the Roer River. We crossed the Roer River on a pontoon bridge. We also crossed the Rhine on pontoons at Cologne.

I got four Bronze Stars. I got one for helping a lieutenant and the other one was when I was knocked out of the tank. I got another one, well, what was the other one? I don't know what the hell I got them for but I got another one. The other two I got in different battles.

I got one when my tank got knocked out and I ran up this hill and jumped into a ditch. We were in a ditch in a forest, trees all around. Here came a tank in at an angle to our front. It was a little light tank and it was Patton's personal tank. I was crouching in this ditch when an old buck Sergeant, infantryman ran over and jumped into the ditch with me. He was a dirty, filthy guy, looked like he'd never taken a bath. He was carrying a bazooka and when Patton's tank came in, he drew up on that tank with that bazooka.

I hollered no, no, that's ours that is ours! The hatch was open in this tank and Patton and four other officers, no less than a Major, got out of the tank, came down to us and asked us how things were going. There was old Patton himself. I told him okay and we were doing all right. He said well, you guys are doing a wonderful job, just keep it up, and then he turned around and left in his tank. Anyway, why I got a bronze star then, I don't have any idea, but I think we all did that were in that ditch. But I have no idea.

When the war ended we were just a few miles from Berlin. I remember that. Shorbergen or Westerode, I can't remember the name of that town.

But I know we raced across the country there fighting away to get to Berlin and we were up on this damn hill and we — I don't know maybe eight, 10 miles from Berlin and ready to go into Berlin and they stopped us and let the Russians go in.

I remember that the war was ending. When they held us up and let the Russians go in first, some of us cussed them and some of us were happy that we didn't have to go. Tell you the truth about it, I was damn glad. I was damn glad because I couldn't be lucky all my life.

I came back on a troop ship. Going over to England, we went over on this little troop ship and I remember we were zigzagging all over the ocean and everybody was getting sick. I was told was the reason that they did that was we were being chased by a German sub. And I know we were all over that place. Everybody was sick except me and I never got sick.

And on the way home, I was on that same troop ship again and it was crowded, could barely move there were so many of us. I was Sergeant of the Guard and I was not sick. Everybody else was sick. I went down these steel steps into the hold down there to check on the guard and I turned around to come back up and I started up the steps. The guy up above got sick, and threw up all over me and then I got sick. I said many times I felt like shooting him.

When I was in the Army I was married and had a daughter. I came home on furlough after I was there for a while and my daughter was just maybe six, eight months old, something like that. And then I left and I didn't see her then until nearly two years later. None of my family or my wife knew where I was. We didn't know where we were. I returned back to New Jersey from Europe and I think from there I went to Camp Chaffey, Arkansas.

That's where I discharged out of the Army at Camp Chaffey, Arkansas. My wife didn't know I was on the way home. I came home on the bus. I got on the bus at Camp Chaffey and made it to the old bus depot on 6th Street in Topeka. My wife's brother was a fighter pilot and he came and got me at the bus station.

I was a machinist's helper for Santa Fe when I went in. When I came back, I automatically got set up to be a Class B mechanic machinist and then a couple months later after I was there, I was set up to be an A machinist. I didn't like it so I took over a lease for a Standard Oil filling station. The only way I would make a living in that station was painting used cars for dealers. I eventually opened up the Jayhawk Body Shop. I have five children four girls and a boy.

I went through my dad's home town in Germany where his dad's two sisters still lived. Oh, God, this town was just absolutely blown away. And his Dad, my grandfather, his two sisters lived there at the time. And they were writing letters to my Dad but after that town was blown away, he never heard from them again. Nobody knows what happened to them.

My grandpa emigrated here from Germany. My grandpa hid on a cargo ship in Germany. He was 17 years old and he got to New York, and he met my grandmother in a New York German settlement. They were married and they finally made it to Topeka where they homesteaded that place on Rochester.

When he ran away from Germany, grandpa's job was working on the river where they would pull a boat down river or across river by pulling on ropes. He was one of the guys on one side of the river who pulled the ropes and towed the boats down river. He was still alive when I went to Germany and came back. He asked me a few questions about Germany and the war and I talked to him a little bit but not much.

The thing I remember the most about the Bulge, was that German tank blowing the whole right side of my tank off and putting me out of commission.

Jack was awarded Four Bronze Stars, a Purple Heart, the American Service Medal, and European African Middle Eastern Service Medal with 3 Bronze Stars.

Kansas GI's witnessed the destruction of their parents and grandparents hometowns. Nuremberg, Germany. *National Archives, ww2-185*

Harland during the war.

Harlan in 2010

Place of Birth:
Keats, Kansas
Active Duty Date:
August 17, 1944
Unit:
1st Plt., D Company,
774th Tanks
(supported 83rd
and 78th Divisions)
Location:
Longueville, BE
Arrival ETO:
January 14, 1945
Rank:
Corporal,
Tank Bow Gunner

HARLAN HENRY

Edited by Carol Lacer

The 774th Tank Battalion was designated a separate battalion September 12, 1943. They arrived at Utah Beach on August 25, 1944, and attacked east arriving in Luxembourg in October. In early December they moved into the Hurtgen Forest in support of the 83rd Infantry Division attacking southwest to the Roer River.

During the Battle of the Bulge, on December 17, 1944, the battalion was put on alert to search for German paratroopers and they patrolled roads and set up road blocks for over a week. They spent much of early January 1945 in the area of Jeneffe, Belgium, with the 83rd Division due to very icy roads, deep snow and land mines.

On February 3, 1945, the battalion was attached to the 78th Infantry Division and moved to Rotgen, Germany, to cross the Roer. On March 9 elements crossed the Ludendorff Bridge at Remagen but the remainder of the battalion had to be ferried across. In April, they were in the battle for the Ruhr Pocket and ended the War in Bergen, Germany.

Here is his story. . . .

I was born in the small town of Keats, Kansas. My dad ran a gas station there. He had been a farmer, but he went broke and had to claim bankruptcy. We were poor people for the rest of my childhood.

We moved to Wabaunsee and then moved to Junction City when I was a high school junior. I graduated from high school in Junction City in May 1944. Two months later on August 17, 1944, I was drafted into the Army. The pay was $50 per month and $7.80 was taken out each month for a life

insurance policy, with my father, George Henry, named as the beneficiary. They soon raised the Army pay to $95 per month.

They sent us to Ft. Leavenworth for a week, and then I took a train to Ft. Hood, Texas, for what was supposed to be a six month training program. But they had cut it short to four months, because they needed us in the war over in Europe. We had a cadre that just worked us to death. It was so hot. We would go on 10 mile marches carrying 80 pound packs. I hated that place. I was a pretty strong person but that just about got to me. I wasn't used to that kind of work and I was ready to go any place, even Germany, to get out of there.

I went over on a ship called the Queen Elizabeth as an infantry replacement on December 10, 1944, from New York. The ship was designed for 2,000 passengers, but there were about 4,000 of us aboard the ship. It was likely the biggest ship available at the time. We had a room to sleep in for 24 hours, and then for the next 24 hours we had to stay out on the deck. We took a zigzag trip, because German submarines were trying to sink as many troop ships as they could.

Coincidentally, my uncle was on the same ship at the same time, but neither of us knew the other was aboard. He was 35 years old and got drafted. He was in the States for a long time and they decided to send him over because of the losses in the Bulge.

We landed in Glasgow, Scotland, on December 16, 1944. We got on a train and rode to Southampton, England. We crossed the English Channel to Le Havre, France, and we were there for one day. Then we got in boxcars, and rode about two days on a train that took us to the front lines.

When we got off the train, we were infantry soldiers in the snow with machine guns on the front lines. We joined the units already there fighting. I was assigned to the 83rd Division. It was colder than hell, I mean it was miserable. I was in the infantry for about a month.

Tankers were being killed, and they needed replacements, so they were taking soldiers from the infantry battalions. I was getting tired of digging a foxhole every night to try to sleep in. The ground was frozen and there was snow, so you couldn't even dig a hole to sleep in while in the infantry. You could try to dig a hole in the snow, and maybe pull a log over you for cover — it was all you could do, because all you had was a miniature shovel. It was so freezing cold. All of that made the tank sound pretty good to me.

They asked if I knew how to shoot a machine gun and I said, yeah, that is what I do. So, I volunteered to be a gunman on the tank. I thought anything was better than sleeping in all that snow. I was only 18 and didn't think I would see 19. It was hell.

I found out the tank wasn't much better than being in the infantry. There was no heat in the tanks. I was in the 774th Tank Bn., popularly known as the Blackcat Tankers. They sent us to all the hot spots. They had radios attached to tell us where there might be a German canon that needed to be taken out, and they sent us out there to do it. I was just 18 years old. All I did was shoot when they told me to.

For the rest of the war I was in an M-5 Stuart tank. I began as a bow gunner. They sat to the right of the driver. That is where I learned to drive a tank. You could drive the tank from either side. Later I was a driver.

I remember going through the Ardennes Forest. The first of February, we were supporting the 78th Division after we had moved through the Siegfried Line defenses toward the Ruhr River Dams. We moved through Rotgen, Germany, toward Schmidt. Our objective was the Schwammenauel Dam that held back a large artificial lake as part of the Ruhr River system. For about a week, we had a horrible battle with the Germans. They were in bunkers and the tanks were called up to blow them all up. I remember we were able to keep the Germans from blowing up the dams and eventually able to cross the Ruhr River on March 1. Things were really tough there.

Our tank had a 37mm cannon which was kind of a pea shooter compared to other guns. It was one of those tanks that didn't have much armor but it was fast and light.

There was lots of infantry when I got across the Ruhr River. That is the reason why we were able to get across so fast. They drove some heavy tanks across there too. We got quite a few people on the other bank but we couldn't get more people across until they built a pontoon bridge.

The battle went on until we arrived at the Rhine River. You had to come over a hill to get to the approach to the bridge, and I looked at it in awe. It might have been the only bridge left on the Rhine that had not been blown up. They had tried to blow it up, they dynamited it and whatever they did didn't work. It might have all turned differently if they had blown it up.

I went over on the Remagen Bridge while it was still standing. I was in a light tank and I think that is why they sent us across. I don't think a heavy tank could've gone across.

We crossed on March 9, I was among the first across, and the Germans blew it up the next day. We could see through the holes in the bridge and see the water below. It was barely wide enough to get the tank through.

It was dangerous. It is a wonder we didn't fall into a hole. The infantry was on the other side trying to secure a foot hold there. They had pushed the German infantry back some away from the bridge. I remember the German planes coming over trying to bomb the bridge.

That was scary! There were four of us in a tank, and we had one guy on guard all night long. We'd take turns. We used periscopes to see outside the tank, but it was so cold they would frost up outside, so you couldn't see out of the tank. You almost had to keep your door open to see, so your head was sticking out because you couldn't see through the periscope. Even if you wiped it off, it would immediately frost back up. The periscopes were worthless.

I watched the German dive bombers as they came into the bridge dropping bombs and our guns were shooting at them. But all a German soldier really would have had to do was get close enough and toss a grenade into our tank, and he would have killed all four of us.

There was a .50 caliber machine gun on the turret of the tank that the tank commander operated. He had to be out in the open but we left the lids open most of the time anyway. I would watch him shooting up at the planes, tracer bullets flying, but he didn't appear to be very close to hitting any of them.

It was daylight when we crossed. There was about 50 yards between each tank. They didn't want too many tanks on the bridge at one time. They set up a pontoon bridge real quick after the bridge fell in. Only light tanks and infantry could get across the bridge.

The Germans had a bunch of young kids doing the work. There were a bunch of fourteen year old kids doing a lot of the fighting. After that they were giving up in a lot of places and still fighting in others. You would see a lot of dead soldiers, theirs and ours lying all over the roads, it was a mess.

I was lucky; I kept the same tank the whole time. My tank commander was an "old man" at 30 years old from Nebraska. He knew the war was coming to an end. He had started in Africa and had been through a lot. He saved my life because I did what he told me to do. He didn't take very many chances and he wasn't afraid to tell the Captain or Major what he thought. I remember at the Bulge before we got to the bridge we were ordered to take out the ground mount 88. My tank commander told them our 37mm isn't going to take out an 88; and there was no way he was going up there. We didn't go. He had their respect because of his experience. The light tank was phased out after the war and the new tank had a 76mm gun.

The 37mm wasn't a very big round, about the size of a bottle of wine. We always said out goes the bullet, in goes the wine. There was a big round turret inside and you could just grab one out and put it into the gun. We didn't go up against German tanks. Their 88 would go right through our tank if we ever got hit. They came close quite a bit.

We came awfully close to getting hit one time by an 88, so close that it took some of the duffle bags that were strapped on outside the tank turret. If it had been one foot more to the inside, it would have killed all four of us.

The Germans moved out quickly as they realized we were approaching, civilians as well as the soldiers. They left in a hurry, sometimes leaving a meal on the table. We would go in and eat the meal. The Germans were pretty good cooks. I also got into the schnapps. I remember the kartofel (German for potato) schnapps; they could make that in a few days. I don't know how they did it. I got to where I kind of like that stuff.

I don't remember having many clothes at all because if you wore yours for 30 days they'd bring you up a new set and you'd just throw the old ones away. You never did take your boots or socks or anything off. You just had what you wore and tried to keep warm. It was colder in a tank than it was out on the ground. Oh, it was cold. And the tanks don't have heaters. The transmission warms up a little bit in the middle there, but that's only when you're moving. When you're sitting there, it's like sitting in a refrigerator.

I got hit by shrapnel on March 15, 1945. It wasn't bad. It happened when I was gassing up the tank with five gallon cans from a trailer load that had come by. The Germans were shooting screaming Mimies at us. I heard the shots go off. It was a terrible noise. They shoot six at a time, and they land all over and explode when they land. One landed about 25 yards away beside the tank. I was laying on my stomach on a track, there was already a guy underneath it, and a piece of shrapnel hit me in the butt. The shrapnel went through the three pairs of pants I had on and it stung like crazy. I felt back there and felt that it was bleeding. The medic told me I had better get in the jeep and to the first aid station, so I did. We drove not too far; they put a patch on it and told me it wasn't too bad, you can go back to work now.

I was in D company, the light tank, and the big tanks were A, B, and C companies of the 774th Tank Battalion. They attached us to different divisions. I was attached to the 83rd, the 78th but I was close to the 1st Division.

I remember the prisoners. There were thousands and thousands of them. One time three Germans were surrendering to me. The first thing you do is search them. Your hand gun in a light tank was a sub machine gun. What I had was what they called a grease gun. If you just touch the trigger it would fire, and it shot a lot of rounds fast. I hated that gun but that is what I had to carry when I got out of the tank.

The prisoners were supposed to keep their hands up, but one of them was trying to show me something in his pocket. I didn't speak German so I didn't want to see what he had. I just wanted him to keep his hands in the air. I kept motioning him to keep his hands up. I didn't know what he was reaching for. My gun was pointing down toward the ground and it accidently went off, and three shots hit the ground. One of them hit the center of his foot and he jumped up in the air as far as he could and started running. I yelled halt over and over, and finally he stopped. One of the other Germans was a medic, and I told him to fix the man's foot as well as he could. It was an accident I really regretted, I sure didn't do it on purpose.

I am half German myself and sometimes I wondered why I even had to be there, but Hitler was trying to rule the world and he needed to be stopped.

We were in Grassau, Germany, at the end of the war. I had been in tanks from January until the war ended. We fought for thirty-some little towns as we pushed through Germany until they surrendered.

When the war ended there was a group of SS Troopers who were not going to take unconditional surrender, so they didn't give up. We were sent to that area for about a month before they finally gave up. They were doing their training and we were doing things on our hill. People were doing negotiations with them in a building. I was standing guard on one side, checking our people in and there on the other side of the road was an SS Trooper checking his people. We weren't shooting at each other but it was a tense situation. Nothing ever happened, but it was a group of higher SS Troops who were not ready to give up. They sent our tank battalion down there. This all happened after the war was over. I was scared standing next to him. We never talked to each other.

In Germany, the name Henry was short for Heinrich. After the War, I was over there for a year in occupation. It was nice, I enjoyed that. I got to travel to Switzerland and a lot of countries and if you were speaking to a German, they would call me Mr. Heinrich. My grandfather came over from Germany and one came from Sweden.

I was put in supply and we were in the southern part of Germany. I learned to ski at Hitler's Eagle Nest on the German-Austrian border. I didn't have enough points to go home. I remember I was on orders to go to Japan until the bomb was dropped and it all ended. I can thank Truman for that.

I came home on the George Washington on June 12, 1946. It was an old ship, built in Germany in 1908 and a lot smaller than the Queen Elizabeth. We landed in Camp Kilmer, New Jersey. I remember I was in a bar in New Jersey and they wouldn't sell a beer to me. I wasn't 21 yet. I had been able to drink it for a long time and it made me pretty mad. It was hard to believe that I came back from the war and all I had been through, and I still wasn't old enough to have a beer. I was a corporal when I was discharged.

When I got home, I bought my first new car, a Dodge convertible, and that was pretty impressive in those days. I drove a cab for a couple of months. I had done that before I went into the Army and the money was really good. Ft Riley was really booming and things were good. But when I got home, it was all over with, because the war was over. I worked a few weeks driving a cab, but I knew I was not going to make any money there, so I bought a truck, and I started trucking. It was really hard work so I sold my truck after a year. I sold it for more than I paid for it and then I sold another old truck. A local dealer was amazed, and he asked me if I could sell cars for him. I said I didn't think I would be any good at that.

He offered me $50 a week, plus I got the government GI training allowance of $50 a week for on the job training. Then he said he would also give a commission on each car I sold. That was a hundred a week, and back then it wasn't bad. I sold Dodges and Plymouths for about a year and a half. Then my brother started selling Oldsmobile's and Cadillac's down the street. He came down one night and took me for a ride in an Oldsmobile. After that, I couldn't sell Dodges anymore. I felt like the Olds was so much more of a car than the Dodge. The Dodge had a fluid drive and a six cylinder motor and the Olds had a V-8. It was a new dealership in town and the guy was glad to hire someone who had some experience selling cars. I worked there 32 years selling cars that I loved. I had the whole Dodge crew there with me before it was over with.

Throughout my life, and even more after retiring in 1988, I enjoy fishing, hunting, working in my yard, cutting wood and growing a flourishing garden every summer, which I share with many friends and neighbors.

My first wife died at a young age. I met my second wife, Jan, while I was selling cars. We celebrated our 50th anniversary last year in May of 2009.

What I remember about the Battle of the Bulge was the misery. One guy would stand guard all night and you would wake up in the morning and you would see dead people lying all over. You just let the medics take care of them. It was horrible, you cannot imagine. It was something you just don't want to talk about. You never had a bath for six months. Every once in a while they would bring up clean underwear and socks, but I never had a bath for six months. Of course it was cold and you didn't sweat much. The weather was wicked that winter, one of the worst ever I am told.

The thing I remember most was getting through the Ardennes Forrest, reaching the Rhine River and driving our tank across the Remagen Bridge before it collapsed. I will never forget that.

It was a terrible war. Every day you'd wake up and think this will probably be your last. Some of us were just lucky.

Harlan was awarded a Purple Heart, Good Conduct Medal, EAMET Ribbon with 3 Battle Stars, Victory Ribbon, and Army of Occupation Ribbon.

Kenny, Fort Riley, KS

Kenny, 2010

Place of Birth:
Evansville, Indiana
Active Duty Date:
July 27, 1941
Unit:
2nd Plt., C Company,
89th Cavalry,
9th Armored
Division Location:
German-Belgium
Border
Arrival ETO:
October 25, 1944
Rank:
Staff Sergeant

KENNY LUIGS

The 89th Squadron was the reconnaissance unit for the 9th Armored Division. They many times were an autonomous outfit with the companies and platoons spread out gathering intelligence and taking prisoners. During the Battle of the Bulge units were in both the St. Vith battle area and the Luxembourg area. The fought as dismounted infantry, eventually participating in breaking the siege at Bastogne. After the Bulge, they scouted into Germany toward the Rhine River and discovered that the Ludendorff Bridge was still intact.

The 9th Armored Division landed in Normandy in September 1944, and headed for the Luxembourg-German border. When the Germans launched the Ardennes assault, the Division was in heavy combat in St. Vith, Echternach and Bastogne. On February 28, 1945, they crossed the Roer River and headed for the Ludendorff Bridge at Remagen. Before the bridge could be blown by the German engineers, the 9th rushed across, forming a bridgehead on the east side of the Rhine. By War's end, the 9th was in Leipzig and moving toward Czechoslovakia.

At the end of the Battle, an After Action Report by Commanding Officer Lt. Col. Fiore, reiterated the fact that recon teams are not suitable for dismounted combat. The cost of replacing both equipment and manpower is prohibitive. They are not trained to be infantrymen but are trained to gather intelligence. Obviously, this was not an option during the chaos of the Battle.

Here is his story. . . .

I was born on September 9, 1918, and I grew up in Evansville, Indiana. I joined the Army in Evansville in 1941 and went in to keep from starving to death. I was married to Helen on June 20, 1942. I did my training at Fort Knox but never had any basic training because I went into a brand new division just starting up, the 1st Armored Division. They put me in the 82nd Field Artillery and since I didn't take any basic training they used me as a truck driver. It was about two and a half months before I ever got a uniform. After that they sent me down to motorcycle mechanic school at Ft. Knox.

They made me a drill sergeant but I still never had basic training. I taught guys how to ride a motorcycle for about 15 months and then transferred to the 5th Armored Division at Ft. Knox.

Kenny at Ft. Knox, KY 1940

At Ft. Knox I was made a staff sergeant when I was a motor sergeant and they made me the platoon sergeant. We had a guy who came into the Army and the only clothing he had was a pair of overalls. He walked in and said, who do I see to join this place? The recruiters talked to him but they thought he wasn't very smart because he was a hillbilly. The next time I saw him, he was walking down the street in the 5th Armored Division as a master sergeant. He was a whip. He was a mechanic to end all mechanics so he was a master sergeant for the battalion motor pool.

They were going to transfer all of us guys because they disbanded our unit. I had to find a new job and they asked if I wanted to be the motor sergeant for Service Company. So the next day I was a motor sergeant in Service Company of the 5th Armored Division.

When I reported in they said they could not make me the motor sergeant and they were going to drop my rating and give me a different job. Finally a guy went AWOL and they busted him down and gave me his job. So that's where I got to be a staff sergeant.

Later I was transferred to Ft. Riley, Kansas, and was assigned to 2nd platoon, C Company, 89th Calvary, 9th Armor Division, and stayed there in the 89th Calvary until I got discharged. It was a reconnaissance group. It was a good job, better than the infantry, you know? We transferred to Leesburg, Louisiana, in October 1943, and went through all kinds of training and evaluation with a lot of field exercises.

We left Leesburg, Louisiana, at midnight on August 9, 1944, by rail for Camp Kilmer, New Jersey. I remember the day we left Camp Kilmer it was very sunny and we were carrying our duffle bags and packs as we boarded ship in the dark. We were sailing on the Queen Mary and she was painted a drab war time gray. I slept on the deck one night and then I was in a bunk the next night because there weren't enough sleeping racks. Everybody put their duffel bags in the toilet so we couldn't use that. You had to go find a place where everybody could go at the same time. We went over in five days because they went faster than submarines.

We had five different units on the ship. There was a whole hospital crew and we had a bunch of WACs. We landed in Edinburg, Scotland and went down to some place in England, and then got ready to go on to France.

We stayed in England long enough to get all of our equipment. We had tanks and armored cars, trucks and jeeps and we got it all ready. The English Channel was a very rough ride. On September 27, we landed on Utah Beach in LST's, boarded trucks and went to Luxembourg.

We went right through France and it took five days to get to Luxembourg. We traveled all day long sometimes seventy miles or more. We went through quite a bit of Franc and arrived at Schönefeld on October 24. We moved to our Company area near Clervaux and had our first combat experience.

Things were real quiet and there wasn't much shooting. We were in Luxembourg and we stayed in a four-story hotel. We would get upstairs and turn the chairs around and use the backs for gun rests. We got in the middle of the room so they couldn't see us and we opened up all the windows. They were across the road where they had a slit trench. We would catch the Germans with their pants down and then we would shoot at them. You should have seen them try to run with their pants pulled down around their ankles. The Germans took down three floors while we were there with artillery fire.

We had a General at Bastogne *(VIII Corps Commanding General Lt.-Gen. Troy H. Middleton)* that we reported to, exclusively. Once, after we got through reporting what happened he said, hell, I already knew that. The General they had there, he was a pretty good general but he didn't seem to appreciate us. He had reconnaissance men and told us his men could have told me what we told him. He said you guys are dumb; it took you a long time to find out about something like that. He always ate you up about something.

So, we figured the Germans were coming anyhow, we didn't care. Down in the area around Luxembourg and kind of north of that, there wasn't much going on at the time so they broke our outfit up and sent me with one platoon over on the Saur River bank. There were Germans on the other side. It was quiet, nobody was doing hardly anything. Every now and then they'd shoot a little bit or something like that. But when the attack came, then we dropped everything and started shooting again.

This has been so long ago, it all comes as bursts of events, no sequence to much of it. All I can tell you is this is what I remember happening.

On December 16, 1944, we were around Berdorf patrolling and had just returned from a patrol. Everything was real quiet and we were just enjoying Luxembourg. We were sitting along the Saur River with the CCA group and the Germans were on the other side. We were like that way for a week and nobody shot at anybody, nobody did anything. All of the sudden the first time they saw us walking around out there, they started shooting at us, so we fired back. It looked like trouble was coming so we moved to an area close to Haller. They were attacking all along our sector from Beaufort to Echternach.

I first realized that the attack by the Germans was big when we got run off from that river because we just had 28 men on one side and Lord knows how many they had on the other side. They ran us back to four different towns before we ever got to stop. Every time we'd set up some place, the Germans would run us out. That was when I knew it was big and serious.

Anytime the Germans attacked us and started firing at us, we would get orders to move back to the next town, but when we got to the next town the Germans were sitting there waiting for us. We went back four towns and they were there, every one of them, but when we got to the fifth town we got orders, hold or die. I said to myself, you are not getting away from here because you can't go any farther so you're going to have to hold right. The town was Clerval.

We were moved to Beaufort but the Germans had already taken some hills there. The artillery and buzz bombs were just everywhere. We made an attack toward Beaufort to help out A Troop and the 60th Infantry but after a while the Germans held us back. We had to withdraw as the artillery and machine gun fire was just bad. For a while I thought we were surrounded.

All during this time it got colder and colder, sometimes it would snow and there was lots of fog. It made it really hard to see and our patrols were something. We would be on top of Germans and not even know it.

My platoon went out on a recon *(reconnaissance)* patrol to try and find the Germans flank but we hit some tanks and infantry. We held them off for six or eight hours, they were just relentless. Everything was all messed up, confusion, nobody knew what was going on.

We had men just disappear, some captured by the Germans. Around 10 PM we couldn't hold out, too many Germans and too much artillery, I didn't think I was going to see the next sun rise. We were surrounded and we all felt desperate but not really afraid, we just didn't know anything. We set up again in Haller and in the early morning of December 18, I think, we set up defense west around Savelborn. Then they started coming, hundreds of Germans, but we really let them have it. There were dead Germans all over and we held them for a while that day.

We had three light tanks and an antiaircraft half track *(quad 50 caliber machine guns mounted on a half track)* group who got lost from their company so they fell in with us and we used them for defense. We used them to support our ground troops and the Germans were hunting that antiaircraft half track and they wanted it bad. They tried everything in the world to get it.

We stayed for a long time at Clerval. The Germans would make an attack on us from down in kind of a valley and we were up on the side of a hill. We just jacked this sucker up and got it going and we started shooting antiaircraft 50 caliber rounds down through the gorge, and we were just knocking the hell out of them.

On December 22, we had been in combat only six days but it felt like a lifetime. One afternoon two German assault guns moved into Savelborn. They were slow armored vehicles, like squatty tanks. They came in from the east but it was a surprise anyway. The German infantry attacked our antiaircraft (AAA) guns, they really wanted them bad. One of the other Sergeants saw that they were gaining momentum as they came into the town square, like a crossroad and headed for our AAA guns. He ran over to our armored car and loaded the 37mm anti-tank gun and fired five or six rounds into the rear of the assault gun and disabled it, blew a hole in the back. The driver tried to get out but he was able to get him with his carbine.

The other gun tried to get away, back where it came from. Several of us ran after the gun and shot the commander and gunner before they could get away.

I rode in an armored car or in a jeep. We had one armored car and two jeeps for every platoon but the biggest part of that time I was riding the first jeep up ahead. We were doing reconnaissance work. Once we got a mortar and put it on the back deck of our armored car and we fired off its back deck.

Our armored car was filled with food, C-rations. The armored cars had chain containers on the side and we never did put any chains on it so we would fill it full of C-rations. Both sides would be full and if you were passed by a bunch of infantrymen, they always asked you if you had anything to eat. We would open that drawer and in they would go like a bunch of dogs going after something to eat. We had stuff all the time.

Once, a guy's sister sent him a box that looked like spam. None of us really liked Spam but that evening he got it out and we all agreed to eat it. He opened it up and he shook it and there was a fifth of whiskey in it instead. His sister knew a person who worked in a meat packing plant. They filled that Spam up with a bottle of booze. When we found that we all stood around and drank it all. He wasn't too happy because everyone told him how good it was but unfortunately he didn't get any.

We were the only reconnaissance outfit in the whole division, and if they needed any intelligence gathered, they would call on our outfit. They would break our company down into platoons and send us out to see what we could find, and usually we ran night operations. Usually we would get what they wanted but sometimes we would have to come back and tell them we couldn't find what they wanted. Then they would send somebody else out but as a rule we usually got it. It all depended on how the guys in the platoons worked out and usually they worked out pretty good.

Sometimes our company broke up. We had three working platoons and sometimes you went out on patrol and when you come back one or both of the other platoons were gone. Before they got back, you were gone again. Our company was all over during the Battle of the Bulge. Part of it was up around St. Vith.

On December 27, we went on the attack along the Neufchateau-Bastogne Highway headed for Bastogne. The 101st was surrounded and we were sent to try and get them out. C Troop was sent to the Bastogne area and patrolled the area around Morhet. Our platoon was right along the German lines, southwest of Bastogne about five of six miles. We got in around midnight. The fighting was constant, the German 88's and machine guns all day and night, no let up. We moved toward Sibert to recon *(reconnaissance)* it for an attack by the 19th Tank Battalion.

They sent us out pretty close to Bastogne and we set up in the woods. The Germans fired in a whole bunch of those trees burst, shells coming down through the trees and exploding when they hit the branches and man, I mean if you were standing up, they would get you. They stationed us in the woods for a while, so we dug foxholes with overhead cover to get away from the tree bursts. They sent guys from the different platoons to find out where the artillery fire was coming from. They found them and knocked them out with our artillery and the tree bursts stopped.

One time we found three German tanks sitting down in a valley waiting in ambush for somebody to come down the road. An American unit was coming down the same road and every man had a rifle on his shoulder and had a cover on the rifle. All the guns were still covered on the tanks. We just got back from seeing the Tiger tanks, sitting down there in the valley waiting. Our recon unit was just coming out of the valley and I came upon the Colonel. We had a discussion about his unit's combat readiness because it was obvious that some of them had not been in combat yet.

The Colonel was sitting in a tank and he said, well I've got a tank gunner here and he said he can take them out. He fired three shots and knocked out three German tanks sitting right down there.

We finally broke through to Bastogne on New Year's Eve. Our CP *(Command Post)* was in the 101st Airborne's CP temporarily. We had beaten the Germans back and opened a lane into Bastogne. It was twelve midnight when we headed back to Petite Rozerie. We were absolutely worn out, constant combat for weeks. We finally got back with the Troop on January 4, 1945.

I remember another time that an antiaircraft half track was sitting along a road. Germans were down in the hold of some tanks and the half track started shooting at them and we had fire flying and everything else. The Germans sent two light tanks up to destroy the antiaircraft guns. All of the sudden here came the two tanks up the street right through the town. They were spread out and you had one to the left and you had another one on the right. The one on the left had to turn around a corner toward us.

Lt. Cleaver and I were standing there looking at them and he says, come on Luigs, let's go. I said, what are we going to do? He said get a couple of hand grenades and come on with me. So we both got a couple of hand grenades and up this street toward us came these two German light tanks, chug, chug, chug.

We went into an alley right in behind the houses. When we got into position, he says now, this last tank, when it comes by, I want you to jump up on the back of it and throw two hand grenades in that tank turret and then sit down on the hatch lid so they can't get out.

He said then I'll get the one in front right here but after you get the back one. Everything thing worked out, I got mine and he got his and they didn't bring on any more of those tanks. The rest of the company took care of the infantry and we got the tanks out with two hand grenades down the turret. I just sat on it, closed the lid and sat on it, they couldn't raise the lid and then it went boom.

That's where, right there at that time, the gliders came in. They were towed by two-motored planes one and then right in behind it was another glider full supplies and they were flying low, I mean really low. *(This was called "Operation Repulse" and occurred on December 26 and 27, 1944. The glider pilots who brought in badly needed ammunition and medical supplies at a critical point in the battle. They had to fly through unbelievable enemy fire and are one of the little remembered acts of heroism of the Battle of the Bulge.)*

I don't know if you know what kind of shape Bastogne was in. When we first went in there, there were guys hanging everywhere, the Germans shot them and killed every one of them.

We were not part of the group that actually broke into Bastogne and helped break the siege. We were on the reconnaissance end of it. We went in and out, in and out. We didn't stay there and fight the Germans like these other guys did. But there were some infantrymen already in there.

Of course we were in and out and back and forth and here and there and back and you didn't really know what was going on until you talked to somebody running the other way. You never got much from headquarters anyhow. They never told you what was going on and I'm not sure they knew either.

We went out on patrol and captured five Germans. When you were ordered to go find a prisoner, you just went out and looked for one. One time, we went out on patrol and it was dark. I mean dark and we went into a town and we saw a bunch of Germans standing in line so we turned and got in line with them. We figured, well, we'll get in line with them and then walk away with one of them. See that's one way you could do that. You could always do that in the dark.

This particular time, come to find out they were in a chow line. We didn't have any mess kits or anything to eat with and I could talk a little German, not a whole lot but some. All the sudden, I said to a couple of the Germans, I forgot my plate, come on, go with me. So they went with me and we just walked right on out of that line without being noticed. But otherwise we would have been hung up in there.

When we would go out to capture Germans we would catch them walking. German boys were the guys you wanted, not the older soldiers. You walked in behind them if you got the chance and put a gun right in their back. "Hande hoch," that means hands up high in German. They would get their hands up and we would walk them on in.

You could bring them back if you were out at night, the Germans surrendered easily. One time we caught over 100 Germans and never fired a shot. They were all old men and kids. You know the Germans there at last, used the old men and the kids. We would take their guns away from them and beat the barrels over a tree or around a tree and ruin the gun so they couldn't shoot them and send them straight to the prison fences on foot by themselves. We would radio back there and say we were sending a bunch of Germans back there.

When they had you as a prisoner, boy you were desperate, but you got out of there as fast as you could. Some of our guys that were prisoners gave up right there and they paid hard duty for that.

We saw some Germans that were dressed up like Americans and directing traffic, they were doing a lot of things. They would report in, you know and they could speak good English but they couldn't pronounce Ws and the Vs. But you just asked them a few things with a W or a V in it and when they couldn't pronounce it you knew. Some of the guys just shot them right there and that was what you were supposed to do. If they ever had anything American on them, it made no difference.

Some guys said they got their pistols off of these Germans and kept them but some of that can't be right because they didn't give up their guns that easily. I had a Luger and I had one of those P-38s, both. And boy, I had the sweetest set of dueling pistols you'll ever see in your life. I got them out of a museum. But somebody stole them. I sent them to the post office but I never did get them.

What do I remember most about the Battle of the Bulge? With everything that was going on, we didn't have enough of anything. I remember walking into Bastogne and we reported to that General. He told us get me two or three Germans from out there. He said the only way I can find out what is going on is to get some prisoners. We rode the biggest part of the time but every now and then we walked. When you went in on anything concerning reconnaissance, you walked.

We got captured one time but we got away. They got our guns but we got their guns before we finally took them prisoners and then got out of there. We caught them, just beat the hell out of them and took their guns away. In other words, it was a fist fight. That's about all it was. You were desperate then in that kind of situation.

I was walking down a real long road, trees on both sides, I was in the middle, and snow was all around on the ground. The platoon was on patrol and one of the guys fell and he broke his ankle. I had to take him back to the aid station. Well, I got him there and I wanted to stay back there until I knew how he was. It didn't work out that way. Just as I got him back, word came to come back to the company because they needed me right now so back I went. Everyone had foxholes dug by the time I got back and I didn't have one. All of a sudden the Germans fired a bunch of tree shots in there and I dove into the nearest hole headfirst. The guy who dug the foxhole, was standing outside and he couldn't get down in there because I was in his way. He was raising hell, he had me by the feet trying to pull me out of the hole and I had him by the feet trying to pull him in with me.

The weather finally cleared and the American fighters came. I was astonished to see so many of our planes and happy to see them. The airplanes came in with all the white stripes on them. I saw every bit of it and just stood out there and watched them fly in. Vapor trails, they left a bunch. They just turned the whole sky white everywhere. Boy, everybody was ready to see them.

The P-47 fighters were made in Evansville, Indiana where I was from. They provided close air support. They made a lot of them too. I read something about it not too long ago that they were turning out about 15 of them a day. They were supposed to be real good airplanes and they made them out on Highway 41.

We also had an ammunition plant that made nothing but 45 caliber ammunition. They made a museum out of part of it and they wanted a GI uniform so I gave them mine.

When we saw the white streaks up there that really it was the beginning of the end of the Bulge. They started dropping bombs and we watched where they put them, pretty close to Bastogne. We just sat outside our foxholes hollering at them and shaking our fists at the Germans.

We had a big radio in our armored car on one side and a radio on the other side. This radio kept you in contact with the jeeps out in the field. You could also contact headquarters and as far as I know they were only in armored cars. The ones in the jeeps were small ones, just a small box, but this big radio was used to call the jeeps because they had greater distance. We acted like a relay between the jeeps and headquarters. They were very good radios. Once we were sitting up on a hill in England and I got a radio station in New Jersey. It had a lot of up to date music on it.

When we would call in a fire mission for artillery, sometimes we would talk directly to the batteries after they finally got them set up but that wasn't very often. Usually we contacted headquarters with anything that we had going on.

We never used the radios at night because you couldn't see who was standing outside listening to you so you just turned them off. We did a lot of night work. At night you would be right among the enemy and vice versa. You would sometimes be walking right along with them. If you are trying to catch some prisoners, sometimes here the Germans came along talking to each other real loud. They never were quiet and a lot of times we fell right in there with them, walked with them for a while.

We had it all planned in advance what we would do. They were doing the same thing we were, but we had it all cut and dry. We would assign a German to each person, and then we just fell right in with them and next thing you knew, we had a gun on them and we took them back down to headquarters.

At night we would travel in a convoy using blackout lights *(small triangular shaped lights for night driving to see the vehicle ahead only)*. You couldn't read with them and there was no way we could have any lights on at all at night so we could not use a compass or a map. But you had little bitty lights on the back of the vehicles if you were riding in a convoy.

We were riding in a convoy one night and all the sudden nothing, only black. We were following two taillights in front of us, no tail lights and all the sudden bang, bang, crash. The road turned real hard and there was a barn right there on the side of the road and we drove right into that barn.

I captured two prisoners that night too without a gun there at the barn. We spent the whole night in the barn and we had two jeeps with us. So when morning came, I went up into the loft and there were two Germans sitting up there and they said comrade. I yelled to my guys that I was sending them some company sent the two prisoners down to them.

There was always somebody finding something to laugh about. Once we went back to a little town that had maybe 15 buildings in it. There was a big barn and in that barn yard was a big red bull. A girl came down the road leading a cow with a rope. She put the cow inside of the barn yard and we sat and watched that cow breed with that bull and laughed. But I know one thing; we had a lot of guys killed there. And there's nothing funny about any of it. Some things I can laugh about now but it wasn't funny then.

After the Battle of the Bulge, we crossed the Siegfried Line. It had concrete and stone obstacles and pillboxes. They came by with tanks with big scoops on the front of them and they just scooped up the concrete and the pillboxes and threw them over in a pile. They opened up a road real quick through there. Germans were smart but they weren't smart enough because we got through there too.

I remember the Roer River but not too much. We went on doing recon work east through Germany toward the Rhine. We were moving through pretty fast, not much resistance and we were at Bad Neuenahr about 5 or 6 miles west of Remagen on the Rhine River. We didn't expect to see the Ludendorff Railroad Bridge still up because we heard that it was blown by the Germans.

We found the Remagen Bridge as we were going along on a reconnaissance patrol toward the Rhine. We were watching a road and we finally moved our position over a hill. We couldn't see it from where we were had been and as we got to the top of the hill all the sudden there it sat. We radioed back and said there's a bridge up here and it's all in one piece so come and take a look. Even old Patton got excited. He always showed up with his flags and sirens and he always seemed to draw fire.

We dug a big foxhole right where we were. We were in contact with the company with our handheld radios and called in to them and told them what we saw and boy they said stick tight, don't move, don't do nothing, don't talk to nobody. You just sit there and wait until we call you.

The command post called back and said let me know where you are again so we told them where we were. The CO asked if that bridge was still there. I said it sure is. And then about that time we had another radio crew over there a ways and they said, yes, it is there, I can see it from here. Then we had another radio crew say it's still here but you better hurry up and get up here because they're going to blow it up themselves and boy I mean that's when things got hot.

We could you see the German engineers crawling around underneath it trying to set the charges. If we hadn't seen them setting the charges, they would have blown it up before we got there. We were up in the hills right back along the river. There was a railroad track that ran along the river on the other side that came across the bridge.

Everybody drove right over the tracks. I didn't see that they put anything on the bridge as we drove across. A lot of them walked. The crossroads we were calling artillery into was not too far from the bridge. It was along the river on our side of the river.

I was out on the top our armored car when I got blown off the bridge. I don't remember what happened. They fished me out of the river and when I came to, I was in an airplane on the way to England. It doesn't take them long to get you moving to medical help. They moved them around and quick.

I was in Piccadilly Square when the War ended. I was walking through Piccadilly and I bumped into the arm of a guy I grew up with. It was the first time I had seen him in a long time. I didn't even know he was there. So I went back to his outfit with him and he talked to his first sergeant to let me go on a B-17 airplane ride. We left about 8:00 o'clock in the morning and we flew all morning. I even got to sit in the copilot's seat for a while and you could see everything. We flew over a lot of areas I was in and you can't believe how much damage was done there.

I had a lot of points because I went in the Army in 1940 and they used points starting from when you came into the Army. They started to send me back to my outfit but they said I would be one of the first ones to go straight home. However, instead, they sent me back to my outfit. We rode a 40 and 8's boxcar for I don't know how many days. They had no roofs and soot would fly from the engine and everything else. When we rode alongside a river the engineers would stop every now and then and everyone would jump in. Three toots meant you better get back on because the train is leaving.

When I came home, I was able to call Helen when I was in the States. We come back on a ship and landed in New York. Then they sent us on a train to Newport News, Virginia, and on to Indianapolis where I was discharged.

I had a Luger. The guy standing out in front asked if I had a gun I wanted to sell. So I sold it to him. He needed it worse than I did.

There was a lot of animosity toward us vets when we came home because people were afraid we would get their jobs. Evansville was all manufacturing and I couldn't get a job any place. I finally got to work with the ironworkers. It was a job moving the production equipment used to make the P-47's out of Whirlpool to make room for something else.

Tad, Joe, Pal, Jerry Ritter

I finally got a different job but I tell you what, it was tough for Vets to get jobs for a while. Lots of returning vets were able to get their jobs back that they had before they went into the Army but I didn't have any job when I went in. I tried to reenlist and go back to Germany. They said they would send me to Japan and I said no way.

We had reunions almost every year and in some nice towns. We had them in Maine, out west and all over and Helen had a good time when we went to them. While I was overseas, Helen stayed with her folks and took care of Joe, our son. Joe was born when I was in California on desert maneuvers.

Editor's note: Joe and I were good friends growing up. I had no idea what Kenny had been through. He was just one of the great Dad's in the neighborhood.

Kenny was awarded the Good Conduct Medal, American Defense Medal, European African Middle Eastern Service Medal, World War II Victory Medal, Overseas Commemorative, and Expert Marksmanship Badge.

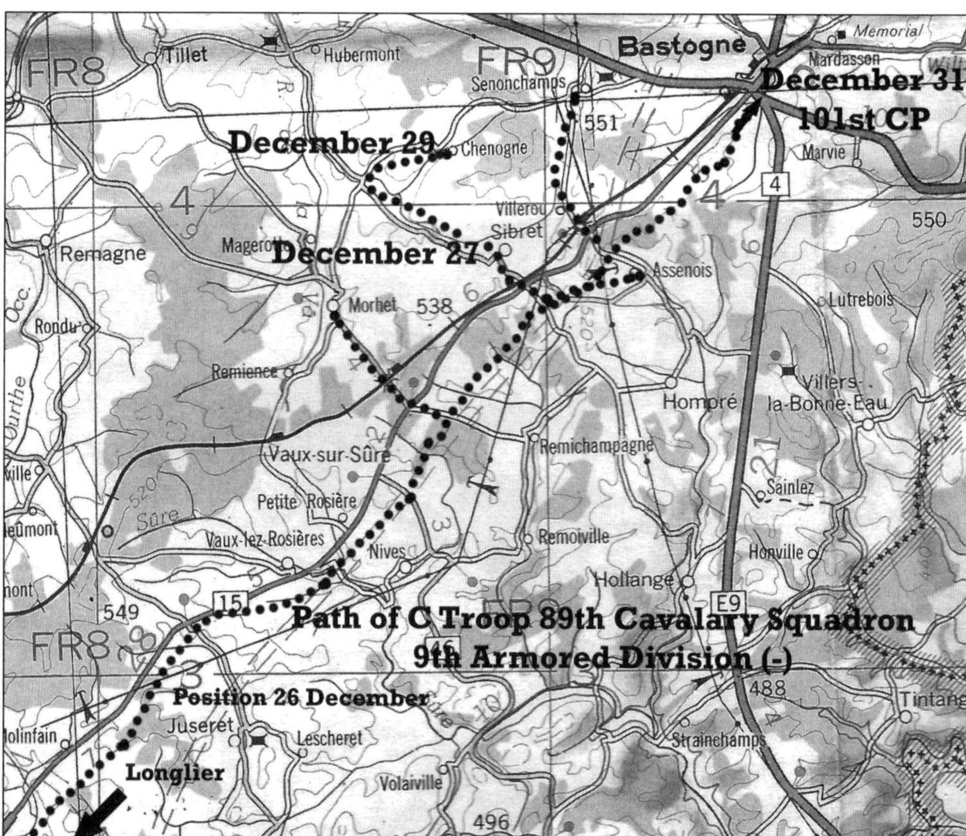

Corporal Carl Shell

Carl Shell, 2010.

Place of Birth:
Virginia, Kentucky
Active Duty Date:
October 31, 1942
Unit:
A Company,
14th Tank Battalion,
9th Armored Division
Division Location:
St. Vith, BE
Arrival ETO:
October 23, 1944
Rank:
Corporal, attained
rank of Command
Sergeant Major

CARL SHELL

The 9th Armored Division landed in Normandy in September 1944 and headed for the Luxembourg-Germany. When the Germans launched the Ardennes assault, the Division was in heavy combat from St. Vith, Belgium to Echternach, Luxembourg. Elements of the Division were instrumental in breaking the siege on Bastogne in late December, 1944. On February 28, 1945, they crossed the Roer River and headed for the Ludendorff Bridge at Remagen. Before the bridge could be blown by the German engineers, the 9th rushed across, forming a bridgehead on the east side of the Rhine. By War's end, the 9th was in Leipzig and moving toward Czechoslovakia.

The 9th Armored Division was known to the German soldiers as the "Phantom Division" because they seemed to be everywhere and nowhere. They were the first armored division to fight in the Battle of the Bulge and the first to cross the Rhine.

Here is his story. . . .

I was born November 17, 1925. I grew up in Virginia, Kentucky. My dad was a coal miner in the Eastern part of Kentucky in around Hazards, Kentucky. I went into the Army in October 1942 at age 16, just one month before my 17th birthday. I volunteered I went over to Norton Virginia. I was sent to Fort Riley and they were forming the 9th Armored Division. I was tank gunner. As you know we were trained to fill all of the positions in a tank in case something happened I could drive, I could load, I could operate as gunner. I was a corporal during the Battle of the

Bulge. I was in Company A. 14th Tank Battalion. We had three tank platoons and I was in third platoon.

In 1943 we went to the desert for desert training. First we had an M-3 tank that only had a 60° traverse on the main gun and then we got the M4's. We came out of California trained but not knowing where we were going overseas. We didn't know if we were going to North Africa even though that is why we trained in the desert. From California they sent us to Fort Polk Louisiana for swamp training. This was in case we went to the Pacific. I trained there in July 1944. We boarded the Queen Mary in New York on August 26, 1944, and they shipped us to Europe and we landed at Hampton England. We were put in the Tidwell barracks. We drew the tanks, cleaned the Cosmolene from the guns and test fired them. Then they put us on LST's to cross the channel and we landed at Omaha Beach and of course the fighting was not on the beach at that point.

We started driving up through France and made contact just beyond Paris. From there it was just the regular combat through Belgium. From there we went on to St. Vith to the town of Ligneuville, about 13 kilometers northwest of St. Vith and that is where we were when the Battle of the Bulge broke out. We were right along the Western Wall and were sitting in defensive positions. The 106th Infantry Division came through our position so they could get to the front line and relieve the 2nd Infantry Division, another infantry division a few yards ahead of us. So we just sat there, on watch, taking a few indirect rounds.

Carl Shell and his tank crew.

I was on guard at that time when the Germans began to attack. Off to my left I saw this machine gun tracer fire and it kind by passed us more or less. But then they came into our positions. Every fifth round on a machine gun was a tracer bullet so you can mark your fire and tell where was going. That was my first indication that the Germans were attacking. I was on guard and I saw the Germans coming across toward me on foot. They were about a hundred yards away. I told everybody the Germans were attacking and some of the crew was sitting in the tank asleep. So I jumped on the 50 caliber and burned up a belt of ammo before I could stop. They had already come through the infantry. There was no preparatory artillery barrage or anything.

I did notice one thing, every night while we're staying there a lone airplane, an enemy plane, an old boxcar kind of a plane, seemed to be taking reconnaissance pictures of our position and we named him "Bed Check Charlie." The plane flew over every night, I wanted to the fire so bad to knock him out but they said a hold your fire, you couldn't even smoke a cigarette, you couldn't let them know where you were. I wanted to blast that sucker out of the air.

It was just survival after that, firing, keeping alive. We got the order to move. On December 23, 1944, a couple of tanks were hit. One section of our company had a German patrol attack their tanks, climbing on the back of them trying to get on top to drop grenades down the hatch and knock the tank out. The driver told me later that he had the gunner elevate the gun as high as he could. We had to close up our turrets.

We did not see any German tanks until the next morning after the infantry attack but the next morning we are making a withdrawal to get a better position, we were not retreating, and that's when I got my first German tank, a mark IV. I felt pretty good about it and glad it didn't get me. We kept pulling back and getting a better position, defending ourselves, it was really chaos I'll tell you. The Germans had an anit-tank weapon called a Panzerfaust, a bazooka and as matter-of-fact we had a couple of tanks knocked out by their bazookas. I was surprised that they did not have more bazooka teams coming through. We just kept withdrawing, defending, withdrawing we just kept right on going through St. Vith. That was a mess. The Germans had gotten in there and turned the signs around and had sent patrols in and they had infiltrated evidently before the attack started.

The Germans were dressed in their regular Army Gray and in white snow camouflage. I don't recall them carrying heavy packs like our infantry did but they were in, occasionally you would see the white camouflage. At St. Vith it was chaos. In addition to us, elements of other units like the 106th were coming in also. I was sitting in the tank and a tank commander had a general idea of making him a sketch map but I didn't know where I was half the time. I could only see through my scope, we were in defense and setting up roadblocks.

I did not have the same tank the whole war. I had a couple of them knocked out.

Combat Command A and Combat Command B was around the Bastogne-Malmedy area. We were part of Combat Command B and we were all around the area around Bastogne. Then there was a combat command in reserve. Combat command was set up as a fighting force, for instance Combat Command B would have a battalion of the 27th armored infantry and some engineers maybe a platoon of recon and a battery of artillery, under one command. It was like a brigade.

I had no idea where my tank was as we were in the defense and attacking. About a month later they pulled us in around Verdun to draw the tanks and equipment and replacements. I think we were around the Liege area when the Bulge ended.

After we got replacements and so forth, we went on the attack across the Rhine and on into Czechoslovakia and there was fighting all along the way. After we crossed the Rhine it was fairly light contact as we would run into some machine guns and some tank fire. The 9th Armored Division was not always part of the third Army we switched between the third Army in the first Army. At the beginning of the Bulge we were in the first Army's area of operation and under their control.

You never knew where you were, what was happening, at least it seemed that way with me, I was in a tank and I couldn't see. We seemed to be surrounded quite a few times. We had seven tanks left at one time, and Captain Sumas, our company commander, we were on some high ground setting up defensive positions and a German recon vehicle came around the Hill and our tanks fired at the vehicle at the same time and blew them up into the air. We found half of them up on the electrical wires. One of the guys ran over to see what he could find and found some maps and so forth and he came back with a German pipe sticking out of his mouth. He came back to the company commander with his pipe in his mouth and the officer's P 38 pistol. We set there overnight, we had a few infantry with us and they dug in around us to protect the tanks and we found that we were surrounded, we got orders to head for the rear and the Company Commander knew which way to go so we just lined up and headed for the rear, we are surrounded. Every tank got through except the last one and they got him and we ran right through a German roadblock. We kind of surprised them. They were facing the other way and we came up from their backs and they looked up and saw our tanks barreling into the roadblock. We had six or seven tanks left and there are 17 tanks to a company, 16 plus the command tank. I believe there was a platoon in another area.

We were not able to get out of the tanks after several days. We had a good supply of C rations and K rations that would last us for a few days. I remember getting one hot meal in those 30 days. One rest period I was able to take my boots off and my feet just puffed up because of the cold weather and I was not able to get my boots back on so I found some galoshes, put those on with two pairs of socks and wore those for a few days and put my boots back on.

My position in the tank was inside the turret. In the body of the tank was the driver and the assistant driver who also the bow gunner and had the coaxial gun. The driver was down front, the assistant driver to the right and in the turret next to the gunner were the loader and the tank commander almost on my shoulders. We had a good supply of fuel and ammunition really; every time we stopped someone would come up and take account of how much ammunition we needed and to fill the tanks up.

I'm not sure I ever saw any of the Germans dressed as Americans because they were in American MP uniforms. And there is a lot of traffic going through St.Vith and a lot of confusion. I heard that the Germans dressed up as American MPs turned road signs to confuse our traffic but I never had any personal knowledge of it. We fought with the 101st at Bastogne.

After the Battle of the Bulge, we went to Verdun, France. At Verdun we cleaned equipment, bore sighted our guns and training recruits. I didn't even get out of the vicinity of the tank. We're kind of in a barnyard. One lieutenant got a pass but I never got a pass.

We crossed the Siegfried Line but don't remember much about it. Most of the Dragon teeth were cleared by the time we went through. March 7 was first time ever seeing the Remagen Bridge. Remagen was our objective and when we got there we found the bridge was intact. Lieut. Grimbec had the first platoon and he called back and said that the bridge was intact and so forth and we finally got the word to cross. We set there quite a while, a few hours, then the infantry from the 27th AIB crossed first and to clear out the machine guns on the bridge, then they sent the tanks across. I went across the bridge in my tank, the bridge was pretty well damaged and they had to patch it up and put up a engineers tape it was a

white tape, to mark holes some so big you could fall through into the Rhine River.

It was nighttime when we crossed. It was dark. We had blackout lights to follow the tank in front of us in the dark and after we got across a couple of tanks went to the left a couple to the right, to set up a defensive perimeter and the next day or two the Germans were continually bombing and shelling the bridge, trying to knock out the bridge. We were across the bridge; I couldn't tell whether the bridge was solid or creaky. The bridge fell in later.

We came across one concentration camp. As we came across one just before we got to Leipzig, there was a concentration camp or prison camp one of our tanks went in. The guards had left so we just broke the fence down and had to continue on and let the infantry and MPs behind us take care of the situation. We couldn't get out of our tanks; we had to hold our position and our tanks, so we did not go into the camp.

One time I got really chewed out. We came into a town and we were advancing somewhere along there and German bazooka teams were really tough on tanks and I saw this three-man crew come out of a ditch and run into a little building as we were going into a town so I jumped out of the tank got my submachine gun went after them and brought them out and they were three young kids about 14 years old. They were in the uniforms, they had ammunition, and they had everything that could knock you out in a minute. So one got really mad at me, I reached over and just rubbed my hand on his cheek because he didn't have any fuzz on his face, making fun of his. I sent him back as a prisoner. The Tank Commander really chewed me out. He said don't ever do that again, you stay in your tank, you could get yourself killed in there but I didn't those little rascals getting away because they had a bazooka and had about 3 or 4 rounds.

You had to be real careful when we got to the German side of the Rhine because you never knew for sure who was going to shoot at you. It could've been little kids. It could've been old man. You just never knew. The German people put out white flags as went through the towns, some would waved at the column and most had a doubtful look on their face wondering what we're meant to do. They had heard some bad things about what the Americans would do to them that our soldiers were bad, but we gave the kids candy bars.

I saw a number of German buzz bombs and saw a number of planes, come in trying to blow up the Remagen Bridge and I burned up quite a few 50 caliber rounds myself. I don't think I hit any planes but I felt good defending myself. We had a 50 caliber outside on top of the turret and a 30 caliber inside coaxial with your 76mm. If the two button triggers are down they both would shoot. I hit the wrong button sometimes thinking I was shooting a machine gun and would throw and HE into a German. I fired the 50 caliber, a 30 caliber and 76mm. The tank commander used the 50 most of the time.

Oh, boy do I remember the day when the Air Force and the P47s showed up. Yes, it cleared up and all of a sudden it was like hundreds of them overhead, going into the front and you hear the boom, boom as they're blowing up the Germans and it was wonderful because we had not had any air support in a week. It felt great. I thought here they come. You could see the jet streams behind them and we had not seen that for quite a few days.

I wish I could tell you more detail of some of that stuff. As you know combat is combat. I do remember once when we were down in Remagen and sometimes things are pretty quiet and you want to go have a BM by yourself so this one fella named Lively jumped out of his tank and went into this house and we started getting incoming artillery. While he was in there a round came in and hit the top of the house and here he came running out with his pants down around his ankles just as white as a sheet, covered in chalk from the ceiling and walls that fell. He was white and he was cursing the Germans, "a fella can even go do his chore without the Germans interfering." We got such a kick out of it.

We had reached the outskirts of Carlsbad in Czechoslovakia when the war ended. We got the order to cease-fire and hold the ground we had. There was very little resistance. First platoon of D company, a light tank company, fired the last round of the war. A Company fired the first round at the start of the combat.

When the war ended we cleaned up our equipment and then they started rotating people back to the states according to how many points they had, their age, if they had dependents, but I had no dependence. But I was one of the first to be shipped out to be a replacement in the Pacific. We were actually on the ocean in a ship headed for the war in the Pacific when they dropped the atomic bomb on Japan. We had been in the states for a 15 day leave and then sailed for the Pacific. They were planning on a large invasion of Japan. When the bomb was dropped the war ended and it saved a lot of lives I believe. I was going in there as replacements. After about 11 months of all that in Europe.

When I was discharged and went back to my parent's home. They did not know I was on my way home. I walked in the house and they about had a heart attack. My dad worked in the mines. I just walked in the house with my duffel bag on my shoulder as I was walking home; I was checking the window to see how many had wreathes in the Windows that showed me who had lost relatives in the war. Worley Hughes was a friend of mine a couple doors down and he got killed Worley was somebody that I grew up with. I worked for about 30 days in the coal mines. I told my dad, dad I would rather go fight wars than go back down inside of that mountain, coal mining was not for me and there's nothing else there for me to do. So I went to Norton, Virginia and reenlist.

I volunteered for overseas again so I got Europe and was assigned to the First Infantry Division and I made rank pretty fast in the infantry. I stayed in the infantry most of my career after I reenlisted. The last two years I spent back in armor because I was an E-9 because there is only one Sergeant Major in each battalion. So I was a Sergeant Major and I went back to the armor group. I spent 15 years of occupation duty in Europe. I married a German girl in 1954 from Düsseldorf named Catharina. We were married for 46 years she passed away in 2001. We adapt two little German children they have their own families now at Fort Hood. I got remarried two years ago to Laura.

Jelle Thys. *Carl Shell photo*

We traveled quite a bit. We've taken two cruises. She's from Montana so we go there quite a bit and we traveled to Ohio and Kentucky down to Texas. I've been back to the Bulge are and while stationed in Europe for a long time. I have been back to Bastogne and St.Vith. A young 15-year-old Belgian boy named Jelle has been corresponding with me. He sent me some snapshots. There is an M4 tank set up in his hometown. We liberated his town.

What do I remember most about the Battle of the Bulge? The things I remember the most about the Battle of the Bulge it would be extreme cold relay number one and the intense fighting and the initial attack was kind of surprising, we were not expecting that.

Carl rose to the rank of Command Sergeant Major in the Army, the highest non-commissioned officer rank in the Army. Only a few attain this rank, so few in fact that there is assigned only one per battalion. Carl currently resides in Junction City, Kansas with his wife Laura. Carl received many personal awards along with the ETO Campaign Ribbon w/ 3 battle stars.

General George Patton's grave in Luxembourg. *Author photo*

Author photo

Ludendorff Bridge today is a museum to the battle.

Nadine and Herman

Herman in 2010.

Place of Birth:
Farmington, Missouri
Active Duty Date:
November 11, 1942
Unit:
HQ Battery,
490th Armored FA,
11th Armor Division
Division Location:
Rechrival, BE
Arrival ETO:
December 17, 1944
Rank: Sergeant

HERMAN WILLIAM WESTMEYER

Edited by Rex Westmeyer

The 11th Armored Division was activated on August 15, 1942, and landed in Normandy on December 16, 1944. They were rushed to the Meuse River where they were assigned a 30 mile defensive sector from Givet to Sedan. During the Battle of the Bulge, the Division defended the highway to Bastogne around Neufchâteau and spearheaded an action that helped pinch off the gap in the Bulge on January 16, 1945.

After reaching the Rhine, the Division turned south through Andernach and crossed the Moselle River at Bullay. They attacked east to Linz, Austria and ended the War in that area.

On December 23, 1944, The Battle of the Bulge is entering its second week when the 11th Armored Division first appeared on the 12th Army operations map. The Division had landed on December 17 and had been in route along with the 75th Infantry Division to the Lorient Pocket in southwestern France when their path was suddenly changed.

Tanks, half-tracks, armored cars, trucks of the 490th Armored Field Artillery Battalion scurried toward Bastogne and by Christmas Day their Division HQ was in Rethal, France, still 140 km southwest of Bastogne.

Here is his story. . . .

I was born on November 15, 1912, in Farmington, Missouri, south of St. Louis, Missouri. I was one of eight children in the family. We lived on a 300 acre farm near the edge of town. One of my earliest memories was the end of World War I. I was helping my oldest brother Walter drive cattle from one pasture to another, and suddenly church bells and storm sirens started going off in the town of Farmington. My brother says, 'the war has ended!' I was six years old.

I finished college at the University of Missouri in 1936, and had been working for years as an Agricultural Extension Service County Agent in Barber County, Kansas by the time World War Two began. I was drafted into the Army on November 11, 1942. It was really a surprise when I received my draft notice, because in 1941 my draft board chairman (who was also a doctor) told me that I didn't need to worry about going to war because of lung problems as a boy. One lung was smaller than the other. On the basis of his comments, I got married and accepted a job in Harper County, Kansas, for more money. Eleven months later I was drafted as a private in the Army, (Chuckle) and that was the deal.

I was stationed in Camp Polk Louisiana. The Army had started a new armored division called the 11th Armored Division. I was one of the first in it and spent two years in training. None of the men I trained with ever went to boot camp or recruit training. Enlisted men and officers who had prior military experience or who were regulars before the war started taught us how to wear our uniforms, march and do our jobs. From the start I was assigned to Headquarters Battery, 490th Field Artillery, 11th Armored Division. In those days I had extremely good vision and my lieutenant assigned me as a forward observer. I never went to a school to learn that, it was all just on-the-job training.

In 1942, the 11th Armored Division left Camp Polk, Louisiana, and traveled by train to Camp Barkley, Texas, and other points west. I remember training in Camp Barkley, Texas, Needles, California, and Camp Cooke, California. In each place we practiced field maneuvers and tried to fire our artillery and tank guns, but the weather didn't always cooperate. From Camp Cooke we loaded trains again and traveled to the East Coast to ship out to England.

On August 15, 1942, the 490th Armored Field Artillery Battalion was one of three self-propelled, 105mm howitzer battalions activated at Camp Polk, Louisiana in support of the 11th Armored Division. The Battalion was composed of Headquarters and Headquarters Battery, three firing batteries of six M7A1 howitzer sections each, and Service Battery.

In the armored division we had tanks, self-propelled artillery and mechanized infantry units. Each artillery battery had several tanks. Some tanks would call and spot artillery fire for the mechanized artillery. That was what my job was. The other forward observer tanks would be attached to the tank companies. Their job was to lay artillery fire down ahead of where the tanks were going during the assault.

Our tank was the same as all the rest. We had a stinking little 75mm and the Germans had the larger 88mm guns. That was not good but we had basically the same firepower as the other Sherman's. We tried to stay out of the Main line of battle so they wouldn't get us and we were continually directing artillery fire.

We arrived at Liverpool from New York on an English ship. We went into a camp somewhere in England and just sat for a couple of months. We didn't even do any training.

We loaded onto LST's, in the middle of December, 1944 and landed in Cherbourg, France, in direct support of Combat Command, B (CCB). CCB consisted of the 21st and 55th Armored Infantry Battalion. We were supposed to contain the channel ports in France which were still German submarine bases.

When the Germans attacked in the Ardennes on December 16, they moved us up to join the 17th Airborne Division in defensive positions along the Meuse River. CCB *(including the 490th)* went through the Arc de Triumph in the center of Paris.

We were with the Third Army with General Patton but I never saw him. The 11th armored was late getting in the fight and I sure didn't spend much time with the unit. We came into France around 15th or 16th of December and we got into the frontline on Christmas Day. On New Year's Day I was evacuated off the line after being wounded. But I saw it all, the Infantry with their wrapped up feet, a guy that was killed by a sniper while I was looking at him. From Christmas Day on when we hit the front lines we had only C Rations to eat. When you get right down to it they weren't too bad.

We wore steel helmets with hearing gadgets in them so we could talk on the radio. In that weather a set of winter coveralls was a prize. The five of us in my tank got coveralls. Most didn't get them. They made a whale of a difference. We carried our personal belongings in the tank and we slept in the tank. I did not have any white camouflage or equivalent or any snow gear. We never ran into any deep snow. But among other things we never slept on the ground overnight. We always slept in our tanks. There were no heaters in the tanks so it was cold.

We carried carbines, little old rifles that wouldn't stop anything, wouldn't kill a hog. It was a gun but the old .45 caliber Grease Guns would have worked better. We never got in a position that we had to use them. The best gun then and now is the 50 caliber machine gun. They were easy to clean and would always operate. It had long range and they still use the same old gun. We had that on top of the tank."

> From the 490th Battalion After Action Report: the 490th moved across what was soon to become icy, slick roads, especially torturous for tracked vehicles.
>
> 25 December 1944
> Battalion completed road march across France, closing in bivouac, an apple orchard south of Jonval, Belgium, at 2150, after clearing Laon at 1630. Distance traveled 59 miles. Roads good, weather cold.
>
> 26 December 1944
> Battalion remained in position, becoming strategic reserve in defense of that sector of the Meuse River between Mouzons on the North and Verdun on the south. Batteries did not occupy firing positions but reconnaissance by ground and air took place for suitable routes of approach, bridges, positions, and observation posts from Jonval to the Meuse River and five miles east of the river.
>
> 27 December 1944
> Reconnaissance continued. Battalion remaining in previous position.
>
> 28 December 1944
> No change.
> Results of Action 23 - 28 December: Battalion completed march from Cherbourg started 22 December, was in position to defend against further enemy penetrations towards the Meuse.
>
> 29 December 1944-11 January 1945 – Engagement at Chenogne, Rechrival Valley
>
> 29 December 1944
> Preceded by reconnaissance parties, Battalion cleared bivouac south of Jonval, Belgium at 0700, marched to Respelle to bivouac, closing at 2100. Reconnaissance parties, preparing positions south of Freux, came under artillery fire. Distance traveled 74 miles.

Now one of the interesting things for me was that we didn't have any idea what was going on. I know some of the top officers knew what was going on, but we had no idea. We were heading in one direction and were told to halt. We stayed in place for about an hour and were told to head out in an entirely new direction. The Battalion got on a paved Autobahn or major highway to move faster. We were told to keep moving and if we broke down, to pull off the road, get the tank fixed any way we could then catch up. It was only later that I found out we were reinforcing other units for the Battle of the Bulge. It was ice and very slick. You don't handle a track in the ice and snow very well and my tank driver got snow blindness and his eyes rolled up and I had to crawl down and inside the tank and drive it.

We went up and down hills that could kill you because you could look down into a gully off the side of the road as we went downhill and when you tried to keep the left track straight and close to the side we were ready to slide into a ditch. I could look down a gully and I wondered how deep it really was, the treads were slipping terribly and I was able to catch a few bare spots in the pavement that caught a little and got us out. We started going uphill and of course we had the same trouble with steering going uphill but at least I didn't have to look down into that canyon below. That was an interesting deal.

> *(On 30 December, the Battalion fired its first combat rounds at Remagne, Belgium in support of CCA's attack to destroy elements of two panzer grenadier divisions which were pressing the 101st Airborne Division, besieged at Bastogne.)*

> From the 490th Battalion After Action Report:
>
> 30 December 1944
>
> 0300 the Battalion cleared bivouac at Respelle and occupied firing positions south of Bougnimont.
>
> 490th AFA Bn was in direct support of Combat Command A (CCA), consisting of the 63rd Armored Infantry Battalion, 42nd Tank Battalion, with elements of the 41st Cavalry Recon Squadron, 56th Armored Engineer Battalion, 5th AAA Battalion, 133rd Ordnance Maintenance Battalion, 602nd Tank Destroyer Battalion, and 81st Armored Medical Battalion.

CCA jumped off from vicinity of Rondu, objective Remagne, as the Division began a drive towards its objective of securing the L'Ourthe River line.

0700, the 490th fired its initial rounds in combat, a volley by B Battery, closely followed by A and C Batteries. Target was tanks in wooded area.

At approximately 1000, 2nd Lt. Hadley Neff, a forward observer with a medium tank company, was killed in action by mortars on a ridge NW of Rondu. Beginning at noon the Battalion displaced by bounds to new position 2000 yards NW of Bougnimont and continued firing. Toward the afternoon light counter battery fire was received. Results of day's action: attacking elements gained high ground NW of Rondu, and then withdrew; one officer killed.

(The 11th Armored Division had regrouped during the night of 30 December with the object of consolidating the entire division for a drive north along the Rechrival valley. CCA moved over the icy roads south of the Bois des Haies de Magery, which had separated it from the rest of the division, and onto the Neufchâteau-Bastogne road, where the columns ran afoul of those of CCB, 6th Armored causing a massive traffic jam.

In the morning CCA turned off the highway and assembled around Morhet. The new scheme of maneuver called for CCA to attack in the center of the valley, emerge from it at Rechrival and Hubermont, then capture Flamierge.)

We were hesitating in this small area where there were houses but not a town. The Colonel was there trying to figure out what to do next when a soldier brought up a prisoner who was a sniper. The guard brought him by and asked the Colonel what to do with him. He told him to take him to the prisoner holding area and he started out away from us and one of the Lieutenants just up and shot him. The Lieutenant acted odd, then went around a house, got in a tank and started shooting into a haystack yelling that someone was in that haystack. He went berserk. Plumb berserk, I don't know the process but they got him out of there.

We moved toward Houffalize from Neufchâteau, attacked the Germans east of Rechrival on New Year's Day. We were trying to relieve the pressure on the 101st Airborne's position in Bastogne.

(Near Rechrival the march into Bastogne was interrupted by an enemy screening force that had been deployed by the 3d Panzer Grenadier to protect its western flank. They were equipped with antitank and assault guns and were quickly reinforced by the 115th Regiment of the 15th Panzer Grenadier. After a few attacks that were beaten back, the 115th used the Panzerfaust to put a number of tanks out of commission. Soon a flight of P-47 fighter bombers appeared out of the clearing skies to assist in pushing back the grenadiers. After considerable reorganization the march resumed. Rechrival was found empty and the command set up positions there for the night. The snow was now knee deep.)

31 December 1944 490th After Action Report:

0030 the Battalion displaced to Le Prey, moving into firing position at 0408 after 20 miles over slippery roads. Weather cold.

Fire in support of continued attack by CCA was opened at 0500 as leading elements pushed forward from vicinity of Moret towards Lavaselle and Rechrival, Belgium. Results of Day's action: CCA gained initially; Batteries firing in support entire day.

(General Kilburn decided that the main effort by the division on January 1, 1945, would be an attack made by the CCB and coordinated with that of CCA, 9th Armored. The CCA, 9th Armor was held up at Senonchamps to the east. At Chenogne, on New Year's Day morning, two medium tank companies made the assault after all the division artillery including the 490th had worked the village over. Thanks to the artillery, this time the village was taken and occupied with ease. However, the retreating Germans were covered by a few of the Mark IV tank destroyers hidden in the haystacks and they destroyed four Shermans.)

I was wounded at Rechrival.

(CCA, directly supported by the 490th, was slow to get started on 1 January and had just begun to assemble on the road when it was surprised by a German counterattack apparently seeing this as a good opportunity to take the armored columns by surprise. In any case Remer did achieve surprise by circling into the Bois des Valets, then bursting onto his enemy. The Germans had their 88's dug in along the forward slope of a hill covered in camouflage netting and straw. Snow was dumped on the straw, completely hiding the artillery. The Americans were advancing over relatively flat land below and were in an exposed position for over 1000 yards when the Germans opened fire, the ferocious 88 rounds ripping through the turrets of the American Sherman tanks. Willie Peter (white phosphorus rounds) burned the straw exposing the positions and the 490th artillery blasted the positions, finally knocking out the 88's.

In the ensuing three hour battle, German tanks and assault guns knocked out twenty-seven American tanks but thanks to the 490th Armored artillery and the fighter-bombers, the counterattack was crushed. At 1530 CCA was moving again. The battle had softened the German resolve and the American tanks pushed ahead rapidly, reaching Hubermont at twilight. The armored infantry had not kept pace with the tanks and fearing another counterattack, the CCA retired and set up its' lines around Rechrival and Brul.)

1 January 1945 490th After Action Report:

Battalion continuing direct support fires. While abandoning an observation post, 2nd Lt. Robert W. Dunaway was struck by a machine gun bullet in the leg. Sgt. Herman W. Westmeyer, his assistant tank commander, dressed the wound under fire, but was seriously wounded himself and evacuated. In the Battalion position area, T/5 Charles R. Smith, HQ Battery, was slightly wounded by counter battery fire. Dunaway and Smith treated and returned to duty. Results of day's action: leading elements reached Rechrival, Belgium; 1 officer and 2 EM wounded.

When I was wounded we had moved our observation team to the top of a hill. The reason we were on top of the hill was that we didn't know exactly what's going on, we only had our point of view and the front line was right in front of us. Several other tanks accumulated by our tank. Our tank and three other tanks were together all trying to figure out where and how to approach the enemy. We had left our tank at the base of the hill and went up on the hill for a better view. We did not want to bring the tanks up the hill and onto the skyline where the enemy could see us, so we dismounted our tank and went up the hillside to take cover and see what was happening and call in artillery during the attack if needed.

I know we used German foxholes they had abandoned earlier and they had them all marked on maps. All they had to do was to see a guy stand up and they could drop a round in there. I just happened to be looking over toward another foxhole and a guy got up out of the foxhole and a sniper got him. He dropped just like when we slaughtered hogs at home. We would shoot them and they would just drop.

Since our tank was down the hill behind us I was not in communication with it by sight and the hill separated us leaving us with no radio communication either. This would not happen in this day and age. We had radios for the tanks, infantry, for the artillery battery and for the top dog battalion headquarters and we should have been able to communicate with any one of those but could not. The only communication we had was a telephone line so we laid a land line a couple hundred yards.

I got in a foxhole up on the top of the hill and checked in by telephone, communication was fine until we got in the middle of a conversation and it went dead. Somebody ran over the wire between us with a tank or it was hit by artillery.

And our tank did not have a cell phone. We had no idea what was happening so when we wanted to evacuate that's when we got in trouble. We could not tell what was going on in the battle and we really didn't know where things were or what the top officers knew. Everything was crazy, bedlam. Enemy fire was right on the other side of the hill and we ourselves really didn't know anything other than what was happening in front of us. And that was during a time when we had no air cover because the weather was so bad, it was cloudy and was snowing hard so that's how it was.

The enemy was advancing and the artillery and mortar rounds were walking around the foxholes. our position became untenable and we were ordered to evacuate. My Lieutenant was hit in the leg by machine gun fire.

(Herman's Bronze Star citation reflects the heroism that he showed on that day in the confusion of the Battle of the Bulge, just south of Bastogne, Belgium:)

> "While evacuating an artillery observation post which had become untenable due to enemy mortar and small arms fires, Sgt. Westmeyer tanks commander was struck in the leg by a machine gun bullet. Disregarding his own safety and in the face of continued fire, Sgt. Westmeyer proceeded to where the officer had fallen and dressed the wound in preparation for evacuating him to safety. While performing this operation, Sgt. Westmeyer received wounds in the arm and hip.
>
> By his complete disregard for personal safety, he aided in the successful evacuation of his commander."

The Lieutenant didn't want to move and I had to prod him to get out of there. As we stood up they dropped in a mortar shell and that's the one that got me. We just had to get out of there. My Lieutenant walked on to the tank after he got his wits together. He was fortunate, shot through the meat of the leg and after he started moving he could operate very well. Lieutenant Dunaway was killed by a sniper two weeks later when he and a sergeant got out of their tank to heat a cup of coffee. I was in the hospital in England.

I couldn't walk so I had to crawl 20 yards to a tank that was being evacuated; I just crawled on my back over to it. They dropped another mortar and it hit a guy I knew in the head, I don't know if he made it or not but it blew meat out of it.

I crawled under the tank and when I got there the crew said 'maybe we should pull him through the bottom of the tank.' The Sherman has an escape hatch on the bottom so they brought me in through the escape hatch. By the time we got to the aid station I had passed out. How I got out of the tank I don't remember but I guess they just pulled me out of the top.

I was on a stretcher from the tank to the aid station, about 100 yards down hill when I woke up and there was my good old friend my tank driver. How they got down there I don't know, but my tank driver was there and he talked to me as I went down to the aid station. It made me feel better.

It was January 1, 1945, New Year's Day. I got carried down to a Red Cross tent and laid inside on a stretcher one night. The next day I was moved outside onto the side of a hill next to the tent. I wasn't alone. There were many many other soldiers lying out there with me waiting to be shipped from the aid station to a hospital. I stayed outside on the side of a hill for two nights before we were loaded up on a train and taken to Paris. We spent two nights in Paris waiting to get shipped to a hospital in England. From Paris we were flown to a hospital in England. And that was the deal on that.

Mortar shrapnel got me in my left side and tore out a hunk of meat. The doctors took skin off my leg and grafted it to my side. I'm extremely fortunate the fragmentation missed major arteries. The doctor in France looked at the wound when I got to Paris and he said 'you're pretty lucky. There's a scratch on your inner belly and if it had gotten a little deeper you would be in trouble.' I'm sure this day-in-age it would heal faster than it did then.

My wound hasn't caused me too many problems over the years. If I stand too long I get pain. I went back to the VA a couple of times for pain in the early years after the war, but they couldn't do anything so they said to just put up with it. Some of the nerves in my left leg were severed as well. I can take medical shots in the left leg and not feel it.

I was in the hospital for four months with that wound. In May 1945, I was headed back to the 490th at the front in Germany when I contracted pneumonia. A doctor at a replacement depot in Germany sent me back to England. The war ended during the short period I was at that replacement depot. After returning to England I spent six weeks on my back with pneumonia before again heading back to my unit. I finally got back to my unit on the fourth of July 1945.

I was discharged January 1, 1946. They did not bring back the whole unit. When the war was over and I returned to my unit we were transferred to an occupational unit. We were just marking time to get shipped over to the Pacific. People were discharged by points. We were still HQ Battery of 490th in the occupation waiting to get transferred out. The war ended in the Pacific before we got transferred out and I was discharged from the old battery.

I returned with members of my battery all at the same time. At New York we split in all directions. We landed in New York on Christmas Day and I got to Camp Chaffee Arkansas, on Dec 31. We did not celebrate New Year's Eve because we didn't want to drink too much and miss the line up the next day to get discharged. I grabbed a train to Wichita and Nadine was waiting at the station in Wichita.

We had no home so we went back with her folks. She had quit her job in Wichita where she was working at the Boeing Aircraft factory building B-29 bombers, so she didn't have a home set up. We then made a trip back to Farmington, Missouri to visit my folks.

I have not spent a lot of time thinking about it. I just don't want to think about it. I never became too active in veteran affairs or activities.

When I left for the Army, I was a County Extension Agent. They promised me a job when I returned and I had a choice of being a County Agent in Newton, Kansas or Garden City, Kansas. I chose Newton. That is where we lived for several years before moving to Dodge City, where I was the County Agent for Ford County, Kansas. Ultimately, I retired as the Southeast Area Extension Director for Kansas State University after many years in Chanute, Kansas.

Herman was awarded the Bronze Star and the Purple Heart, European African Middle Eastern Medal.

Carl during War.

Carl White today.

Place of Birth:
Seattle, Washington
Active Duty Date:
March 29, 1943
Unit:
A Company, 749th
Tank Battalion
Location:
Simserhoff-
Sarreguemines,
France
Arrival ETO:
December 1944

CARL WHITE

The 749th Tank Battalion was activated on December 2, 1942, and disembarked at Utah Beach on June 29, 1944, where it was assigned to the 79th Infantry division. They fought to the Saar River near Sarreguemines against the German Nordwind offensive in January 1945.

In March 1945 they were assigned to the 71st Infantry Division, grinding through the Siegfried Line and crossed the Rhine at Mainz on March 30. They were in Limbach at the end of the War.

Here is his story. . . .

My friend, Harlan Henry and I were in Europe same time but we never knew each other. We were kind of running neck and neck and never had any idea in the world. After the War, we knew each other very well and we didn't talk about it for over 40 years and I really didn't even know he was in the Army. No, you know, I thought about that. I don't know, when I came home, I left the war over there. I was so glad to get home.

Nobody talked about it until recently. All this stuff seems to be coming out now, when we're getting old enough to die, you know. They want to find out a little bit more about what went on back then I guess. It is just like WWI. I know nothing about WWI and all; it just never even entered my mind I guess.

I was born on December 28th, 1924, in Seattle, Washington, and grew up in the South End of Seattle.

I took basic training with a boy from Junction City, ole J. D. Hale. He got a little mad at some girls here once that were raising cane with him on his date. So he passed out their addresses and I took one of them and I wrote this girl here in Junction and she wrote back.

While I was in basic training my dad died and they give me a dependency discharge. I was in the middle of my basic so I went home and I wrote to her once in a while. I didn't write as often as I did later when I was in the Army but then I wrote her. My mother decided to get married again so I went back in the Army.

When I went back into the Army again and it was 1944, the first time was in 1943. So I started writing to her because I needed somebody to get some mail from and I didn't see her until my delayed route after basic training was over the second time.

She told me where to meet her, down by the high school. She said stand there on 9th and Adams and she would pick me up and I had never seen her. I didn't know who I was looking for. And at noon, here she came with a car and picked me up and we went to lunch and so I stayed. She wanted me to stay out at her folks' house out here in the country. Her father Jed was a farmer.

He farmed pretty hard and long hours and I stayed there two weeks and then I left for Camp Chaffee Arkansas. After that they shipped me out to Fort Ord California to get me ready to go to the South Pacific. We were out there long enough to get a couple of long road marches in with all of our combat boots and everything else. And all of the sudden they loaded us up on trains and took us clear back to Fort Meade Maryland on all the sidetracks you could find. It seemed like it was one delay after another. They outfitted us for overseas and then we went to New Jersey.

We didn't get married until I got back from the War. I got back in May of 1946 and of course I had to go out to Seattle to get discharged. I bought an old '36 Pontiac, loaded up and headed for Kansas and of course her dad put me to work right away. He didn't waste any time and I was out in the field before I could turn around. I got back in May and we were married in July, July the 7th.

So I didn't have time enough to get acquainted with anybody let alone talk about the war. When we got back here, there was nobody to talk to about it. Or you didn't talk about it.

Nobody was either interested or didn't have any idea about it or something; I really don't know what the trouble was. But, of course I was amongst all the strangers here in town. Heck, I didn't know them. Like I say, I was out in the field all the time except when I'd come to town there to work.

I never really was in Ft. Riley. When I reported in I went to Fort Meade for more training. I remember we had new combat boots and they put us on a 32-mile road march.

I was put in armor and I took my basic in Fort Knox. So I was in armor. I went to Europe on a Liberty ship. It was one of those small ships that listed 42 degrees they told me. I arrived in Europe in December 1944 and was assigned to A Company, 749th Tank Battalion. I was in the big tanks. It was just a small unit, tank battalion. We were kind of a little outfit but whenever they needed us somewhere, we took off on the road. That's the way they used those small tank battalions.

I was a lowly loader. I had that job 'til we shut down. I'll tell you that was a job that you never had any idea where in the world you were going or where you were or anything else. We had the 75mm gun on it and then finally we got a tank replaced and it had a 76mm with a longer barrel.

The old tank had the old Wright whirlwind engine in it, a nine cylinder radial engine in it, like an aircraft. You could hear them coming a mile away and you never could shut it off. If you shut it down it would get a hydrostatic lock. The oil would pass the cylinders down there on the lower end and when you tried to turn it over, it'd just lock up. So they kept having to keep them idling all the time. The new tank had Ford engines in them, I believe they were. The light tanks had Cadillac motors.

The medium tank would roll right along if you had the right driver. We had a guy that used to drive tanks, and then he made Tank Commander. His replacement was an old guy trying to drive a tank. He would get on the road march and first thing we were clear at the tail end of the road march. That's the hardest place to keep up. And we used to call him Squeaky. The Tank Commander told the old man, he says pull this thing over. I'm coming down and it wasn't long, we were in place in the column.

We had a crew of five. I went over as a replacement because the unit was already in Europe. When I got there, they loaded me up in a 6X6 and I don't have any idea where I went or anything else.

I was surprised I went to a tank unit. They just hollered out my name I think if I remember right and told me to get on that 6X6. I can't even tell you where I ended up but it was some little town somewhere in I guess France somewhere and joined up with the tanks. I don't remember the exact date.

When you are 20 years old you just do what they tell you to do. You had no authority whatsoever. I couldn't even see where I was going. I was busy loading the guns, you know? I couldn't even see out of the sides of the tank and if I could have I wouldn't have known where I was anyhow. I didn't see any road signs or anything.

I can't remember whether they talked about the Battle or not. Usually just the tank commander would come and tell us we were on the right road and where we were going, but I didn't have any idea in the world. I think that's the way we won that war. Everybody was so confused they didn't know where we were.

Well, you didn't get out of the tank too often, I'll tell you that. If you took a leak you did it in a shell casing and pitched it out the pistol port. You didn't get out and mosey around any. And sometimes we'd pull back and get ammo and gas and everything else, that's about the only you got out for. If we got back far enough sometimes we'd find a house somewhere and get us a few nights sleep or a few hours' sleep in there. I know we didn't use the big gun near as much as they do the machine gun fire.

We heard that Germans were dressed up like Americans. I remember that we heard that they had small bottles of sulphuric acid cigarette cases, to throw into faces of Americans.

I remember one time I was so damn scared I couldn't pee. When the infantry was pulling out or moving ahead, they didn't have enough people to keep the line tight during the night. And so they set our tanks side by side, quite a ways apart but then side by side and couple of us stood guard at night. One guy was at one end of this tank and another guy at the other end of the tank. Hell, you couldn't see. It was darker than black there, you know? They told us that there was a bunch of Krauts coming that spoke just as good of English as we did. I think that was the worst night I ever spent over there. I didn't sleep while I was out there. I don't think I ever saw any of the Germans dressed up like Americans but I sure looked for them.

I could hear talking. I could hear talking back in the background but I couldn't make up my mind where it was coming from and I didn't know whether it was us — our guys or the Krauts. I didn't have any idea in the world. The rest of it's all kind of a fog. I remember a river there but I couldn't tell you what it was. Of course I really don't know. I remember going over a big river on a pontoon bridge but like I said before, I was in the loader's seat and didn't get a chance to see out or anything else.

When the Battle of the Bulge started we had just taken Rohrbach and Petit Rederching and we attacked a Fortress on a steep hill called Simserhoff close to Bitche France that was along the Maginot Line. The main Offensive during the Battle was north of our position. But German units did attempt a drive toward Strasbourg south of us and through us. We spent the whole Battle in that area supporting the 44th Division.

Then we moved 20 miles west to Sarreguemines and set up in defensive positions to protect the area south of the main battle. We were under attack constantly, horrible time; I don't think I left the tank once. In February we were attached to the 70th Division and for weeks around Fohbach we were in constant artillery and mortar fire.

When the war ended we were in a defensive position around Limbach, Germany and had been there for quite a while. After the war was over, they set up a military government in a small town there in Germany and then they finally sent the tank battalion home as a unit and I didn't have enough points to come home so I went to the 360th Engineers in Graves Registration.

After the war was over, they sent me down to Lehar I guess it was or Marseille, one of the two and left with the 360th engineers. It was the greatest outfit. I drove a wrecker up and down the Autobahn and we were hauling caskets up there to get the dead guys ready for permanent burial. I stayed about a year after the war and came home in May of 1946. I didn't have enough points to come home. I remember I was on orders to go to Japan and then they dropped the bomb and the war ended. I wasn't too happy to think of going to Japan after what I had been through.

There's another little incident that I remember. First day I went in combat the .30 caliber is coaxial with the .75 and you aim with tracers bullets with it. The gunner went to fire the .30 caliber and it ruptured the round and the tank commander says well there's an extra barrel out on the deck. I had to climb out the damn tank and go out on that deck and I'm telling you, I didn't take too long getting it either. But then I had to reassemble that .30 caliber while we were moving and keep the 75 loaded and as far as knowing where we were going, I didn't have a clue and I didn't care. We didn't use the big gun as much as the machine guns. There wasn't too much thought with some of that stuff like changing a hot barrel on a machine gun.

For a side arm I had one of those grease guns. If you just looked at that trigger it would go off. And try to find a place in the tank to fit that thing

A company 749 Tanks at Grosblieberstroff. *National Archives*

where it was out of the way. It wasn't long that I picked myself up a .45 in the holster and clip.

We kept our belongings hanging out on the deck. You wore what you wore for quite a while. You didn't change every day. I don't remember having many clothes other than what I had on because if you wore yours for 30 days they'd bring you up a new set and you'd just throw the old ones away. You never did take your boots or socks or anything off.

We ate mostly C rations. Sometimes we would eat K rations; it's kind of what you could scrounge up. It was that after things kind of settled down to where we was normally coming back at night and loading up with ammo and gas and stuff, we'd raid anything we could find that was edible. Sometimes we would eat German food but of course that's the first thing they told you, don't fool with any of the German's wines or food or anything else because they could poison it or whatever, you know?

Well, that was the first thing you hit anywhere you were but we had quite a few C's. We'd throw them on the exhaust manifold on a tank and warm them up that way, ham and eggs and stuff like that. Leave it there too long, it blows up. You can tell we had a lot of fun. Yeah, it's fun to laugh about now, right, not so funny then. No bath for six months.

We crossed one river, I can't tell you which one it was now. I absolutely don't remember but we had to wait for the engineers to put up a pontoon

Memorial at Bastogne marking the battle lines. *Author photo*

bridge. So a bunch of them got out and wandered around this little town and found a champagne distillery. Well I don't remember going across the river at all. Nobody else did I don't think. Lord it was just so nice to have a drink. Some people found schnapps but we never found any lose schnapps anywhere other than just around some of these houses or something.

Joan and I have been married for 41 years. We had two daughters.

I actually worked in the same dealership as Harlan Henry. We worked together in two different places. He started selling Dodges. Then I went to Olds Cadillac after they closed up the Dodge place.

I joined the American Legion many years ago. Old Fred Bramlage had a big drive years and years ago for the American Legion, membership drive and then I joined then. I can't even remember when that was anymore. Fred's been dead for a long time. I never did go to much like that, didn't want to remember. I went to a pancake feed or something like that occasionally.

As far as Europe goes, I have never been back and don't really want to go. My memory of the place, it was just in shambles by the time we got done with it. When you would come into a village, the first thing we'd do is blow off any tall buildings for observation or anything and then by the time we left there, it wasn't much left of it. It just — what we didn't tear up with a gun we tore up with the tracks on our tank. Yeah, you drive through — you look at the town and it's a shambles, you know? Just absolutely nothing but a rock pile. I don't think I had a periscope where I sat. Of course the tank commander had one, the gunner had one and both assistant drivers had them. But other than that I don't think I had any way to see out. I do remember that it was so cold that they would frost up.

One of the things I remember most about the Battle of the Bulge was that was about as cold I think as I've ever been in my life over there in December. God almighty it was cold. You couldn't find enough clothes to put on, gloves or anything else. Standing guard was so cold. Oh, it was cold. And tanks don't have heaters. The transmission warms up a little bit in the middle there but that's only when you're moving. When you're sitting there, it's like sitting in a refrigerator. The tank itself, oh, yeah, that thing, it absorbs the cold like you couldn't believe and just hangs onto it.

Well that stuff is just kind of off in the past somewhere.

Carl was awarded the European African Middle Eastern Medal w/three Stars.

Armored Infantry moves through Wernberg, Germany. *National Archives*

Chapter 7

The Artillerymen

IF THE INFANTRY IS THE Queen of Battle, the artillery is the King of Battle. In Europe, the artillery operated directly behind the front lines in many cases, providing direct support to the Infantry. Its role is to destroy the enemy by any number of methods including cannon fire or rockets. They had to provide the coordination of all the direct support fire by understanding the location and situations of the front line troops.

Much of their time is devoted to training and becoming proficient in operating their gun. They had to understand what each other's job was and how to do them all efficiently. They would train endlessly on immediate actions and drills that would require not only loading and firing but also the map reading and technical expertise to provide support. Artillery soldiers have to closely rely on each other for both support and technical expertise.

"Our training was continuous and a lot of the training had to do with the officers and enlisted men," related Elmer Blankenhagen. "In my gun section we trained continuously and we trained on weekends. We trained in evenings because there was a war going on and we had to use techniques that had never been tried before."

As the war progressed, like the other branches of the Army, the artillery began to modernize. "This unit that I joined was the last unit that got mechanized, we went from a French 75's horse-drawn artillery to the truck drawn 105's. When the war first started they were guarding the coast that would have been Half Moon Bay, south of San Francisco with the French 75's and they kept their horses at the Bay Meadows Racetrack," said Don Huse.

Communication was essential and they relied on radio as well and wire communications. "Each battery had a communications group in it," remembered Don Huse. "We took care of everything that had wire, wire between every telephone in each gun position and to the commanding officer so that there was constant contact with all the guns. We set the switchboard up somewhere and we had to run a wire to headquarters and all of the smaller units had to go to a bigger unit."

The artillerymen also were victimized by the German imposters. Harold O'Malley remembers one night when a German dressed as an American tried to come through the wire. "I remember Sgt. Ryan who was in our battery, was on duty and someone came through camp, messing with the barbed wire and he told them to halt and give him the password. But they didn't know it. He noticed that they had on German soldiers' pants and boots so he jumped behind a tree just as they were trying to shoot him. He fumbled with his rifle yelled for help and the Lieutenant came up and the Germans ran off."

ELMER BLANKENHAGEN

I was raised in Iona, Kansas, and I was attending Iona Junior College when I signed up with the Kansas National Guard which was being federalized into regular service. I signed up for a year and a day. The reason I signed up was that the war clouds were coming up and I had some college, nearly a year and a half, and wanted to get my military service over so I could finish college.

I was stationed at Camp Robinson, Arkansas, and we were sent on the 1941 Louisiana's maneuvers where Eisenhower was the Colonel and Chief of Staff for the Third Army, and then was a Brigadier General. We went to the Louisiana maneuvers and the greatest thing we got out of there was tactics. We didn't have any equipment, we had wooden guns, our anti-tank weapons were orange flags, but we did learn tactics which proved very successful down the road.

We got back to Little Rock, Arkansas, and we got some of our first selective service group, right off the street. I was assigned a group of them and was the drill Sergeant to train them. We wanted them to join the rest of the Regiment. Pearl Harbor came on December 7, and I was supposed to be released on December 16, a year and a day from when I signed up. But we were alerted and President Roosevelt said everybody in the military will be in for the duration of the war and six months. The next thing we knew we were on a train, destination unknown. The next thing we knew we were in Kansas City and going across the plains of Kansas. We got to El Paso, Texas, and we went blackout, were issued live ammunition and a set of machine guns. We still did not know where we were going.

Next thing we were passing LA. After we passed LA they called us together and they told us we would be debarking at Watsonville, California, because we had these big 155mm howitzers and there are a number of bridges that would limit us. We had to get off the convoys and go on to Fort Ord, California. We never got live ammunition in our barracks when we went to the field to protect the West Coast. Our division had the zone between Santa Cruz and Santa Barbara and we were right in the Monterey Bay Area. I spent the most miserable Christmas other than the Battle of the Bulge in a foxhole in 1941 in Monterey Bay. Try to eat Christmas dinner with the fog and the rain. I will always remember that night.

We were in the 35th infantry division, and we had four regiments and four regimental combat teams. Then they were square divisions with four regimental combat teams. Later, they changed to three Regiments. As we went to California we were under the VIIth Corps and we had to protect that wide shore.

Our training was continuous and a lot of the training had to do with the officers and enlisted men. In my gun section we trained continuously and we trained on weekends. We trained in evenings because there was a war going on and we had to use techniques that had never been tried before. So we trained and carried live ammunition wherever we moved. We had live fuses, we had all of our powder and powder kegs and we just practiced and practiced.

When we got to California they triangularized the divisions, which mean they changed the form for combat teams to three combat teams. And the 138th Combat Team, which was part of 35th Division, was sent the Aleutian Islands because the Japanese were taking over that area. And they shipped part of the division over to camp Saint-Louis Abyss, California and we went over there and trained with the Marines. The second Battalion of the 1/27 field artillery Battalion became the 1/95 field artillery. It was the first 8 inch howitzer Battalion in World War II.

When they redesignated the second Battalion to the 1/95 they shipped out all the old soldiers, to Salt Lake City, Utah, and the Santa Anita racetrack to guard the Japanese-American internment camps. They put us all together into a cadre to train the 1/95. I was with the one 27 1/95 until the last of July 1942. I was at the theater at Fort Ord, and was paged to report to regimental headquarters. There at the headquarters, Col. Candy said you will be on that train to go to Fort Sill officer's school. So I left at the end of July 1942, for officer's training school at Fort Sill. I was in class XXXVIII it was a class that you marched in double time between all classes. When I was halfway through they stopped doing that because they were losing people in the heat and it got to everybody. They were trying to weed you out and making people jog around between classes in the heat was a way to do that. In your barracks, your bunks had to be perfect. We had little cards with names and buttons and the button had to be where you were.

I bunked with two Black soldiers, an attorney name Bonetti, he was my bedmate and was bunked with another at Fort Sill. We got up at five in the morning and 0530 we had our physical fitness, then we went to breakfast. Everything was attention, attention at the table. It was run like the Military Academy. And so it was.

I was selected go to Fort Leonard Wood, Missouri to the 77th field artillery brigade and I was a spokesman for this group. I reported in to this general who had just returned from North Africa and he was assigned brigade commander. I remember when I reported to him he said, I have one project for you; this Battalion had flunked the Army ground force test. I want these units to pass the armored ground force test as soon as possible.

I was assigned the second Battalion of the 1/77 field artillery Battalion in November 1942 the last of November. Immediately I was assigned as the battery executive officer to train Battery A 1/77 and it was just below the headquarters and the old general used to come down all the time and watch our guns drill. One morning he walked into the Battalion headquarters and told the Col. I want Blankenhagan in for promotion, so that's where I got my first promotion from Second Lieut. the First Lieut. But I can tell you I worked very hard to get that. I had been the chief of guns before and a Sergeant. In the artillery and knew what they had to do and they respected me. Shortly after the battery commander was moved to the Battalion staff, I was assigned battery commander. In January they sent me to Ranger school.

I was in Ranger school at Camp Force Tennessee and it was a precursor school for some of the Ranger and jungle warfare centers. We were a test group on just how much a human being could handle. We spent a month in the forest and my section started with 35 and 17 graduated. The only way anyone got a second chance was if you fell out flat on your face which meant you were putting all out. After that I went back to 1/77 and I was assigned as the PT expert for the whole Battalion. So I got them ready to go to the desert but I still was battery commander.

I took them to the desert in California, went to desert maneuvers in the Mojave Desert. Then we were alerted for overseas movement, September 1943, and we were sent back to Fort Bragg with the 82nd airborne in 1943. In the meantime I was sent down to the XII Corps staff at Fort Jackson to test two divisions for overseas movement. We tested the 106th division and we turned them down. General Jones was the commanding general. He got mad on the stage and I was with Col. Lewis, who was chief umpire for the artillery and I remember him telling General Jones, your division is not ready. As you may recall, when they went to Europe and were in the Battle of the Bulge, the 106th and the 99th were overrun.

I was still a First Lieutenant because you had to have two years to be a captain then. The old man called me up one day and said that they had orders to send a cadre of battery commanders to activate four brand new artillery battalions with the 155, the newest howitzers, the newest tractors, and you will get a promotion pretty fast. We were likely to stay with them, and they brought in some captains, so we went.

We were sent to Fort Sill, Oklahoma, to a new unit officers course where cadres of the four battalions took the officers they trained together and became acquainted and two weeks before we were to go to Fort Jackson, South Carolina, to activate the units they changed our orders to Camp Hood, Texas, and that's where I met Roxy who would become my wife.

On February 21, 1944, we activated these battalions. And I activated battery A 666 field artillery Battalion, we had a cadre. At that time they had what they called Army Specialized Training Program where sharp guys went to college and they got rid of that program and we got a lot of those guys and they were really sharp. We were activated on February 21, and we got orders in August to ship out. We went through all the Army ground force test and old Army maneuver tests and we were ready to go. We were just waiting the word. A sister of ours, the 667 Battalion, and if you remember the Missouri Orange Bowl team that had the Morse twins, the ends, they were our sister Battalion.

In October 1944, a train came in and we got on. Both of my units that I commanded were in the D-Day landing. Had I stayed with them, I would've been there. In August 1944, we were alerted and received what was called the POM, preparation for overseas movement, a train slipped in the camp 1 October and we boarded.

We headed to Camp Miles Standish, boarded a ship, the "New Amsterdam," in Boston. It was a new high-speed luxury liner and we went over by ourselves. The ship had the new radar and the ship's captain took us up on the bridge and showed us the new radar and explained to us why we are not going in a convoy because the new radar would be able pick out the subs.

The 666 was a Corps Battalion we supported numerous divisions. With 155 and the high-speed cats we could move, we could move with speed. We worked with armor divisions, we worked with airborne divisions, and we worked with infantry divisions.

We were in England at Wolverhampton when things started developing over in Europe. We were immediately alerted; boarded LST's and took off for France. The first position we occupied was right around Malmedy, where they annihilated and machine-gunned our artillery forward observers as we went into combat. We worked with the 82nd airborne, the 35th division, 83rd division and the 17th Airborne Division, the 29th division. The Corps artillery worked with the Corps Commander J. Lawton Collins. I always remember when we first occupied our position was gentle Collins. He was an outstanding leader and well respected.

We had some tough spots in the Battle of the Bulge, we had some tough spots where we were attached to the Combat Command A of the 2nd Armored Division and went to Julich and our target was Cologne. We headed toward Cologne and we did a left flank, met the British north of Bonn and captured thousands of prisoners on that March. In March, we began encircling the enemy and were engaged in many gun battles in that area. And that's where I got my bronze Star. I pretty near lost my life but my team accomplished its mission.

We went through three campaigns. When the war was over we were in the Ruhr pocket. We had helped clean up the Ruhr pocket. We were actually in what became the British sector in Germany and we were assigned military, government duties. I was assigned to government work. Then we were shipped down to Austria and we did military government work down there. Then our work was completed and they started shipping the high point men out and that was a problem. I was assigned to the 9th Infantry Division as a battery commander of battery A in the 26th field artillery. Then I was sent to Munich, and I was in military government work in Munich. That was an area that challenged us a lot.

We had orders to prepare this unit for overseas movement. We trained again just like we did before but we got all these guys in there, and we didn't get a chance to train them. They kept pushing because they wanted us to go to the Pacific. When they started moving them out, there wasn't much we could do. We just marked time a lot, but we had jobs to do. Security, political, military government, stuff like that. We are in cigarette camps and my job was to escort troops from Brussels and be the "parents" of the high point guys. The only way to keep them busy was to do things like escort duty and that was in the summer and fall of 1945.

I started home in July of 1945, and I got back to Iona, Kansas, the night before Christmas 1945. The reason we didn't get home quickly was because of all kinds of strikes. There were coal miner's strikes, a ship strike and it all delayed my homecoming.

I was sent home as a battery commander of the 1/41 Field Artillery Group. We were on the "Joseph T Hooker" ship, which was a liberty ship and we had 700 troops on there. It was a slow boat. We had a Sergeant from Atwood Kansas who came down with appendicitis, and we did not have a surgeon on the ship. We had medics but no doctors, so we sent out an SOS for a hospital ship. We were supposed to sail from Le Havre but couldn't get out. One sailed from Marseille and she sailed through the Straits of Gibraltar and was passed the Azores. It was December, very rough seas, to get to the hospital ship we had to turn around and go back toward the Azores. We were going to transfer a little boat to the hospital ship, but unfortunately the little boat sank. So they ended up shooting a buoy from the hospital ship to our ship and transferred a surgeon over to our ship. He was a surgeon from University of Pittsburgh. He said he would not perform surgery on the ship but would use all the conventional methods that he could to pull the guy through.

Four days out. We have high seas in December and our ship started taking on water and the pumps couldn't handle it. The ship's Officers called us together and said prepare your life preserver stations and shake your lines, check preservers because we don't know what's going to happen. The ship would roll and roll and shift up and vibrate when it would move. So the Captain said he was going to send out an SOS.

About midnight we looked around the whole ship and saw lights. The sea was very, very rough. The Captain said were we were going to ride it out, he said if we abandon ship we would lose a lot of troops not through drowning but through hypothermia. But the Lord was with us and the seas began to calm down and by daylight we could look all around at the ships, cruisers and all kinds of ships, ready to pick us up. We were close to New York. They assigned the USS *Honolulu* cruiser escort us in to New York. That was December 1945.

I had to clear the ship because I was asked to sign my life away. It was never planned to see that many people. There were radio people and newspaper reporters because we sent out an SOS and everybody was there to cover the story of the landing. From there I went home.

We were supposed to go to Leavenworth to be discharged before Christmas but they couldn't do it because Leavenworth was backed up. So the next choice was in Missouri but it was full as was Camp Chaffee, Arkansas, so they sent us to Tyler, Texas. Of course, Roxie was in Texas. So I went to Dallas and then got on a train to Brownwood to see Roxie. I left Dallas on a train and I got to Iola day before Christmas 1945.

I married Roxie in February 1946, we were engaged, but I did not want to be married going overseas. So we dated by mail while I was overseas. We were engaged in September, I was promoted to Captain, and we did not get married until I came home to Manhattan, February 1946.

I was very proud of the Bronze Star that was awarded to me. I am also very proud of these other metals that I got that my granddaughter prepared for me in a shadow box. She organized and presented them to me.

In January 1946, I enrolled the Kansas State and graduated in January 1948. We had tri-semesters at K State. I went to work for Swift and Company in Chicago. I trained in Chicago and I did that for a period of time and then I went to work for Kansas State Agricultural Extension in 1950. My professional life was with the University of Kansas doing USDA and agricultural extension work. I went to school using the G.I. Bill.

The General wanted me to stay in the service but I wanted to get married and went to college for my degree. Those are the two major reasons I didn't stay in the Active Duty Army. I was in civil affairs military government work for 20 years in the Army reserves. I was assigned to the Pentagon for 10 years and civil affairs military government. I carried six-day pocket orders which meant I could be called up within six days.

I would go back to do my duty and they always had a project for me in my area. I would have to prepare the project and brief the general and his staff. I might be out in Dodge City and I might have a telephone call from the Pentagon wanting my opinion on some issue somewhere. I commanded the Kansas mobilization detachments here in Manhattan for 10 years, which was a civil affairs military government attachment. All the people attached were over civil affairs military government, military occupational specialties.

My son is a career Army officer and he married a career Army nurse. They have two children a girl and a boy.

If I had it to do all over again, I would probably do the same thing.

This is taken from an interview by the Riley County, Kansas Historical Society, 2003, Manhattan, Kansas, and available through the Manhattan Library.

Don in WWII.

Don and Nona Huse.

Place of Birth:
Manhattan, Kansas
Active Duty Date:
1943
Unit:
76th Field Artillery Battalion, 3rd Army
Location:
Mont, BE
Arrival ETO:
August 11, 1945
Rank:
Staff Sergeant
Communications

DON HUSE

The 76th Field Artillery Battalion was transferred to France after completing shore defense duty in California. They landed on Utah Beach on August 11, 1944, and proceeded with the Third Army toward Belgium. From December 1 to December 24, 1944, the Battalion supported the 8th Infantry Division and the 39th Infantry Division and was stationed west of Germeter, Germany, in the Hurtgen Forrest. On December 24, the unit was moved to the area of Mont, Belgium, in general support of the 30th Infantry Division and was also answering calls from the 1st Infantry Division. During February and March 1945, the Battalion supported the 1st, 2nd, and 8th Infantry Divisions and was attached to the 78th Infantry Division before crossing the Rhine River on March 20. They finished the War providing direct support for operations in Czechoslovakia.

Here is his story. . . .

There are not many Huse around. We're mostly from the Manhattan area, some from south of Manhattan. My great-grandfather came out from Illinois, homesteaded out there. I grew up with in Manhattan. I was born February 2, 1925, Groundhog Day; the weather was 20° below zero that day in Manhattan. My dad told me that. I finished high school when I was still 17; I wasn't 18 until February, so I had to wait quite a while before I was drafted. It was 1943.

I got my greetings from the president. It was a letter, "letter of induction." Our local draft board was in the National Bank building, and they sent me this letter. I wasn't surprised when I got drafted, it was coming, I was categorized as 1-A, everybody that was 1-A was going. So we went to Fort Leavenworth where we were inducted after that I went to Camp Roberts, California. It was a great big Camp similar to Fort Riley it had two story barracks, they were pretty good barracks. It was close to Pasa Robos, California, on Highway 101 it was about 250 miles north of LA and about 200 miles south of San Francisco. The ride was about the same to both places.

I feel like I was very lucky the whole time I was in the service because Camp Roberts was basically an infantry post, but they also had an artillery section there and I went to the 105 section, and I was in wire communications of artillery. We dealt with switch phones, phones, got everything set up laying communication lines. That's where I took my basic training in. I thought there's no place worse than Camp Roberts. I found out later that I was wrong.

I was there this summer 17 weeks doing our basic specialty; they sent us out to the Mojave Desert where we rejoined our unit which was the 76th Field artillery Battalion. It was still a battalion unit and there were still a lot old-timers in there when I joined them. This unit that I joined was the last unit that got mechanized, we went from a French 75's horse-drawn artillery to the truck drawn 105's. When the war first started they were guarding the coast that would have been Half Moon Bay, south of San Francisco with the French 75's and they kept their horses at the Bay Meadows Racetrack.

But then it was when we were sent to the desert, that they converted us to mechanized and that is when I joined them. We did everything we were supposed to do to get prepared to go overseas. These were all drawn by trucks. They were pulled by deuce and a half truck, we call the prime movers. The 105 wasn't a very big artillery piece compared to some of the others like the 155, 8 inch and on up.

We did all the training that we were supposed to do there and then I happened to get a furlough to come to Manhattan. While I was there my unit got shipped to Camp Polk, Louisiana, in preparation for overseas movement. So then after training at Fort Polk, we went to Camp Shanks outside New York City, and we got on our ship it was a pretty good-sized ship, it was a ship that was captured in World War I from the Germans, and it was called the "Amerika." We had renamed it and the British were operating at. We got aboard that and there were about 5,000 of us and I thought the next morning I would get to see the Statue of Liberty and everything in New York. But we got up the next morning to see what we could see, we already sailed under cover of darkness, they took us out of there during the night and we were all by ourselves out there and we didn't see anything.

We sailed on and I woke up the next morning there were about 35 ships all around us in a big convoy. We did not have artillery with us, we drew new guns when we got over there.

The convoy zig zagged back and forth to avoid German submarines they came back down from the north and we landed at Liverpool, England. One of the Queen ships was in harbor, it dwarfed our ship and made it look like a small sailboat as we sailed by. It was so huge. I wasn't sure whether that was *Mary* or the *Elizabeth* it was one of the Queen ships. What a ship!

We went to Wolverhampton which is South East of Liverpool somewhat, and we lived in a house there that an old Lord had donated to the military, named Bishton. Part of it was a new house and part it was an old house that was over 200 years old. We stayed there for maybe a week or so until we got our new equipment, then went down to Wales to get it calibrated. From there we went to Southampton, got onto an LST, and sailed across the channel and landed at Utah Beach.

Of course this was after D-Day, and we had captured most of the Cherbourg, Peninsula, at that time, so we didn't have any opposition getting on the beach. We landed August 11 at Utah Beach and were originally assigned to the third Army, Patton's Army, and he went south. We were immediately changed to the first Army after we landed. Our battalion was what was called a Corps artillery Battalion and we were not assigned to any division, we supported the whole Corps. For instance we were attached to several infantry divisions, the 1st and 2nd, 9th. We were attached to the airborne units were attached to other artillery units; we were attached to wherever the Corps needed us. At the end of the War we were attached to the First Division in Czechoslovakia. That was our last assignment.

When we went through Germany, we didn't have a lot of resistance. We did get into some really tough battles in the Hurtgen Forest and also the battle around Aachen. It was right before the Battle of the Bulge started. We were next to the 106th Division and we watched as they just got annihilated. In fact it was their first contact with the enemy in combat.

After we got through the hedgerows at Utah Beach, the hedgerows were rough. It was just a matter of erosion over years, there were several acres of flat ground and then there are great big walls of dirt with hedges that held the dirt together and they were about 20 feet wide or so and there was a series of those like small farmlands inside there.

The 82nd airborne was in there. When they landed they got the heck shot out of them; the Germans were already in the hedgerows. So we went through there but we didn't have any opposition there until he got to the place called Argentan-Falaise, France just shortly after the hedgerows and they captured a lot of prisoners there, many more than they expected us to capture.

Then we went through France so that fast that they took our trucks away from us and they started hauling gasoline and things. Even the C-47's were hauling gas to keep the tanks going so we ran through Paris and their liberation there. There wasn't much going on, just some small arms fire but you know Hitler had a general who was in charge of Paris and he ordered him to destroy Paris but he didn't do it. And he saved the town. Just outside of Paris and in Paris there were just ricochets of bullet holes without much damage through the city. We weren't there very long, and while our truck was sitting ready to move, a guy came walking down an old army fatigues and coveralls talking and by golly it was Ernie Pyle the correspondent coming by talking to us. Later on he went to the Pacific where he was killed. I got to talk to him but we didn't know who he was, but soon we figured out.

We left Paris and took off toward Belgium, went through Liege and as far as we could go just east of the Liege. And that is where we got into the little towns on the border up there, we were always farther north and never as far south as Luxembourg.

One of our first assignments there was supporting the battle in the Hurtgen Forest. I remember the beginning of the Bulge, it was overcast and we didn't have any air support. That's why they were able to get so many tanks in through our lines. I remember the day, the first sunny day after the Battle of the Bulge, we were moving to a new position, and our convoy was stopped, and I could hear our planes. I could hear the German anti-aircraft fire, and a lot of machine guns shooting in the air. At that point we could hear them and still couldn't see them, and all of a sudden here came a plane, smoke coming out of it and it appeared to be heading right for our truck and so we all bailed out of the truck, then as it got closer and you could tell it would go on overhead and as is a went by we could look into the German plane and see the German pilot's silhouette as he sailed by, and you could watch his silhouette as he tried to handle the plane. He crashed right there next to us. A P-47 was following him and shot him down as he followed him to the ground.

I also used to watch with my field glasses as the bombers came over. We couldn't hear them and if you couldn't see the vapor trails you would not know they were up there. But I would watch them and sometimes I would see just a big ball of fire explode, you knew that one had been hit.

I really felt for the guys. They looked to be B- 24's. There were nine crewmen on the bomber and you knew that with that ball of fire they all died. As the planes came near the first thing they got was AKAK, then after that the German fighter planes to take off after them and that was before the P-51. The German fighter planes shot a lot of the 24's down. The B-24 was the most produced plane in the war, and they were used all over Europe and Pacific, and that is why so many of them were shot down.

After they go through the AK AK's fire and the fighters, they flew over the cities and more fire. After they flew 25 trips they got to come home. That would've been some pretty harrowing experiences. Many of the guys returned to religion because that was all that they had to look forward to. Of course, we all had a fatalistic view of what we were doing and you would do the best you could to protect yourself, if it is going to happen, it was going to happen.

I don't remember when I first heard that the Germans had attacked during the Battle of the Bulge. We thought the main thrust was coming through at Elsenborn. It was so small you couldn't see it on the map. There was Elsenborn Ridge.

I was a buck private when I first went in, and then I became a corporal. They have what they call a Battery Agent, and he's the guy who hand carries the real important messages back and forth, so I became a corporal and was the Battery Agent. Later on I was promoted to staff sergeant and I was in charge of the radio, and survey and wire communications for the battery.

We had portable radios and we also had, our headquarters battery had all the information coded and messages sent. I worked in records center after they activated my battery and we had to decode all the morning reports that the first sergeants sent in. We had a code called the Sylvax Code, a secret British code, it was a card with a lot of letters on it, strips with a lot of little letters on it and you had to have a key to set it up. You would go through the message with the key, code it and then send it in.

Headquarters battery had a fire direction center. We would set up our position and we would have to survey our guns and register our fire on something we could zero in on. That information would go into the fire direction center and in those days the guys had to use slide rules to figure it out. Then they would figure where the target was and establish a base point of fire, then from that base point mark the target. They would figure that out and then send word to the gunner, the guy who will pull the trigger, and then they would have to adjust and shoot again. But everything would have to go through the fire direction center and then to the batteries. I remember some of the officers we had were killed when they were up with the infantry. They were forward observers who were calling directly into the fire direction center, adjusting artillery onto targets.

We had a situation and I don't know how it happened. Our field artillery battery where there were 3 batteries in a battalion and four guns in each battery was running short of 105mm ammunition so they decided to give us some British artillery pieces. We didn't know how this happened. They were called 25 pounders. But they only used two six gun batteries, 26 batteries total, so they didn't need as many batteries and they did away with B battery which is what I was in. They could only have two six gun batteries. They brought over some British gunners a show us how to operate them; I didn't have anything to do with the gunners since I was in communications.

During the Bulge there was lots of units right next to each other. We were supposed to shoot down any of our own liaison planes if we saw them flying because the Germans had been flying our planes. If we saw them flying there was no way to tell so we were supposed to shoot them down. We had 2 Cubs in our battalion for observation.

We had two British sergeants and we had a sign and a countersign and you had to know that, and if somebody asked you a question about baseball or something any American would know. These two British soldiers got stopped. And they couldn't answer any questions because they didn't know so they got arrested and hauled them off to a POW compound. They finally got them back after a while but it was edgy situation.

Field artillery was not on the front lines line with the infantry so we didn't see everything that they saw. I was lucky that I was in the artillery instead of the combat infantry. Those guys are the ones that had rough combat. Our biggest problem was that we would get hit with a lot of German artillery and once we'd shoot they would register the position and retaliate with fire against us.

When the bulge began, it was peaceful up to about the 24th. We were about 1 mile west of Germeter, Germany in the Hurtgen Forrest supporting the 8th Division, and on the 24th of December, the day was clear and very cold. We repulsed a German attack at Malmedy. Then we moved on close to Mont, Belgium in support of the 30th infantry division. We moved every week or so from Mont to Chodes, then Geromont, and Faymonville. For the last week of the Battle we were part of the 1st Division Artillery.

After the Battle of the Bulge was pretty well settled, we headed to Germany toward Bonn and the Rhine River and crossed the river on a big pontoon bridge there next to Bonn. The Rhine is really a big River as big as the Mississippi. Then we went through Germany and there was hardly any opposition and we went on clear to Czechoslovakia. When the war ended we were sitting in a position close to Hundback. Hundback is close to Cheb. At Cheb, the first infantry division has a very nice monument to the men who were killed in the Rhineland campaign.

There sure are some nice people there. The Czechs were so happy to see us and we were there after the war for over a month and I lived in a home with a couple of people. They took me in and I slept in a nice bed, really nice. The women all worked over there and they would cut the grass with those scythes. Back then they had a socialist government; they were given 10 acres to make a living on. They wore those old wooden shoes, and looked very primitive in their work. Everything was done by hand. They would flail the grain off the wheat with a stone and they all owned ducks and geese.

Our colonel was the first one killed in our unit. He was some place he should not have been, he was out looking for a new position. We were already really close to the infantry because we didn't have a lot of distance to our guns when we would shoot, and he was looking for a closer position in a Jeep. The Jeep driver, the colonel, and executive officer were in this Jeep and the right front window took a round and blew the colonel out of the Jeep about 75 feet. It broke the Jeep driver's leg; it didn't hurt the major who was in the back. He finally came back to the unit after being evacuated. After they gave us those old 25 pounders, one of the gun crews got killed by muzzle bursts. The shell exploded almost as soon as they got out of the barrel. It got so bad that they would try to protect themselves by piling sandbags in front of them and they would put a longer rope on the lanyards so when they fired them they could be back behind a tree or something.

The 105 used flashless powder and at night you can see a flame of blue come out but with a 25 pounder the whole area was a big bright orange and they really got a lot of retaliation from the German artillery. We only used those British 25 pounders during the first part of the Battle of the Bulge and the Hurtgen Forrest.

I remember going into the Hurtgen Forrest. It was a horrible haunting place and the Germans knew that area very well. It was their homeland.

Each battery had a communications group in it. We took care of everything that had wire, wire between every telephone in each gun position and to the commanding officer so that there was constant contact with all the guns. We set the switchboard up somewhere and we had a run a wire to headquarters and all of the smaller units had to go to a bigger unit.

And Headquarters would go to an infantry unit and there would be an attached infantry officer they were working for. And then you set up your guns. Because those artillery positions were all scattered out maybe a half a mile from each other in the quarries and things, they weren't all just four guns in one place. The batteries were spit up and the guns were all in different places. Each of the guns acted individually. They had their own kitchens and everything. A battery was about 100 people.

We were along a river we were close to Malmedy but we moved around some. I remember once our bombers bombed our own troops. We had an old chicken coop just outside a Malmedy and all of a sudden I heard a boom boom boom boom everywhere. Our bombers had dropped bombs very close to us and in Malmedy. You could see the bombs walking up the side of the hill toward the town.

As they needed assistance they would call in a request to Corps and that's how they decided where we would go. We operated under Corps artillery. In fact toward the end of the War we were attached to the First Division in Czechoslovakia. That was our last action. When we went through Germany we didn't have a lot of resistance.

The war ended May 8, and we stayed back in Czechoslovakia and finally left December 19, 1945, from Marseille France on an old liberty ship called the *John Fiske*. I was a staff sergeant then, and we had civilians on board. Some of the guys had bunks in the holds, but we had better bunks because we were NCOs. There wasn't enough weight on a ship to hold the ship down and we were really bucking. Out by the Azores it was really rough, the ship would bang and shake. It was really something, like an earthquake. About three days out the ocean was as calm as Tuttle Creek.

We came back to the states and landed in Virginia, got passes and I came to Manhattan. I got discharged at Camp Chaffee in Arkansas, the separation station for this area, and came back to Manhattan. It is really quiet. It seemed, like not much going on, all my buddies were coming back, the ones who did not get killed. We had a lot of parties.

When I got out I didn't really want to do anything, I just got out of high school, didn't have any trade, and there was hardly anything going on the building area. So I worked for a contractor remodeling some houses, then I went to work for an electrician, and learned the electrical trade. Later on I got into refrigeration and air-conditioning in the commercial business, so I was working on the big systems.

I also was working part-time for the vo-tech school and finally I put a lot more time there and I retired from the vo-tech school. And then after that I still work there part time, so I haven't done any industrial work in a while. But at that time there wasn't much going on right after the war.

K-State was moving old barracks in for student housing and everything was really temporary. There were a lot of people going to school on the G.I. Bill, I intended to go but never got around to it.

My wife's name is Nona. We have one daughter, Donna, who lives in Council Grove. My family, great grandfather came from Illinois started homesteading south of here at Ashland. He was a lay preacher and built the church and built a stone house across from it. My dad had two sons. I am the last one of that era.

The thing I remember most about the Battle of the Bulge was I almost got killed. We had a small concrete building that the Germans had started but was not completed next to our position. It did have a concrete roof on it and concrete blocks for the Windows and we fixed it up for us to sleep in. When they dissolved my battery I went to the message center and headquarters and I worked there and I was standing on guard one night, it was after dark. We stood there their carbines upside down and you could swing them up if you needed to. An artillery shell landed less than 50 feet from us and a big chunk of shrapnel hit the stock of my gun. I don't know if that would've killed me or not, but it would have ripped a hole in me. That carbine saved my life. I took it back to the armory and he couldn't believe what happened to my carbine. That was really close. The guy who ran the message center had just left, and he got some shrapnel in his ankle and they had to take him away.

At night, we couldn't sleep because all the shells were coming all the time and every time we would doze off, in would come more shells, whistling in and someone would say, hear that one? Then the buzz bombs would harass you and you couldn't sleep. I don't know if they knew where we were or not but you could hear them coming, just screaming, and then you hear them hit boo-boo boo-boo boo-boom. They would have accidents with those things and their engines would quit and they would go over you and then would crash. Some would fall in the trees, and we really felt a lot better when you'd hear them go on by.

We were guarding the Remagen Bridge and that was the first time we had any connection with a lot of the German jets. They just came in on the latter part of a war, I never did see any of them. I knew the guy who crawled out on the Remagen Bridge and removed the charges. I didn't know it at the time that he was the one that did it. It ended up that he was the guy who put the stonework on the front of my house. His name was Gene Dorland and he was from Chicago and received the Silver Star for that.

The thing I remember most about the Battle of the Bulge was a lot of confusion, and you never knew what was going on, you never knew what was happening, we are out in the boonies. We knew they were launching a big attack on us, and everybody, even the cooks, made sure they had a carbine. They never had to have one before, but they better have their gun with them because you might need it.

Don was awarded the ETO Ribbon with 3 Battle Stars, American Defense Service Ribbon, and Good Conduct Medal.

Harold in WWII.

Harold in 2010.

Place of Birth:
Wabaunsee, Kansas
Active Duty Date:
February 5, 1943
Unit:
335th Field Artillery Battalion,
87th Infantry Division
Division Location:
Bertrix, Serpoint, Freux, BE
Arrival ETO:
December 2, 1944
Rank: Sergeant
Section Chief

HAROLD O'MALLEY

The 335th Field Artillery Battalion 87th Infantry Division was activated on December 15, 1942. I was assigned to the Third Army and landed at La Havre, France, on November 28, 1944. They arrived in Belgium on January 12, 1945, and supported the 87th in the areas of Bertrix, Serpoint, and Freux. They breached the Siegfried Line on February 26, 1945, and crossed the Mosel River on March 16 capturing Koblenz. The Division crossed the Rhine River near Boppard and ended the War at Plauen near the Czechoslovakian border.

Here is his story. . . .

I grew up in Clay Center, Kansas. My Dad was a farmer but lost the farm during the Depression and moved to town to be a carpenter. There were five of us kids and I am the third one in line. I graduated from Clay Center High School and went to work for the Union Pacific Railroad and worked for them until I retired. I was a station agent.

I was married in 1943, and we had two children, one boy and one girl. My wife, Georgine, died three years ago at 85 years old. We were married while I was in the Army at St. Ann's Church, Walker, Kansas.

I was drafted in February 5, 1942, and went to Fort Leavenworth, then to Camp McCain, Mississippi, then to Tennessee for the Tennessee maneuvers then to Fort Jackson.

We did our advanced training there and then went on to our POE and headed to Europe. We sailed out of New York November 4, 1944, on the SS Louis Pasteur and I remember seeing the Stature of Liberty as we left. We landed in Liverpool on November 13.

We boarded LST's on November 28, and headed for France and landed at Rouen on December 2. I have a map here showing the route through France. Lots of homes were destroyed and lots of farm animals were lying dead in the fields.

We were part of Patton's Third Army. I was in the 335 Field Artillery Group, 87th Infantry Division, and our job was to support the infantry with artillery fire. We had 155mm howitzers that were pulled by prime movers. We had 13 guns in our section. I was a gunner and didn't want any part of being a forward observer.

We fired our first shots in combat at Metz at two old forts built to defend Metz. The Germans held the forts. We followed the infantry east and on December 17 we were in Obergailbach, about 95 km east of Metz. We were in Metz when the Battle of the Bulge started. I learned about the Battle through scuttlebutt but finally the Officers told us about it, that is how I found out. We had Christmas Day off, the only day, and then we started up toward the battle. We towed our guns with M-13 prime movers.

From Metz to the battle area we ran into ice and ice storms and the tracks slipped everywhere. Someone got the idea of welding tips onto the treads so that they would dig into the ice and it worked really well. We stayed at a convent and I remember the Nuns meeting us at the door. They gave us a place to sleep and breakfast, then wished us well as they sent us on our way.

We set up our first position in the battle around Bertrix, Belgium, and began firing on enemy positions. I remember it was so cold and snowy. I remember seeing Germans dressed up like Americans. We were in position, had the guns set up. We took turns being on watch. Each guy would do his watch, then come in and wake another guy up and we would rotate. We slept in tents and never got far from our foxhole. We dug them every night after we moved to a new position.

I remember Sgt. Ryan who was in our battery, was on duty and someone came through camp, messing with the barbed wire and he told them to halt and give him the password. But they didn't know it. He noticed that they had on German soldiers' pants and boots so he jumped behind a tree just as they were trying to shoot him. He fumbled with his rifle yelled for help and the Lieutenant came up and the Germans ran off.

We talked directly to our forward observers and to the wire section too. We had four gun emplacements with a telephone and the FO had a radio. We also used aerial observers. There was a Lt. Bounds, our FO. One time his plane got out in front of our big guns and his plane was hit and blew up. We also had so 105's and some 75's. Of course the Germans had the 88's. I remember the first time I got in one of their barrages. They sounded like whistling, screaming jets that slammed into the ground. We could use direct fire with our 155's just like the 88's. We would lower the barrel down and kick it off.

We used timing devices on our HE rounds (high explosive) with what they called quick fuses. They gave a quick explosion. Sometimes we would use air bursts with timing devices. My job was to make sure that our section was set up right and using the right rounds and fuses. I was a Sergeant in charge of the 1st Section, A Battery. It was a section of 155mm howitzers.

We were set up just west of Bastogne and were supporting an infantry platoon that was surrounded in an old farmhouse. We were laying rounds within 20 meters of the house, holding off the Germans for hours. It was so accurate the Lieutenant thanked us later. On January 16, we moved on very icy, slick roads into Luxembourg, then a few days later back up through Bastogne. It was a wreck. Then we moved on up just outside of St. Vith and on to Schonberg.

After that, we spent most of February supporting the assault on the Siegfried Line in Manderfeld, Belgium. I remember the mud, so bad that we had a hard time getting supplies and ammo had to be parachuted in to us by airplanes.

In early March, we were around Koblenz along the Rhine. I remember firing at targets across the river. After that we went across the Rhine at Koblenz.

Hundreds of prisoners were everywhere. It looked like the whole German Army had quit. Toward the end of the war, I remember we took in a lot of prisoners. I remember being part of a task force heading across Germany toward Czechoslovakia. We would find something to shoot at, unhook, blow the hell out of it, hook up and set out again. Some of the German roads were like turnpikes and we were right behind the infantry who was making great time traveling over these. None of the towns had any signs so I never really knew where we were.

We knew the end of the war was getting close. Some people had bets on when it would end. The last city I remember was Plauen. We had fired nearly 24,000 155mm rounds at the enemy when the war ended. When the war ended we bivouacked in Czechoslovakia for a while, rested a while. Then we began to return home to get ready to fight the Japanese.

Our entire division left as a unit. I went over with the unit and came home with the unit. We went to Camp Lucky Strike, one of the Cigarette Camps at Le Havre. We got a 30 day pass and I was ready to go to Japan.

The war ended with Japan while I was on the 30 day pass so we didn't' have to go. So we went back, the division broke up and I was discharged. I fully support Truman's decision to drop the A-Bomb. It kept a lot of us from going to Japan.

I sent back a rifle and some other things. I was lucky that I got those back. I had a pistol, a P-38 and some other things in a duffle bag and a flag, another rifle, it was all stolen. I found the rifle in a safe. Another guy and I were going through a town and stopped in an abandoned house to spend the night and there it was. It had 5 gold bullets in it but it disappeared on the way back home. It seems like everyone had things stolen and I think people in the Port of Embarkation stole it. They feel something heavy in your duffle bag and they steal it.

I lived in Clay Center where I grew up when I got home. I worked in several of the stations in the Kansas City area and Topeka. I donated some of my railroad stuff to the station museum in Topeka, a lot of Union Pacific memorabilia. I donated some pictures too.

I have taken two trips back to Europe and once we went back to Koblenz. We also returned to Luxembourg and Trier, both places I had gone through. I remember the castles along the river. There were lots of vineyards. We also went to Metz and some towns in Luxembourg.

I have a friend I served with from Minnesota and we tried to find everyone. We would go to his home in the summer, 25-30 people, would come to reunions and we had a great time. Our old Captain came once, it was just great.

The thing I remember most about the Battle of the Bulge was lots of dead Germans. We dug a big trench, piled them in and covered them up. It was a mound of Germans. And I also remember the destruction of homes and buildings.

Harold was awarded the Good Conduct Medal, American Service Medal and the European African Middle Eastern Service Medal with 4 Bronze Stars.

Chapter 8

The Airmen

AS THE WAR PROGRESSED FROM D-Day to Germany, the fighter groups moved along behind the front providing close air support for the Infantry. As the battle line moved east, they would construct temporary airfields. Many times they were exposed to the same artillery and aircraft fire as the front line troops, scurrying to find cover as the first sound of incoming fire. The pilots and crew lived in tents.

One of the most vivid memories was the day the weather cleared and the Allied planes were in the air.

Bob Knight remembers the exuberance when they first saw the planes. "We didn't have close air support, but suddenly the skies cleared and huge groups of B-17's came over. We all cheered their arrival, you know."

"I remember the beginning of the Bulge, it was overcast and we didn't have any air support," remembered Don Huse. "That's why they were able to get so many tanks in through our lines.

The path to the air was not always obvious. Art Holtman boarded a troop ship in August, 1944 as a member of the 2nd Armored Corps and disembarked in England assigned to the XVIII Airborne Corps. Art found himself converted from a tanker to a glider pilot. "The 82nd Airborne and the 101st airborne made up the XVIIIth Corps Airborne," said Art. "I was with the Headquarters Company of the Corps

Bill Wright photo.

and we were not paratroopers, we were glider troops and we got about two months of glider training in England."

Air ground coordination was relatively new during WWII. The team of coordinators were assigned to divisions and traveled just behind the front, coordinating attacks from the air. "We would have to get our antennas set up exactly right. We would turn the antennas with a wheel that was marked out in degrees." Wally Jeffery related. "When we talked to a pilot we would get a fix on his voice and listen to all of their conversation in the air. We would say give us a fix and the pilot would say this is "Bluebird 5, I need a fix," then he would say aahhh for a few seconds and you took a fix on the voice."

Kenny Luigs probably summed it up for the rest. "The weather cleared finally and the American fighters came. I was astonished to see so many of our planes and happy to see them. The airplanes came in with all the white stripes on them. I saw every bit of it and just stood out there and watched them fly in. Vapor trails, they left a bunch. They just turned the whole sky white everywhere. Boy, everybody was ready to see them."

Wally in WWII.

Wally in 2010.

Place of Birth:
Keats, Kansas
Active Duty Date:
November 13, 1942
Unit:
327th Fighter Control Squadron (Supported 1st Infantry Division)
Location During Battle:
Malmedy, BE
Arrival ETO:
June 7, 1944
Omaha Beach

WALLACE JEFFREY

The 327th Fighter Control Group was activated on March 20, 1943. They were scheduled to land on D-Day with the 1st Infantry Division on Omaha Beach but the beach had not been advanced enough and were postponed one day to June 7, 1944. Not only did the Group control fighter attacks but also was involved in controlling B-25 and B-26 operations. They followed the Infantry and during the Battle of the Bulge they were positioned around Jamoulx, Verviers and Liege, Belgium. They proceeded into German going through Marburg and ending the War in Weimar, Germany.

Here is his story. . . .

I was in the 327th Fighter Control Squadron and headquarters were usually several miles behind the infantry. We landed on Omaha Beach the day after the start of the Normandy invasion and were on the front lines in eastern Belgium during the Battle of the Bulge. The 327th Fighter Control Squadron also participated in the Rhineland Campaign and was part of the American Occupation Army in postwar Germany.

My Dad was born March 13, 1882, in southern Missouri somewhere near the town of Nevada. They lived on a farm, and I think he had two brothers and a sister. I think they were about twenty years older than he was. I know very little about my Mother's early life. She was born in August 1897 and was the oldest of seven children.

My Dad was hit very hard by the Depression and trying to keep the family together was very difficult. As a child, I lived with relatives while

Dad and my Mother would work various jobs that required them to be on the road. I never had my own bed until I was in the Army.

After the first of the year in 1940, my brother Jim enlisted in the Army at Ft. Riley. He took just a short period of training, and then they sent him to Hawaii, where he was in a Military Police unit.

I met my future wife Geneva, at a dance and after a while we were dating regularly. One day, I decided to quit my job and go get married. My brother Bob and a lady friend stood up with us. I can't put into words how proud and happy I was.

I was doing hauling for a contractor when the draft board sent me my greetings and salutations notice. Geneva was just about ready to have our baby, and I got a reprieve of about 3 or 4 weeks. I reported to Wamego, and was put on a train for Leavenworth for my physical, etc. on November 13, 1942. I was sworn in the same day, and we got to go home, reported back November 19, 1942, and were issued Army clothes, etc. and waited to see where I would be shipped.

They loaded a bunch of us on this troop train at Leavenworth and went west. We were not told our destination until we arrived at Fresno, CA. After the tests there, I was told I would be in the Army Air Corps and to choose three occupations where I had interest. They sent me to radio school and I boarded a train for Chicago. We would split our days between radio mechanics and Morris code. We studied radio mechanics all morning and then marched to the coliseum for code instruction. We had Army instructors and civilian instructors.

Out of our class, they picked a few of us to go to an advanced school at Tomah, Wisconsin, and we started classes the next night. Our purpose was to study all the radio equipment and all the methods of maintaining and operating the equipment that would be used. All the separate pieces would together make up a Control Net System, and we were to be a part of that system to control our fighter planes, and sometimes some twin engine bombers. We were to be assigned to the tactical air group. Each one of us studied each facet of the course so that we would be capable of maintaining and operating any component.

When we first trained on the equipment, we were supposed to string wire and we had EE8's handsets. We got five months of training in Chicago in radio school. We took both radio mechanics and Morse code. Then they sent a few of us Tomah, Wisconsin where they had the equipment that we were going to use. We studied the equipment that we would be using. It was made by Bendix and it was called a CNS, Control Net System made by the British. They had designed it at the beginning of their part of the war, 1939 or so for their Spitfires. We had Bendix make the Equipment and all of it had the royal crown for the British invention. We used the same equipment for the whole war.

All the radio equipment we used, as well as the Control Net System, had been designed by the British, and was supposed to be highly secret. Our Bendix Corp. made the equipment, but every piece of a complete unit had a red royal crown affixed.

I think that we went to school for six weeks, and when we graduated, I was made Buck Sergeant. We went to an airfield just outside Orlando, Florida, where we were to be stationed. The outfit that I became a part of was the 327th fighter control squadron. As far as I know, we weren't part of any other unit as of then.

We were next loaded on a troop train and were unloaded at Camp Miles Standish near Boston, Massachusetts. We were put in barracks and waited for our shipping orders. We loaded on a small ship called the "Mexico." It was a small wooden ship that had been used in coastal waters to haul bananas to the U.S. from-Latin American countries. We had about 300 men in our outfit, and we stayed on the upper deck that was above water. On our tenth day we could see land, which was Ireland. We steamed into the harbor at Glasgow, Scotland, where we got off the ship.

We were assigned to the 8th Air Corps, but we weren't being used. About all we did was stay around our barracks and wait.

One day they told us we would be leaving by convoy the next day. We went through London, and it seemed like it took us forever to cross that city. After we were north of London, we split up with the headquarters part headed for Colchester and the 3 D.F. station crews going to our different points of operation.

I got word from my wife that she had given birth to a little girl and she had named her Dixie. She said all was well. I can still remember how happy I felt, but up to that point, I was never more homesick.

I couldn't get my mind-off of what was ahead of us. I knew our outfit was trained to be close to the front line, and the job I had called for me to be up front with or near the fighting boys. As the front moved forward, we were to move also.

I never knew where I was. I was on an outpost and I was just given a grid to go to and set up. I never knew of any towns around where I was. I knew where headquarters was but I seldom ever went back to headquarters. I was in the Army Air Corps in the 327th Fighter Control Squadron. We had about 300 men and were part of the Ninth Air Force. We had radio

Communications Van National Archives

truck operators and ground controllers and we guided fighter pilots to their targets and directed them back to base when they were lost or hit.

We never used maps where I was which is why I cannot tell you where I was. The officers would tell us where to set up. We were always out in the country and never understood the relationship to any town or city we were just all out by ourselves, most of the time we weren't even close to anyone else. They picked the spot because of elevation and there were three units of us so they wanted us to stay on vectors where they put us.

When we arrived at our location, we would have to calibrate our vehicle according to where true north was, not magnetic north but true north. The officer would tell us how to set up our equipment accordingly to make sure that we were given the right vector for the airplanes. We knew how to set up for true north but they would tell us what true north would be for each location. They would tell us how many degrees we would have to set up different from magnetic north.

We would have to get our antennas set up exactly right and they were rotating antennas. We would turn the antennas with a wheel that was marked out in degrees. When we talked to a pilot we would get a fix on his voice and listen to all of their conversation in the air. We would say give us a fix and the pilot would say this is "Bluebird 5, I need a fix," then he would say aahhh for a few seconds and you took a fix on the voice. You would turn the wheel and then you depressed a switch and if it became louder you knew you were 180 degrees out. When you turned the wheel and his voice got lower you knew you had his position and vector in degrees. We then had FM radios back to headquarters as to their position. We did not have radar, we had VHF radios, and the radar was back at HQ.

In addition to fighter control, we used to control B-25's and also B-26's later on when the bombers would bomb bridges and things. We were able to talk directly to the pilot but it wasn't under normal circumstances. Officers back at HQ would talk directly to them. With the radar they could pick up enemy planes and they could tell them the vectors so they could intercept them. It worked out swell. Later on in the war, they told us it was pickle barrel bombing because they could drop the bomb in a pickle barrel.

We would stay set up until the infantry would move ahead, we just stayed with them. Of course, they would get further away from us when we were getting set up because we couldn't just move the van. We would have the van with the radio equipment in it and then we would have to raise the antennae and get it set up and then camouflage all that. Then we pulled a generator on a trailer, a PT 99, and we had a portable generator that we used mostly, it was a PE 75 and it is where we got our electricity to operate with. We were self contained.

We ate K rations almost all the way through the War. We did get C rations at first when we went into Normandy but in France, Belgium and Germany all of our units except the HQ unit, ate K rations. The HQ lived in Chateaus, pretty high on the hog. In Normandy, when we were waiting for our equipment to get in, they were in the foxholes with us and we were together until our equipment could get in. Our equipment at that time was highly secretive and they didn't want the Krauts to get a hold of it. We had orders to keep from getting it captured and we had orders to blow it up if necessary.

We went into Normandy on a Liberty ship boarding at Liverpool, England, and then came back up the Channel to Normandy. It took us eight days to land at Easy Red on Omaha Beach. Our radio people were supposed to go in D-Day afternoon where the 1st Division went in, but the guys couldn't advance, so we were rescheduled for D-1. I watched all the fireworks. Our heavy cruiser and other gunships blasted at the German positions. That is a long time to sweat out an invasion to think about with might happened. They ran out or K rations it took so long and we were on half rations.

I don't know how the other guys felt but to me when we went in a lot of it is a blur. I think we were all in a state of shock. Some of the scenes are still very vivid to me now. The most valuable tool I had was my hard hat. I used it to wash my socks and underwear, cook, take a sponge bath.

When we were in Florida we were practicing controlling planes and when we were in England we actually did control planes in combat for the last five months. Our unit was out in the country somewhere but the HQ was in Colchester, England. I don't know where the other two units were.

Finally we were ready to go ashore and the water was still pretty choppy when we climbed down cargo nets called Jacob's Ladders and jumped into bobbing Ducks, a 6 x 6 truck designed to float and go on water like a boat. It was soon my turn. Several of us radiomen had orders to go. We had our Mae West life jackets on and our haversacks, rifle and ammo. I climbed down the side of the ship. It took our group of 30 men right to the edge of the water. We then crossed the beach area, which was still littered with shot up equipment, and there were pieces of men not yet cleaned up. Out of curiosity, I picked up a GI shoe that had the heel blown off. The rest of the foot was still in it. I then fully realized this wasn't a game. We fought through the hedge rows of Normandy and St. Lo. We helped liberate Paris August 25, 1944, and then we moved on north to Belgium.

We didn't stay there long, as the Germans were backing up pretty fast. Our headquarters set up in the Verviers, and our unit set up near the German border. We had all but outrun our supply line, so we were mostly remaining in place while waiting for our supplies. We did have some of our boys in heavy fighting, trying to go north through Holland. At that time we were controlling fighter planes that were supporting them. We were also controlling some English fighter planes. I always knew that our pilots were good, and some of the bravest men you could imagine. At this time, I found out the difference in our pilots and the English pilots. Our pilots were always calm and looking for a chance to combat the Germans.

Sometimes, when one of the planes would get hit, the pilot would calmly say I'm hit, I'm bailing out. His flight leader would then call us for a fix, so we would know where he went down in case there was a chance for a rescue. One time while I was on duty, we told some English pilots that there were some hostile planes near them, and they asked for a fix on them so they could avoid them. Then an American pilot cut in and asked for a fix on them so they could engage them and they did. I felt proud of them.

By this time we were starting to get some Christmas presents from home, mostly cookies. I'll never forget December 16th. It was foggy and just about dark when we could hear the drone of the German planes overhead. The fog grounded our planes. We didn't know at the moment, but they were dropping German paratroopers on us, but we soon found out. They were in American uniforms and could speak English.

We were still operating our radios, even though our planes were grounded due to fog. We were to stand guard at night, and help fight and capture paratroopers during the daylight. This was the beginning of the Battle of the Bulge. The Germans were attacking in full force. They caught us by surprise. I know we had a green infantry division along the line where we were, and our battle hardened 1st Div was back by Verviers where our headquarters was. They were back there getting a well-deserved rest. Because of the many Germans in American uniforms, we had to challenge just about everyone you came across. Since some of the Germans could speak English, we used passwords that were hard for them to say such as "thistle". Sometimes you would ask them something about the USA that any American would likely know. This went on for several days, and I know I was getting pretty well worn down due to weather, lack of rest and wondering if we would survive. The infantry was getting pushed back, and the division where I was had their front line troops walking back, mostly as stragglers.

They were gaunt, hollow-eyed, and looked like they were in shock. I felt sorry for them, as they were cold and hungry, and we had nothing to offer them, so they kept on walking. According to the Stars and Stripes Army newspaper, a lot of the guys had to have their legs or feet cut off in order to save their lives because of frostbite.

We were informed pretty soon after the Battle of the Bulge started. The officers at HQ told us that there were German paratroopers dressed in American uniforms and some could speak English as well as you and me. The first thing they told us was that our planes were grounded but we still had to man our stations. Somebody had to be there and one man could operate it and we would take turns because it was a 24 hour deal, not always, but sometimes. Unless they called you from HQ that we could shut it down because no planes would be flying.

We were ordered to retreat, but I sure hated to. We had always gone forward, and I felt real humble and almost ashamed to give up territory we had already fought for, and people we had already liberated.

So they told us to man the station in the daytime and go out and try to locate and catch the paratroopers. At night somebody was there, and we could just hunker down. It was cold but I was lucky in the unit I was with, we had been issued winter clothes. We had overcoats and boots but we didn't suffer like the infantry did. It was colder than Hell but we didn't suffer like they did. We all had our jobs to do with different degrees of danger. There were times when we were in danger but not all the time.

One time we went down a road. A Lieutenant in a jeep came towards us and stopped. He asked us where the 1st Infantry headquarters was. We told him it was near Verviers. He turned around to go back. I asked my buddy was he thinking what I was. He said yes, so we raised our guns and halted him. We got into the jeep, and had him drive to the Interrogation Center, and told them we thought he was a Kraut. We left, and we heard that they shot him the next morning. They didn't mess with Krauts in American uniforms, and considered them as spies.

You get accustomed to danger. I now had two little kids of my own but I didn't' know if I would ever see them. You just kind of get numb. I noticed when I went into Normandy; there were a lot of green troops. Some had come up through Africa in the 1st Divisions and some were battle hardened. As I was around them, I thought these guys were tough and they were but then I got to realizing that after I had been there that the Germans were tough but I would rather face them than our boys, they were tougher.

When they went into battle they were in battle. The paratroopers and infantry boys were cocky. Some of them got killed but they were still

National Archives
Operations Tents 327th fighter Control

cocky. I made up my mind that if I was in a war I would rather face the Germans than the Americans.

One day, when I was on duty, I heard a buzz bomb engine quit. I stood in the open doorway of the van and watched as it dove toward us. It hit an anti-aircraft gun pit about 100 yards from me. I was knocked out from the concussion. I don't think I was out very long, as there were soldiers just getting to the gun pit location when I came to. On Christmas Day, a bomb fell near our mess hall building, knocking a lot of plaster and dirt in the pots while they were preparing our Christmas dinner. But it was a good meal anyway.

The unit my friend and I had been with a couple of weeks before the Bulge in order to observe their work was captured. I was glad we hadn't remained with them as long as they wanted us to. I didn't know until later how close I came to being captured myself.

I knew the tide had turned when we got those battle planes up in the air again. We just tore the hell out of them. That was the difference. That is the reason they named it the Bulge because they pushed us back. The weather turned better right after Christmas, and the sky cleared. Our planes came out like a bunch of bees attacking the Jerries (as the English called them). Our fighting units were regrouped and reinforced and went on the offensive. We started pushing them back. It was tough going, and conditions were still bad, but I felt better moving forward. It got easier as we went on, with the exception of a few locations. We pushed them back to the Rhine River, and across it. They were blowing bridges in order to use the river as a natural line of defense.

We never knew for sure how many planes we were controlling at once. Our code name was "Sweepstake." So they would call in and say "Sweepstake this is Bluebird, give me a fix." There were several hundred that we were controlling throughout the day. There would be 6 or 8 planes in each flight and they would go all day. We controlled all the planes that supported the First Army and there were several divisions there. We were following where we were needed. They picked out the spot, the altitude and so forth.

Bombs dropping from the bombers were controlled by either HQ or the bomber. The pickle barrel bombing, the HQ officer would control it. The big bombers like the B-17 and B-24's would miss their targets and they even dropped some on us. I remember one night in Normandy a B-17 had been hit and was flying low, losing power. He got on our frequency. I picked him up and gave him a vector to get back to England.

What sticks in my mind the most was the infantry guys who were green troops; I don't think they would have had the Battle of the Bulge if there had been experienced troops on the front. They had hellish conditions and a big force attacking them. The site of these guys coming back is what sticks in my mind. They were gaunt, their eyes were hollow and they looked like they were in a state of shock. They were straggling 2 or 3 at a time. That is what I remember, the way those guys looked. They asked for food and when we didn't have any they walked on toward the rear. It was terrible. They were the most pitiful looking guys.

We kept moving forward. We were just outside Weimer, Germany when they gave up. I saw German civilians cursing German soldiers. I guess that is what happens when you lose a war. After Vietnam, I thought of how we were treating our returning soldiers.

I started on my first leg home, by riding in the back of an Army truck in a long convoy of guys headed home. I didn't know one person in the convoy. They took us to a railroad siding. I think it must have been near the French border. We didn't know where we were or where we were going. We were next loaded into boxcars. Their boxcars were a lot smaller than ours in the US. I don't know how many of us were in the cars, but we were pretty crowded.

Finally one day we got orders that we were to move out. We took another train ride to a port. I think it was Bordeaux. Anyway, it was a port in the south of France. We loaded on a troop ship, and I remember going by the Rock of Gibraltar, and out into the Atlantic. About all I remember about that trip was that it was peaceful and relaxing. The weather was so nice, and the water was like glass. The food and bunk was the best I had since I couldn't remember when. I just lay around in the sun and meditated about the past and the future. I didn't try to start any new friendships. I did miss my old buddies who I had been with. I think we were about 10 days until we steamed into Boston Harbor.

When I came back to the States I did not call anyone. I just came home. I saw my boy for the second time but saw my daughter for the first time. I finally got my discharge and received $300 mustering out pay. I caught a bus from Kansas City to Manhattan but somewhere between Kansas City and Topeka it broke down. We waited for several long hours, and then continued on our trip to Manhattan.

I stopped at Smith Jewelers and bought my wife a pearl necklace and walked to her upstairs apartment. I was greeted by my honey and my two beautiful kids.

After the war, I returned to Europe and I was amazed that they treated me so well. I saw the monument at Bastogne; I knew I was part of the history of it all. The cemeteries are extremely emotional. They are beautiful but sad.

What I remember most about the Battle of the Bulge the green infantry troops who were coming back through the lines, their eyes were hollow and they looked like they were in a state of shock. It was terrible. They were the most pitiful looking guys.

Wally was awarded the Rhineland-Ardennes ETO Ribbon with 5 Battle Stars, American Defense Service Ribbon, Good Conduct Medal, Sharp Shooters Metal, and the EAME Ribbon with Meritorious Wreath.

Arial view of the 18th and 115th Infantry Regiments assault Omaha Beach on D-Day, June 6, 1944.
Footnote.com

P-47 fighter downed on D-Day crashed on the beach. *Footnote.com*

Above and below: Omaha Beach is now a resort area. *Author photo*

Carrol in WWII.

Carrol, 2010

Place of Birth:
Alta Vista, Kansas
Active Duty Date:
January 9, 1942
Unit:
514 Squadron, 406 Fighter Bomber Squadron
Location:
Mourmelon, France (A-80)
Arrival ETO:
April 4, 1944
Rank:
Staff Sergeant

CARROL JOY

The 406th Fighter Bomber Squadron was activated on March 1, 1943. It arrived in England on April 4, 1944 flying its first operational support mission into France on May 9, 1944. The Squadron flew support for the D-Day landings on June 6, 1944 and arrived as a group in France on Omaha Beach on July 20, 1944.

In mid-June, the 513th was selected to be the first P-47 fighters to be fitted with new high velocity rockets that became very important to the close air-ground support for the infantry. By building temporary airfields and only miles behind the front lines, the 513th moved from location to location across France. During the Battle of the Bulge, they provided close ground support for the 101st Airborne as they fought for survival at Bastogne flying from airfield A-80 at Mourmelon-Le-Grand, France. They moved to Y-94 at Handorf, Germany, where they ran operations until the War's end.

Here is his story. . . .

I was born in that house right over there on the corner. My grandfather bought it in 1906 and they moved out here in 1908 from Republic. I was born in 1920 in the house right there on the corner. I did not get drafted into the Army, I joined. The first bunch that got called up for physicals after Pearl Harbor went to take their physicals and went into the Army. A good friend of mine who already had been drafted in June, about six months before, was stationed Sheppard Field, Texas. His brother and I had talked about going down and visiting him about Christmas time, and really thought about going to enlist when along came Pearl Harbor. So I

decided I would go enlist, and I wanted to get in the Air Corps so my cousin and his cousin all went to Wichita to see the recruiting officer. We had to have three letters of recommendations so they sent us home.

This picture was taken in France in 1944 while we were stationed in Mourmelon at A-80. The Eisenhower jacket was British made. We were issued two jackets, one for work and the other for dress. The scarf is a piece of German parachute cloth. Several of us wore them. In addition to the British made jacket, we also had American made Eisenhower jackets.

I was with the 514th Squadron of the 406 Fighter/Bomber Group and we provided closed air-ground support for the D-Day invasion of France from its airfield in Ashford-Kent, England. The Squadron landed their P-47 Thunderbolts in France and continued to march east close behind the Infantry, building temporary runways and living out of tents. Keeping the tactical fighters airborne was a wonder in itself, especially considering the conditions that the pilots and crew endured. Without a doubt, it was a welcome site to all of the Veterans, the return of the Air Corps during the Battle of the Bulge.

They sent us a letter after the first of the year to tell us that we had to go to Wichita Monday morning and from there they bussed us to Fort Riley. We were there all week prior to being sworn in on January 9, 1942, and I got my first shots. I am not too sure when the other two boys were sworn in. They gave us clothes, gave us shots and put us on a train to Leavenworth. At Leavenworth they gave us clothes and the same three shots again. Then we went by train to Wichita Falls, Texas and they gave us shots again. Same shots, three times now!

We took our basic training in Wichita Falls and all we learned there was close order drill, nothing else, no rifles, nothing. I was on guard duty one night in the hanger with five rounds of ammunition and they told me not to put them in the gun but I did. I picked them up and put them in my pocket after the Officer of the Day had shelled them out on the floor.

All we had then was close order drill. Later we went to mechanics school and carburetors were very hard. At the very first, they gave us a book and we went around and looked at them, they were some old obsolete ones. I remember one, a B-18A. Finally, they said, don't just stand around and look at them, and they gave two of us a couple of brooms and had us sweep out the hanger. If you have been in a hanger you know how big they are. We missed most of one day of the school sweeping the hangar.

In training, we trained and worked on the whole airplane. They gave us a few tech *(technical)* orders to read. Each type of plane had its own particular tech orders and covered everything about the aircraft, how to repair and adjust, etc.

After I got out of school they sent me to Santa Monica, California to go to the Douglas Aircraft factory school. There were some of the movie stars who came out to eat lunch with us. From there I went to Key Field, Mississippi, and the unit was formed there in the spring of 1943. We were originally called the 406th Fighter Bombardment Group and it was changed to the 406th Fighter Bomber Group, then they dropped the "Bomber" later on. I was in the 514th Squadron. In the summer of 1943 we were just kind of messing around because we didn't have any airplanes.

We were transferred to Congaree Air Field outside Columbia, SC, and began to get a few P-39's. I got the first plane assigned to the unit, a Douglas A-24. It was made as a diver bomber for the British, a big airplane. The wings were bolted on with lots of bolts all the way around and they kept working loose.

Then we got the Douglas A-24 dive bomber, a version of the SBD dive bomber that the Navy used in the Pacific. It was a good little airplane. When fuel was scare and we had to run 91 octane fuels, we had to change the timing and they didn't run very well. The two 50 caliber machine guns fired through the propeller and every once in a while it would put a bullet through the propeller. So they had to go back to 100 octane fuel. They were good airplanes. At Congaree Air Field we had P-39's for a while then finally got the new P-47's. It was called the Thunderbolt.

When the flights were coming back I could pick my "A" flight out, it was smooth and they were the hottest pilots we had. Trouble was we lost a lot of them. The P-47 was a really good airplane. The A-10 Warthog is also called the Thunderbolt II. I went to an air show in Topeka a few years ago. The crew chief on an A-10 Thunderbolt gave me a 20mm shell casing from the nose gun, one of those Gatling guns.

We left the States as a unit from Camp Shanks, New York on the "Sterling Castle," a British ship. We had boiled potatoes and biscuits that was all that was fit to eat. We were talking to some of the cabin boys who were just kids and they said they brought some German and Italian prisoners back to England from North Africa, they said some of the Nazis would go to the rails of the ship and yell, "I would die for the Fatherland," and the crew said sometimes they helped them out.

We arrived in Liverpool, England, April 4, then we went on to Ashford which is on the southeast corner of England, I think we were the closest group from the English Channel, about 80 km from the French coast. On May 9th we began to fly our first missions over France.

I remember seeing the Buzz Bombs. I saw them going over by the dozens, going right down doodlebug alley. We had A, B, C, and D channels. I remember in the morning in the early part of June, it was still dark and we were pre-flighting the aircraft. My plane had on its running lights and you would think twice before you got close to that propeller. I saw one coming over, not very high, I could see the exhaust coming out of the back, and I thought, "what the devil it that?"

So I turned off the running lights and thought that was what the other guy was doing. You could tell when they were coming over at night, the sky would just light up with anti-aircraft fire so you knew they were coming. We saw a lot of them at night and we would go out there with our carbines and bang away at them, never hit any of course. They landed all around us but never landed on the runways. If one had landed there it would have wiped us out.

This picture is when we were just getting ready to taxi out. When they taxied out we would ride on the wing because the pilot couldn't see over that big engine, we held on to one of the 50's and guide the pilots along. When we got out there we were going to taxi out west and I looked over and here came a buzz bomb about 100 feet off the ground. A British plane chased it, firing at it; it looked like it went over there and then over here and others at the side said it went up and down. We watched the British plane and the buzz bomb kept coming so some of the pilots jumped out of their planes and ran for cover. It hit maybe 100 feet away from the runway. It had a ton of TNT and it dug a hole two feet deep. They just came over by the dozens.

Carrol and his crew

Our planes didn't escort bombing missions much, we were attack and destroy, and that was our motto. We got six battle stars from major battles, a star for each major battle. Ground forces received up to 5 but we also got one for air Offensive-Europe.

We had a pilot who got hit and crashed his plane into a German gun position. Our first squadron commander shot down a Japanese Zero with a P-39 in Alaska and he was the first man of the 406th shot down in a P-47. We had been in England about two months when the D-Day invasion took place and we supported it. We lived in tents and were the first fighter group to operate in field conditions in England. We were part of the Ninth Air Force, the Eighth lived on a base with accommodations, and we were in tents in whatever the weather brought to us. Our runways were temporary, that was like chicken wire with metal rods running through it. In Europe we used planking. It was called Pierced Steel Planking runway.

Our planes were painted with invasion strips so people could tell they were Allied planes. We put them on the night before the D-Day invasion. On the night before the invasion another guy and I had a pass to London. A guy asked us if the MP's had caught up with us yet. He told us to tell them that we were based just outside of town or they will cancel your pass. So just outside of town, the next morning, we were getting on a train and that is when we found out the invasion had started. On the way back to the base from Ashford, the jeep stopped to pick up water for the water injection that was on the plane engines and we hitched a ride back to base. It was as busy as a beehive and was that way for the next few days.

We flew support for Utah Beach because we were the closest squadron to it. An air echelon went over to France in advance of the group to get things set up, and then we went over on a Canadian ship and landed on Omaha Beach. The ground echelon operated in England until the air war got established in France. So we left piece meal, some of the group went to get established in France while we kept the operations going. We never stopped operations at all during this time of moving to France and also all the other moves to new bases. We landed in France on July 28, 1944, and set up at Tour En Bessin, France, just outside Bayeux and about 20 km from the front.

We had external fuel tanks that could hold about 100 gallons of fuel and made of pressed paper. You could knock off the nose and we used those to put in our tool boxes and clothes and everything we could put in. Some of us crew chiefs did that so that our things would get there to France from

England. Then we tied them onto the wings of the planes with bomb arming wire and I guess some of the men didn't do a very good job tying them on because they lost everything they had. They just fell into the ocean. We lived out in farm country, sometimes muddy and showers were primitive and almost nil. I took many baths out of my helmet when time permitted. We were in Tour En Bessin.

We had two of our reunions in England, France, Belgium and Germany and one of the towns had a square named after us. My wife and I attended all but one of the fifteen reunions here in the States. I went to see a P-47 at the Combat Air Museum in Topeka and the name of it was "Big Assed Bird" after one of the 513th Squadron planes. They got that plane down in South America and put it back together. There was the 512th, 513, and 514th in the 406th Group.

I lost my first plane but it wasn't my regular pilot. The pilot crashed somewhere over in France but he came back. It must have been when I was in town. He belly landed the plane called "Bloom's Tomb," named after our CO whose name was Bloom.

My regular pilot was Lt. Lewin Epstein. I remember once he took me up in my P-47 so that I could get a look at the front lines. It was against all the rules, the plane was a single seater and he sat on my lap.

My first pilot was shot down but he was in a "Bloom's Tomb," one of several shot down. His name was Lt. Levitt Clinton Beck from Huntington Beach, California. He was shot down, captured by the Germans and died in a German concentration camp. He was in a dogfight on June 29th, and he and the Kraut had been shooting at each other, and on his way down he managed to get the Kraut and brought him down too. He actually shot the Kraut down before he crashed. Another pilot, Lt. Gaudet, was also shot down in the same fight and never heard from again. Nobody knows what happened to him.

Lt. Beck was flying a plane also called "Bloom's Tomb." He was able to crash land the plane in the French countryside and was helped by a young Frenchman. The French Underground hid him for about two weeks in an attic in a French café in Anet. They were planning to get him back into American territory but some Krauts found him, and he was in civilian clothes so they put him in a concentration camp, Buchenwald. Later in 1944, the Germans transferred the airmen to regular POW camps but Lt. Beck died before he could be transferred.

While he was at the French café in Anet, he wrote a manuscript about his experiences and it is worth reading by anybody. It is called "Fighter Pilot." He had buried his manuscript and Americans picked it up and later gave it to his parents who lived in California. He was a good looking guy and he and I made a good pair, he was a really good pilot. His mom and dad sent me a copy and wrote in the book for me. A Frenchman sent me a picture of the plane later.

I had one plane that I was responsible for and one assistant and an armorer. I took care of the plane and the armorer took care of the guns and bomb racks.

I remember a pilot named Kramer. He was flying another plane, and he came in one day and said, "I hope I don't have to use this parachute. It's not mine and it doesn't fit." They sat on the parachutes. He didn't come back, got shot down. We never knew when we sent them out if they were going to come back. When you saw your plane come back it was fine but when you didn't it did not feel good. I lost three planes in the War.

Most of the time, I had no idea where the front was. Sometimes we were close to the British artillery and I could hear them next to us. According to the plan, the British artillery was off to the side so we could take off and not get in the way of the guns. We supported Patton's Third Army most of the time. That is why we went into France in late July, following Patton's army. The battle of Falaise Pocket was our first big battle and we were close to the front then. It was here we received our first Presidential Unit Citation.

We moved on to A-80 at Mourmelon close to Reims and stayed there until the Battle of the Bulge was over. We were at A-80 during the Battle and it was known as an Advanced Landing Ground because of its "advanced" position. We were in tents and there was a big staging area for our supplies.

We heard on the night of December 16, that the Germans were supposed to be dropping in paratroopers and that gave us kind of a funny feeling to have to worry about enemy paratroopers. This was not true on the scale that we heard but the false information was spread by the Germans to scare everyone. The size of the force was very small.

The 101st was right there next to us when the Bulge broke out. Some of them thought they had won the war but they were replacements for the 101st so they were training, no experience, and they kept calling us "fly Boys." When they got trucked up to Bastogne, we couldn't get our planes off the ground because it was so foggy. So the Colonel pulled an inspection on us. I found a faulty engine and had to replace it. At that time clocks were kind of scarce on those planes and some of them didn't have them. I took the clock out of the plane until it came back with a new engine, and

I "replaced" it in the flight forms after the plane never came back. I still have it.

They put me on guard duty for several days and nights. One night I forgot the password when I was coming up to the Officer of the Day, and we nearly got into a fight over it. By that time I remembered the password but I didn't tell him. We were fogged in so we didn't have much to do. I was a Staff Sergeant and it seemed that they once threatened me to take my stripes because I didn't read up on the latest tech orders.

We never had Germans try to attack our airfield. When we were at our first base in France, A-13, just off the Normandy beachhead, we had a French boy who said the Boche were following our planes back to our base but I don't know what happened.

When the weather finally cleared on December 23, my plane was shot down and I didn't have one until the battle was almost over. After the Battle was over we moved on to Metz, Airfield Y-34 at the end of January. We were there for about a week, then sent us to Asch, Belgium, and put us behind the First Army. On the way up there we stopped in Luxembourg City and Bastogne and on to Asch, Belgium. The road we took had been pretty well cleaned up. I remember an incident in Belgium when another plane had taken a bullet, leaving a bullet hole that had just clipped the wing off the pilot's chest.

The airstrips all had designations. We only made one more move after Asch We ended the war at an old Luftwaffe airfield at Handorf, Germany, Y-94. Our last move from Y-94 was to Nordholtz, Germany, up on the North Sea. We were part of the occupation force. It was from there I left in September to go home.

While we were at Nordholtz, Major de Seversky, one of the P-47 designers, paid a surprise visit to our base. So, I guess we were a highly regarded fighter group in Europe.

My first plane was named "Sheriff of Los Angeles." The last one was "Paula." Several crew chiefs were awarded Bronze Stars for meritorious service. We had at least 75 missions without an abort. I had 115 from the first of May, 1944 to October, 1944. None of my planes ever aborted a mission until the end of the war. I had close to 200 missions at that time.

Our unit, the 406th, was awarded two Presidential Unit Citations. The first was at the Falaise Gap and the second was at the Battle of the Bulge. We were credited with over 13,000 flights or sorties, so you know we were busy, usually sending 36 planes on a mission.

I got out of the Army on October 13, 1945. I rode a train from New York through Chicago to Ft. Leavenworth. I got enough money to call home from Manhattan and I asked the Alta Vista operator to call my folks. They had a farm program now called FSA (Farm Service Agency) and the manager saw my name in the Kansas City paper. That is how my folks knew I was coming home. Our local operator called my aunt and she called my folks who had a phone.

The day after I got out was Sunday. A year later to the day, a friend of mine knew a girl in White City and she had a friend who turned out to be my wife later on. Her name was Edna Mae and it was kind of a blind date. We have four children. I stayed on the farm and farmed with my Dad until he retired. After I was discharged I served another three year hitch in the reserves for a total of nearly seven years of service. I have eleven grandkids and six great grandkids.

What I remember most about the Battle of the Bulge? I was delighted that we could get our planes in the air to give help to the "Battered Bastards" of Bastogne.

Carrol was awarded the Bronze Star Medal, ETO Ribbon, 2 Presidential Unit Citations, Bronze Star Medal, and Good Conduct Medal.

Above: Bill Wright's P-47 at the Battle of the Bulge.

Right: Bill Wright by prop.

Chapter 9

The Medics

FORREST ADAMS BOUNCED IN the heavy waves as sea water spilled over the closed ramp on his Higgins boat as his boat ground toward the beach. His thoughts were drowned out by the roar of the big diesel engine. It was June 6, 1944. Hours before he had climbed down the cargo net of a troop transport, dropped into a Higgins boat and chauffeured to war leaving life as he knew it. His Higgins boat was heading straight toward Hell. Forrest was a Combat Medic assigned to the 116th Infantry 29th Infantry Division, the group of young men in the Virginia National Guard who history would remember as the ill fated Bedford Boys. They were the first wave onto the death of the beaches of Normandy.

Due to a coordination error, his 10th wave was behind schedule and the tide was up and very little beach was showing. He finally stepped into complete bedlam around 0900. He wondered, almost out loud, how did I get into this mess?

Dr. Dr. James McConchie had already been to Africa and Italy when the 77th Medical Evacuation Hospital arrived off Utah Beach on July 7, 1944. They had been staged on the Empire Lance so that they could board their assigned Higgins boat with the least amount of confusion.

The 77th was a unique unit made up of doctors and nurses from The University of Kansas Medical School. The 77th provided hospital support from Ste. Mere-Eglise for the 30th, 29th and 35th Infantry Divisions and were not far behind the front during the battle of the Falaise Gap.

"Divisions didn't mean much to me," said Dr. McConchie. "During the Battle of the Bulge wounded would walk in and "some of them we'd ask, well where'd you get hit? Well right over the hill was the answer and we just walked into the hospital." They were only miles behind the line in the Battle of the Bulge at Verviers, conducting the human repair from the damage of the war.

Dale Smith was another combat medic for E Company, 2nd Battalion, 2nd Medical Battalion, 39th Infantry, 9th Division and came ashore as a replacement for a D-Day casualty. His first combat operation was at Cherbourg, France, his last was along the Elbe River at the end of the War.

"The medics are usually pretty popular guys because you never knew when you needed one or how badly you would need them. I do remember how bitterly cold the winter was and my feet being frozen pretty badly," he said. Medics were seen as angels, one step under God.

Both Forrest Adams and Dale Smith were awarded Silver Stars for their bravery in Europe. Both said they were only doing their jobs and they didn't do anything anyone else wouldn't have done.

Forrest as a medic in WWII.

Forrest in 2010.

Place of Birth:
Burden, Kansas
Active Duty Date:
June 13, 1943
Unit:
1st BN,
116th Infantry,
29th Division
Location:
D-Day to Elbe River
Arrival ETO:
December 22, 1943
Rank:
T/5

FORREST ADAMS

The 29th Infantry Division was called the Blue and Grey Division because it was made up of Army National Guard units from Maryland, Virginia and North Carolina. It was reactivated to active duty on February 3, 1941, and sent to England for training on October 5, 1942. On D-Day, the 116th Infantry of the 29th Division took very heavy casualties when they spearheaded the assault on Omaha Beach. They participated in the battle at St. Lo fought their way north east toward the Roer River. During the Battle of the Bulge, they held defensive positions along the Roer preparing for the next major offensive. In March they were in the Ruhr area and by the end of the War held defensive positions on the Elbe River.

Here is his story. . . .

I am a D-Day veteran but not a veteran of the Battle of the Bulge. *(Editor's note: Forrest is too modest as his unit was just north of the edge of the Bulge.)* Our outfit, the 1st Battalion of the 116th Regiment of the 29th Infantry Division, had just been decimated by fighting west of the Roer River opposite of Julich, Germany, and had been relieved from line duty on December 8. 1944. However, on the day before the Battle of the Bulge began, four of us Medics went via Jeep through that area on a leave to Paris.

I grew up on farms in the Cowley and Labette Counties of Kansas, graduating from Labette County Community High School in 1942. In the fall of 1942, my brother Russell and I enrolled at Ottawa (Ks.) University in a Pre Med curriculum. In November of 1942, we both along with all other eligible male students enlisted in the army because we knew that we would be drafted and felt that enlistment might give us a better choice of an eventual assignment In January 1943 all enlistees were called up except for Pre Med students who were then called up at the end of the school year on June 13.

I do not remember the goodbyes as we left the Parsons Katy railway station for our trip reporting for duty at Ft. Leavenworth. The train trip was uneventful other than the main function of that run was to pick up cream and eggs. Passengers were secondary. When we arrived at Leavenworth we were greeted and escorted to the Fort where we were processed and introduced to a new way of life in which we were made to understand that others would do the thinking for us. It was ours to obey! We learned to pick up cigarette butts from the parade field, how to peel potatoes while on K.P. duty, learned something about marching and were introduced to the Army Manual. It turned out that we were housed in a 2 story barracks building under the supervision of a corporal who had been a student at OU the previous year. After two weeks of processing and the gathering together of a number of other Pre Med students from other mid-west colleges we were loaded onto a train made up of Pullman cars and headed down the tracks. To where? No one knew, but we soon learned that rumors were a norm in military life. We saw a huge distant city, was it Dallas or Brownsville? We finally wound up at a place called Camp Barkeley outside of Abilene, Texas.

The first two weeks of life at Camp Barkeley was pure torture. The Officers and Non-Coms were of regular army types who belittled us new recruits. They did have the challenging job of shaping 200 fresh smart-aleck college boys into a disciplined group that could be educated into becoming soldiers, who would listen and follow orders. After the first two weeks our group was transferred to a Basic Medical Training Battalion where we not only got First Aid medical training presented by medical doctors but were also exposed to what real combat conditions and life might be like and what skills might be required. Later, in combat, I was amazed at how that exposure returned to us as we needed it.

Many of the boys in our Basic Training Company were pre-medic students. After thirteen weeks training and just before Thanksgiving many of us went home on leave. My brother who was 5 years older than me did not get leave. He had been selected to remain as a trainer. I became engaged to my high school sweet heart while at home. I'll never forget the parting when I left home for the unknown.

On returning to base after leave many of the boys in our training group were sent through Camp Shanks, N.Y. to the European Theater of Operation. We marched through the streets of New York with the band playing all the way onto the deck of the Queen Elizabeth (l), right down to E deck with all of our equipment. We were assigned 9 men to a room which must have been about 8 x 10 feet in size. Bunk beds were 3 high. There was little room for creature comfort. The QE (I) got under way and slipped past the Statue of Liberty in the gray morning light.

Soon we were at sea. By ourselves??!!! Usually troop transports went in convoys protected by armed ships equipped to (hopefully) fight submarines. German subs were called U-boats. The QE I was fast enough so that by changing course every 3 minutes the U-boats could not target her with a torpedo. Most of us experienced sea sickness. Chow (English style) consisted mainly of mutton stew and biscuits.

After 5 days, we arrived at Glasgow, Scotland and docked at 4 PM December 22, 1943. We were immediately loaded onto a southbound train to a Litchfield England military personnel replacement depot. All of the boys who had graduated from Basic Training with me were assigned to different units of the 116th Infantry Regiment of the 29th Infantry Division which was a National Guard Unit from Virginia and Maryland. I was assigned to the Medical Detachment of the Head Quarters unit of the 116th. The 29th Division was the second Infantry Division to be deployed to the European Theater in WWII. The First Division had been sent to North Africa and because at that time Hitler was threatening to invade England, the 29th was sent there as a precaution.

Colonel Charles Canham was the regimental commander. At a replacement welcoming he made the statement that the plan was to train, train and train some more because "Some of you boys will not go home and by training you will increase your chance."

About a week after joining the Head Quarters Medical Unit, I was given the opportunity to temporarily transfer to the Medical Detachment of the 2nd Battalion and assigned as a dental assistant. My assignment did not allow me to participate in any of invasion practice runs but did allow me to keep my legs in shape pumping a pedal for the dentist's (Lt. Frederick W. Kraus, a Czechoslovakian refuge) drill. Our assignment was to make sure that there would be no normal dental problems once the invasion occurred.

Combat status of a Battalion Medical Aid Station units consisted of about 30-40 men whose function was to provide wounded casualties immediate care at the front line and then to get them checked by a medical doctor as soon as possible before their evacuation to concentrated medical care at field hospitals.

About May 1, 1944, the 116th Regimental Combat Team went into a pre-invasion assembly area where it gathered in innumerable support units. I was reassigned to the Regimental Head Quarters Medical unit. About the 2nd or 3rd of June we were loaded aboard an Army Personnel Assault ship (APA) in the harbor at Plymouth England.

About 4 PM on June 5, 1944, our ship got underway and joined that gigantic invasion armada. When our ship was a considerable distance at sea General Eisenhower's voice came over the loud speaker telling us that we were actually going to make that invasion on the French coast against Hitler's forces. This was not to be another dry run. It was the real thing! We had all been expecting it but we just did not know when! Ike's speech made us all proud to be who and what we were! We represented the United States of America, that great big melting pot of people from all over the world! (A fairly large number of our men spoke a foreign language that they had learned from their parents at home.) That night on the ship, many prayers were silently said. Sure there was bravado, but there was silence also. Equipment was checked over and over, letters were written. Sleep, such as it was . . . was fitful.

At about 03:30, the claxon sounded for breakfast. As we stood on deck in the morning mess line the supposed softening of the German controlled beaches started. It seemed that every ship carried artillery of some size and let loose at 04:00. There was even small craft that carried batteries of rockets. The sky was a continual flash of gunfire. The bombardment continued until just shortly before the Infantry hit the beach. I later learned that there were many of our special units on the beach during various phases of this bombardment blowing up implanted metal barriers which were intended to keep boats from landing. The bombardments were trying to make it easier for the actual landing parties to get safely ashore. Many of these barriers were fitted with contact explosives.

The Headquarters Medical Detachment of the 116th Regiment like many other units departed from our AP A via cargo nets thrown over the side allowing us to climb into bouncing LCVP or Higgins boats. Waves were 12-15 feet high that morning. This meant that as you climbed down that cargo net you would time the upward bounce of the LCVP and then drop into it. I do not know how many casualties happened in that operation. Most everyone became sea sick even though we had been previously been given the new Dramamine tablets!

Our unit was assigned to the 10th wave and scheduled to hit the beach at H +70 minutes. Our LCVP's order was to rendezvous with the 10th wave at a certain British ship. Because that ship knew nothing of the 10th wave the coxswain of our boat returned to our own APA and got shore bearing. We went in by ourselves and never received enemy fire, probably because we were a lone target in that great big fish barrel. I have always credited that fact for part of the reason that I am here today. As we neared the shore in that terribly rough sea we could see other LCVPs backing out into the Channel with their ramps down! We could not figure why they would do such a dangerous thing but later realized that they had taken direct artillery hits on the front of their craft while discharging their loads and were unable to raise their ramps. They hoped that by getting away from the shell fire on the beach they might be able to repair their craft or possibly be picked up by other craft. Many of these craft sank and I presume many of their crew drowned. The Navy guys were a courageous bunch trying to get the landing parties in to shore as close as possible. Many craft, men and equipment were blown out of the water by explosive metal barriers.

The initial landing took place at low tide. But by the time we got in there wasn't much beach showing because the tide had come in. It must have been about 09:00 when we got in. Where we landed the beach had a rocky shoulder 5 or 6 ft high behind which some of the initial infantry survivors had dug fox holes before they moved inward.

There were dead men and equipment scattered all over the beach. I shall never forget the first casualty that I saw. That young man looked so alive as he lay on the beach but was missing the lower half of his body. Confusion ruled on the beach. Previously laid plans meant little. Our orders had been to leave casualties on the beach to be treated by Navy Corpsmen and to move in land. It was the middle of the afternoon before we moved up that hill where the engineers had cleared a path through a mine field.

I remember there was a neutralized artillery Pill Box on our right as we went up that hill. There was still German 88mm artillery inland that was firing at massed targets on the beach. I shall never forget while we were on that high bluff watching a wave of small craft coming in, seeing men running for the shore and hit the ground as they hear the scream of incoming shells and then not arise . . . just a circle of bodies, dead and wounded!

Able (A) Company of the 1161h Regiment was the spear's tip of the landing on Omaha Beach. They suffered 95% casualties. They were

National Guard boys from small Virginia communities. The Bedford, Virginia, community suffered especially heavily that terrible day.

I have often thought that General Eisenhower must have had divine guidance in making the decision to 'go' on that particular day when there was such a storm in the English Channel.

His counterpart General Rommel decided that because of the storm no invasion would occur and he went to his wife's birthday party 80 km. inland. German troops did not do anything on their own initiative. They had to have orders. The beach defenders followed previous orders and cost us heavily. But because of Rommel's absence and Hitler's orders to his staff not to be awakened until noon that day, coupled with their previous concern that the main invasion would occur elsewhere, their nearby Panzer Units were not released until we had secured a footing. It is entirely possible that if the German forces had been able to act on their own initiative our invasion force could have been defeated. Thank the good Lord for Yankee ingenuity!

On D +1 (Day after D-Day) our Medical unit with Regimental Head Quarters moved through several small towns. After Vier-Ville-Sur-Meir we crossed the flooded estuary at Grand Camp. I remember in one small village there were U.S. parachutes and bodies hanging from church spires. On the morning of D + 2 we received fire from snipers hidden on tree limb positions. It was quickly eliminated. The Infantry line companies were out ahead of us and took many casualties. They had to be evacuated and then replacements had to be brought in to keep the fighting force effective. Regimental Head Quarters and its attached units had fewer casualties.

I remember seeing a forested hillside where the trees were stripped of all foliage including limbs from aerial burst of German artillery. Any time they saw movement of any kind they would open up. All cross roads, gates or other natural passage ways had previously been zeroed in by their artillery. Front line advancing troops faced small arms and machine gun fire and fire from Kraut tanks. Artillery fire would fall on them if the Krauts had time to get it coordinated but certainly was a constant hazard for supporting personnel.

St. Lo (France) was one of our main initial objectives. The 116th made good gain of territory after the beach but was checked from getting to far out in front of neighboring forces who ran into trouble. Whatever caused delay in forward movement in our area allowed the Krauts to regroup, maneuver and resist much greater than previously. The earthen hedge row fences of the Normandy countryside provided good defensive positions for the German infantry and held up our tank movement because of previously zeroed movement locations. After a period of time Yankee ingenuity placed bulldozer blades of the front of our tanks allowing them to cut their own openings through the earthen hedgerows and the German defenses. The German S.S. was with the regular Wehrmacht troops. They enforced the discipline of their captured foreign troops and made them fight. The S.S. were radical, i.e. even mocking surrender with hands over their heads calling out "Commerad-Commerad" approaching Allied troops without visible weapons but hurling hidden hand grenades when up close.

Sometime before the final push for St. Lo I transferred from regimental medics to the 1st Battalion medics and was assigned as a litter bearer bringing wounded back from the front line to the Battalion Aid station. (Front line litter squads were replaced by the medical evacuation helicopters in later wars.) This Aid Station had seen heavy duty all the way from the beach.

The medical officer in charge was a Captain Heffener who had made a name for himself by jumping up on French hedgerow, waving a Red Cross flag and trading a dead German soldier for a wounded GI. He also was responsible for keeping the Aid Station close to the front line and getting radio communication with the front line troops rather than depending on an easily disrupted wire telephones as was called for by the Table of Equipment.

We finally took St. Lo sometime in the later part of July, 1944 but it was brutal time. There were thousands of casualties on both sides. One of our casualties was Major Howie, the former Battalion commander of the 2nd Battalion of the 116th. His body was laid by his troops on the steps of a damaged church there in St Lo after that final battle. A poem later appeared in the newspaper 'Stars & Strips' in his honor and entitled "The Captain of St. Lo".

After St. Lo, our next engagement was for the transportation hub at Vire, France. There always are many walking wounded who did not need to be carried back to the Aid Station. I remember one young G.I. who had been seriously injured in both lower arms who refused to be carried back because he thought that our litter squad was more needed by others.

After taking that transportation hub General Patton's tanks made a rush eastward towards Paris. However there were many German troop westward in the Brittany area especially around the sea port of Brest which was the home of a huge submarine base. The Germans had built massive underground caverns in which to hide and refit subs and also to use for warehousing all types of supplies and equipment. The ceiling and wall of these enclosures were several feet thick reinforced concrete. Fort Montebery, with its long range sea pointing artillery set on a hill above the city of Brest. Its original mission was to protect the seaport base from an ocean assault.

Our eight Division force was expected to encounter only a few days resistance but it took better than 2 weeks before the collapse of Ft. Montebery and the submarine pens. We took more than 5,000 prisoners.

On September 25, 1944, many of us headed to Holland on a WWI French 40 & 8 boxcar train. Others of our unit drove our equipment there. Soon thereafter we crossed the Siegfried Line. It was a series of concrete barriers, built between WWI and WWII, designed to keep any western enemy force from crossing into Germany. These concrete structures consisted of massive concrete posts resembling huge saw teeth. Their purpose was to prevent vehicular passage except at designated areas. Massive pill box bunkers with artillery and other weaponry was meant to defend these passage ways. The walls and the roofs of these bunkers were of reinforced concrete many feet thick. They were usually camouflaged. Many times they were covered with dirt or made to look like an ordinary building to avoid detection from the air.

Fortunately for us these bunkers were stripped of heavy weaponry by Hitler's forces themselves after they had conquered so much of Europe. They thought they would never need it to protect their homeland from invasion by westerners.

After the Division's strength was gradually rebuilt from our losses at Brest the next action was to help encircle and eliminate the German forces at Aachen Germany. The 29th Division relieved the 28th Division in this pincer movement at the suburb community of Wursland. Fighting here was quite different from the hedge row fighting of Normandy or attacking a major fortification like at Brest. Fighting was house to house and street to street. As before our litter squads evacuated wounded German soldiers as well as our own injured. Here we encountered the shells from the huge Kraut railroad artillery. I am alive today because of the existence of a tall brick building taking the blast from the downward arc of one of those shells. It could have been fired from as far as 20 miles away but of course with the help of a nearby observer.

That action would have been in October. In late November we were on the plains west of Julich which is located on the East side of the Roer River and about 18 kilometers west of Cologne. The weather was very rainy and cold, mud was everywhere and deep. The 2nd Battalion was in muddy trenches on top of a hill overlooking the small town of

Kosler, which lay about 3/4 mile west of the Roar River. We met a lot of resistance, especially from artillery located on a similar hill east of the city of Julich. Our Aid Station unit lost one litter squad from an artillery barrage here.

When the 1st battalion relieved the weakened 2nd Battalion, we made a night time attack down that hill into Kosler and relieved the 2nd Battalion's Fox Company. Fox Company had been cut off for 3 days without help or supplies. Our Infantry companies tried to get to a sports arena on the banks of the Roer River opposite Julich, but took many casualties because of the wide open field that they had to go through in broad view of heavy German fire.

On December 6, our infantry companies were again ordered by Division command to make an attack across that 3/4 mile of open field toward that sports platz. Lt. Col. Tom Dallas had been our battalion commander in England before the invasion. He was a good leader. He always said "Come on boys. Let's go". He recognized such an attack was a suicide effort. He went back to Division to reason with them but was told either to order the attack or be relieved by some new greenhorn leader.

Baker and Charley Companies with Able Company in reserve made that attack at 16:00 hours and were badly chewed up. When our 3-four man litter squads and 3 other medics from the aid station got on the field about 1730 hours, which was after dark, there was only 10 members left of Charley Company. Two days later, on December 8, 1944, those three infantry companies were relieved from line duty and their 44 surviving members were transported to the rear in three trucks.

During this rebuilding time, four of us medics were able to get a short leave to go to Paris. The day before the German break through that was later called the Battle of the Bulge, we drove our Jeep through that immediate area. Many of our supply units that we saw on that trip were eliminated the next day. We heard about the break though when we got to our R&R (Rest & Relaxation) hotel in Paris.

After a time of rebuilding our Infantry companies and their training as a unit the battalion was reordered back to Kosler. Because of the German effort in the Bulge their resistance was greatly reduced and our Infantry was able to get into the farm houses and their courtyard arrangements near the banks of the Roer River. There was only spasmodic artillery fire from the hill above Julich during this time of holding and rebuilding and considerable snow. I liberated some civilian bed sheet and made a snow suit. Many others did the same.

I was a litter squad leader but did drive a medical Jeep when it was to our advantage. I remember one extremely dark night driving the jeep (without lights) about 1 '12 miles up to one of those farm houses to recover wounded. There was a corner to be turned and I knew that the German artillery would hear the motor and bracket that corner. We could depend on them to be very methodic. By careful timing we were able to complete our evacuation.

After about a month or six weeks of our retaking Kosler, the German forces had been weakened by their losses in Battle of the Bulge and the Allied Forces were able to resume their eastward push. A large dam up the Roer River had been blown by the Krauts to slow our crossing. On the day of our new Allied attack under artillery fire known as 'Time on Target', which means that all artillery guns could fire as often as they could reload and shoot at a given area of concentration for a given period of time. During this time the Engineering Company attached to the 29th was able to construct bridges across that River. This push was helped because the sky had cleared and for the first time in several weeks we were again able to have help from the Air Force. Following the capture of Cologne we were engaged in numerous small engagements as we pushed onward northeastwardly.

At the end of the war we were 40km. Northwest of Berlin and met the Russian forces at the Elbe River. I do not remember the name of the town of that last battle. We took a lot of prisoners. They had no choice they were trapped in a vise.

I remember most of the countryside consisted of cultivated farmland with housing collected to form small villages. Each farm house in those villages would consist of the dwelling on one side connected in a square with a courtyard in the center. The other three sides of the square would be poultry housing, milking or pig housing and farm implement storage.

I was at the motor pool working on the Jeep when someone said "Hey the war is over". I replied "Hey that's good," and went back to work because rumors were always rampant and we were afraid to hope. Thankfully it was true - we were still alive- but it caused us to remember those who did not make it!! Our three litter squads carried approximately 3,000 casualties from the front line from the time I joined the 1st Battalion aid station to the war's end. The 116th Regiment with attached units lost almost a thousand men on D-Day itself. Our regiment then went into occupation duty in the area near the Bremerhaven seaport.

Our battalion Headquarters was located at a small town by name of Bitterkaese (Bitter Cheese). In September, after the war in the Pacific was over our men started being discharged on a point system based on length of duty, time in combat and other factors. Having joined our outfit as a replacement I did not have enough points to go home with the first group.

When my time came to go home we sailed back to New York, were shipped by train to Ft. Leavenworth where I met a high school classmate who had been in the Pacific. His skin was totally yellow from quinine. I was discharged at Camp Crowder Mo. On October 20, 1945, I married my high school sweetheart on January1, 1946. We had 3 children. She passed away in 1966. I remarried in 1971 and she passed away 2 years ago.

You can call me a jack of different trades. For different reasons I have had to switch jobs numerous times. I have taught vocational agriculture, written service repair manuals for Caterpillar, done office management work, done research work with Kansas Farm Bureau, sale work and had my own small home installation service. I am now retired and keep busy with my own home's maintenance and with volunteer work.

For his service in WWII, Forrest was awarded a Silver Star and two Bronze Stars, European-African-Middle eastern Campaign Medal, Distinguished Unit Metal, Combat Medic Badge and Good Conduct Metal.

Omaha Beach Monument German machine gun bunker. *Author photo*

Omaha Beach National Guard monument. *Author photo*

Dr. McConchie, Captain WWII

Dr. McConcie, 2010

Place of Birth:
Washington, Kansas
Active Duty Date:
May 19, 1942
Unit:
77th Medical
Evacuation Hospital
Location:
Verviers, BE
Arrival ETO:
August 18, 1942
Rank:
Captain

DR. JAMES McCONCHIE, M.D., FACR

In 1942, The University of Kansas Medical Center formed the 77th Evacuation Hospital and began operations in North Africa. It was made up of forty doctors, sixty nurses and three hundred enlisted personnel. They participated in the Sicily and Italian campaigns until they were completed in 1943, and then returned to England.

They landed in France on July 7, 1944, and moved with the troops. During the Battle of the Bulge, they set up a schoolhouse as a hospital in Verviers, Belgium. Through strafing by German planes and hits by buzz bombs, the staff continued on treating the casualties.

On the eve of U.S. entry into World War II, the War Department approved a plan to form mobile military hospital units to serve in a national emergency. Under the plan, certain units would be affiliated with outstanding medical civil institutions. U.S. Army Surgeon General James C. Magee wrote to Dr. H. R. Wahl, Dean of Medicine and administrator of the University of Kansas Hospitals, as KU Medical Center was known at the time. Would KU Hospitals accept the affiliation of the 77th Evacuation Hospital?

The medical center responded. KU faculty and staff joined with School of Medicine alumni and area physicians, dentists and nurses to form the unit. Activated in May 1942, the 77th Evacuation Hospital was attached to Gen. George Patton's 7th Army during the North African Campaign and treated troops in the European Theater, moving to the point of greatest need over a three-year period.

Verviers, Belgium

The Army had two medical units close to the combat zone in the

Battle of the Bulge. They served as army evacuation hospitals, receiving wounded brought from division clearing stations.

Our 77th Evacuation Hospital, a 750-bed unit located at Verviers was one of these.

We were located just behind the center of the new First Army front. The 77th was assisted for a time by the 9th Field Hospital where for about a week we handled most of the casualties from Army divisions trying to stop the German advance. We worked eighteen-hour and longer days to sort, treat, and evacuate the flood of patients.

The 77th had taken over a five-story school making a hospital out of it. The multi story construction created difficulties at times, requiring carrying the patients up the stairs on stretchers. Our orderlies and litter bearing soldiers became so stout and good at carrying wounded that they very nearly ran up the stairs with patients on gurneys.

The buzz bombs were launched right over the Siegfried Line trying to bomb Liege because this was the hub of everything over there. They went right over us and some hit into top of the hill behind us. The nurses were at their positions when the buzz bombs came right over us. They said that they would just leave their windows open on both sides of their building so the bombs could go through. That was a joke then because the bombs sounded so close.

But none of them hit us but the thing is when they came down the line, you could relax and hear their motor, but the minute you didn't hear that motor, you better take for the ground because it was coming down. The buzz bomb was lethal in its blast and it was just made out of a little tin and as that exploded of course, whatever that hit was shrapnel but the major effect was that ruined the lungs. If you had your mouth open, it'd just explode your lungs with the concussion wave of the explosion.

On December 20, a shell blew off a corner of our school building-hospital wrecking a bathroom; damaging the nurses' quarters, laboratory, pharmacy, one medical ward; and mortally wounding a Red Cross worker. I can remember hearing three rounds, just bing, bim, bam, wham and they hit us. The hospital staff just cleaned up the wreckage and continued in operation.

We were close to Verviers. We were right at the line. The casualties were walking in at the end of the Battle of the Bulge from nearby, some of them we'd ask, "Well, where'd you get hit?" "Well, right over the hill," was the answer and we just walked into the hospital. They were that close and we were real close to the first Army headquarters and 8 miles from Liege.

Three days before the start of the Battle of the Bulge my partner, Dr. John F. Bowser, was sent to a General Hospital in France. He went right through the middle of Bastogne, down to the General Hospital in Commerce, France. That was three days before the Germans had surrounded Bastogne. The Germans wanted us to surrender but the General said "Nuts" and that was the gold word for the rest of the war.

When Dr. Bowser left me alone, the Battle of the Bulge came along and I had to contend with it alone with all of these casualties. We had gotten a new replacement radiologist but he fell under shell shock and was sent home leaving me again.

I supervised two X-ray machine rooms. My job was to find and mark the skin where the bullet or shrapnel was located for the surgeons to remove. I had 1,000 and some patients. I used fluoroscope during that period of time. The fluoroscope of that time was not as bright as today's and we worked in the dark to be able to see the details of the image. In order to keep my eyes accommodated to the dark I wore red goggles like the night time pilots when I was out of the scope area. Fortunately I had six technicians, all of them technical sergeants who were excellent. See, they'd been with us since our original training and they knew about as much as I did except for the extreme cases.

If they had a problem at all they would send for me, I'd put out my fluoroscope and solve the problem by flicking up my red goggles without breaking my dark accommodation. But they were able to do all the major stuff, to send it on without my okay, which the other radiologist would have been doing as well. I couldn't have possibly seen all of these X-rays myself to say they're okay. These boys were so well-trained by then that the ordinary type injury, they could do it as well as I, making sure that you got the right angles. If it needed another angle and things like that, then I would stop what I was floroscoping in the next room. They were doing a great job, the three of them that were running that room; they were able to come up with a better diagnosis than I was except some of the little details. I was able to keep working in the other tent across the hall.

A report would go with every set of X-rays, but sometimes I couldn't keep up. We'd get a patient ready to move before I was ready to do any typing on it, and because there were the ones that were a little over the ordinary. All did get reported upon because a surgeon would need to see our X-Rays and the report first before operating on the soldier. The X-Ray department saw 3,660 cases during this time.

I have always been interested to hear what was going on during the German offensive. I didn't see it, but I know we had an awful lot of prisoners, and we took care of them after we'd taken care of our boys. We never touched them before our boys but then they got as good of care as anybody else. And I'll tell you about one instance on that subject, well kind of goes along here and now. Whenever we got air raid warning, we were supposed to go down to the basement at least all of us that could because we wanted to be able to work again. We had two rooms, which had, black out tents and just outside was a big while hall, like schools have, and we had room for all the patients that were lying sideways in the hall.

When we received the casualties they'd already been through a triage procedure, that's the first thing that happened. You'd see the ambulances lined up sometimes back 20 in a row, and as they came in. When a prisoner of war was brought in, the triage captain there would say, well we can't take care of him, take him aside. We'd take care of our soldiers that were there and that we could do something about and then if the prisoner was still going then they would help him.

I was first scoping a prisoner at the time we got an air raid alert. So I just started to talk to the corporal who was outside because a big share of these men couldn't go down and he was out there looking after them. I told him that I had a prisoner on my table kind of keep that in mind. And when I came back up, the prisoner was deader than a doornail. One of these men out here, who could still move, slipped in and throttled him. I'm sure that's what happened. Of course I didn't examine the wound but before I left, I didn't find anything of any significance wrong with him. But that's just an instance when they come along like that.

I usually would work 16, 18 hours a day and then I slept some and ate. In the meantime, they were taking the films as they could and then any problem I would go out and I had this stack of things to do. We'd work together and end up with 16 hours apiece and then we'd both be there for a little while.

We moved a bit after New Year's, the 4th or 5th of January. By the end of December, the battle was beginning to surround our town. The line had started to push in and we were getting really close to the line. Actually the 4th — I think the 4th of January that we stopped and moved to LaLouvie.

It got cleaned up and they didn't need us at that site. Some of our group went to work with other surgery units and that type of thing. But I didn't go myself after that. So I had quite some spell there that I was there, present.

The Division numbers and units mean nothing to me because I had the patients. The patient had a number and that was it. Of course the nurses could tell you all about that because everybody was coming there, stalking around the nurses anyway. It had nothing to do with me.

So I was just working there. I saw 1,186 patients during that period of time that we were in that area and the radiographs that we took were 4,428 making 6,000+ radiographic examinations during this short period of time. It was nothing to have 1,000 patients in a day. And we just put on tents and tents and more wards to have a row of 2,000 patients at a time. And that's a lot of patients.

But that would be coming down to why it was so important for us to get that bullet or piece of shrapnel out of there. Because it came, the bullet in the first place came through dirty clothes and all that type of thing, was filthy, filthy and bacteria and stuff in with it but the main thing was that if we didn't remove it of course I had to determine if it was close to the knee, if it was in too close or this or that or whether it needed to come out from other purposes. And the foreign body in there is foreign and the body wants to get rid of it but it's going to take the easiest way to squeeze it on out and this is going in deeper and deeper and deeper and they can go in right here and they find it clear back here in the middle of the back. That type of situation. And this would implant and make the infection go deeper and deeper and deeper, going into a joint, into a blood vessel and this type of thing. That's why we had to remove them.

There was a group of four nurses that we took care of each other pretty much. We had dances when it was quiet, so forth. But you'd see many evidences in a rural setting coming through early in the day when the weather was so terrific and they apparently were not equipped at all for this because they weren't expecting it. They didn't have winter boots and underwear and all of that stuff. And they just froze. And they came in by the by the 100's and a lot of those 60-man tents added on and on and on until they could take care of them. But I don't know exactly what they did for frost bite, I didn't treat frost bite. I didn't have too much time to piddle around with somebody's toes. If it had a foreign body in it, I was there.

My job above everything else, 1,500 out of 4,400 of the other injuries that we actually treated was getting foreign bodies located. They could be either shrapnel, bullets or whatever. I had to find them on the X-ray first and know that they had something in there and where was it and how are you going to get it out or do you need to get it out. I would march with the fluoroscope moving my arms and things like that. I had a pen like this that had a metal point on it and I could press on the skin and tell exactly where that was because if I pressed down and it didn't move, I was right square over it. If I was just a centimeter off, it would tip it because there's softer tissue around it. A surgeon without my locating efforts would normally go in, see it on the X-ray but they may be just enough away from it and would push it aside and never know where it was. Well I was marking directly above it in at least two planes and they knew if they would triangulate into that, that's where it's got to be.

Years later, I saw a WWII foreign body localization device with all kinds of equipment all over the place to triangulate the foreign body. It was a horrible situation we went through to find one of those foreign bodies and here I was with just my own idea of how to locate it. And I was skilled at it.

I would accommodate myself to the dark because you couldn't see well. The fluoroscopes weren't like they are now. You had to strain to see each detail in the low level image and you had to be accommodated to the dark. I wore red goggles just like the night flying pilots did all the time and I could look at an X-ray by push it back on my head fast enough that it didn't break my accommodation and that's what I would do when they called me to the other X-Ray room. It was similar to placing your feet in one of those old fluoroscopes in the shoe store only you weren't in danger then, the foot was in danger because it was getting an awful lot of X-ray, but it's the same principle, the fluoroscope. That's the thing.

I feel good about the job we did because I've always been good at making the best of situations. My father was in the plumbing, heating, windmill and pump business and I helped him out whenever I got old enough and when I wasn't playing hooky some days. And so I had that touch with those old fluoroscopes. I was also able to take a flat plate X-Ray and make a three-dimensional picture out of it in my mind to know what's going at different depths and I had that quality, that skill.

I lived across the street in the neighboring homes, which was an interesting situation. And in the movie that we had that goes with the book, it has a little bit about my association with my family, the hosts that I lived with during that period of time.

Well anyway, I got to go to boxing matches and stuff before the Bulge, we were sitting up there and having a good time, and the only part of the Belgians spoke French so we talked to part of the French. Then some spoke German or French but no English. I was invited one night when the things were quiet at the time, to dinner with one of the high city officers with the colonel. Now, I don't know why he invited me except to shock you or something and anyway, we'd have about six different kinds of wine in between the dinners between courses. And I asked one of the locals that could speak English, how you came about this good wine when we knew that the Germans had been there before and had taken anything that was any good. Well then I came in with the only actual French I knew was, "the wine behind the woodpile." And that's where they had the good wine hidden. That's the only thing I know in French.

We became extremely close with them. My partner lived in the house across the street and then the hospital was right up the hill. After the war settled down, I got an opportunity to take a weekend off to go back to visit these folks. And I took the old man a bottle of German cognac. I went back 17 years later and I took my four boys and my wife. My wife and I slept in the same bed that I slept in during the war.

In this supper that they had for us, the kids were all there and after we were through, he brought out a bottle of cognac and showed it to me. He said have you ever seen one like this before? And I said, no, he said well that's the bottle of cognac you brought to me 17 years ago. I vowed never opened it until one of the most important days in my life and here it is! We're all here together here tonight.

So that's the kind of relationship that you would get from the Belgians. Here we were, saved them and they and their families took it to heart. They took one big picture of all of the family members, and he gave me the picture of it that day. So it as very, very touching and that was of course just before the Battle of the Bulge. We went from there up to La Louvie and spent some time and that was one of the most enjoyable times.

When I think of that period of the Battle of the Bulge, the thing that sticks out in my mind was that I can remember the day that the clouds left and the sky was filled with our fighters and our bombers and stuff going over. The sky was full of American planes. It seemed to me, I never got them counted but I saw what I think was as many as eight of their planes going down at one time at that time.

Well, things turned around pretty fast when they did because we had the stuff but we couldn't use it before because of the weather. And I can remember Christmas Day when everything was pretty quiet around because that's the time it opened up and the skies cleared.

I saw the German jet planes and I saw them go down. We just had too many planes, and the Germans just didn't have it after that.

The burden on our weary 77th staff eased only in these last days of the year, as First Army hospitals began opening in Verviers.

Dr. James McConchie, M.D., FACR

See additional information.

Medicine Under Canvas:
A War Journal of the 77th Evacuation Hospital

Medicine Under Canvas is a book and a documentary film about the 77th Evacuation Hospital during WWII. The very rare book is 200 pages long and is arguably the most detailed history of an evacuation hospital in the European and North African theatres of war. There were over 40 evacuation hospital units [2] in the European Theater of Operations, but very few have published unit histories.

The first printing in 1949, edited by Max Allen, MD was done for the unit members and only 500 copies were produced. The second edition, printed in 2008 was limited to 1,000 copies. Over 200 photographs are included, which does not include the head shots of the majority of the unit members in a "yearbook" type section filling the last pages of the volume. Early sections of the book include a view of what the injured soldier experienced from arrival through treatment and then until he is evacuated to home, another hospital or back to his unit.

The 77th Evacuation Hospital was formed at The University of Kansas Medical Center in 1942. Preliminary organization began earlier, but the actual unit was not drawn up until the Japanese attack on Pearl Harbor. Made up of 40 doctors, 60 nurses, and 300 enlisted men, the unit shipped out to England in May of 1942. They began active operation in Oran, after the Allies invaded North Africa, and continued there and in Sicily until the surrender of all axis forces in that theatre in 1943. They returned to England and prepared for deployment on the European mainland after the anticipated invasion. On July 7th, 1944, the 77th crossed the English Channel and supported the Allied troops through the end of the war.

The companion DVD film documentary is 72 minutes long, has an original score, over 17 interviews with members of the unit, and never before seen color footage from WWII filmed by Mervin Rumold, MD, one of the surgeons in the unit. The making of the documentary was a four year process for the director, Dan Ginavan.[4] Beginning at the final reunion of the surviving members of the unit in 2004, he began filming the interviews in a small hotel room not knowing if a film would ever be made or not. Continuing on with the project and completing it with a very small staff the film premiered on Veteran's Day, 2008. The film is expected to compete in film festivals in 2009. The film won a Silver Telly in 2009 in the category of History/Biography.

Please visit KU Medical Center for additional information: http://www.kumc.edu/ea/urelations/uvideo/77th_project.html

Crossing the Seigfried line. *National Archives, ww2-116*

Schrobenhausen, Germany

Dale, 2010

Place of Birth:
Liberty Center,
Kansas
Active Duty Date:
1943
Unit:
E Company,
2nd Battalion, 2nd
Medical Battalion,
39th Infantry,
9th Division
Division Location:
Kalterherberg,
Germany
Arrival ETO:
June 10, 1944
Rank:
T-Sergeant Medic

DALE SMITH

The 9th Infantry Division had many firsts during WWII after it was activated in August 1940. It was part of the first U.S. combat force to engage in ground operations on November 8, 1942, when it landed in North Africa and it liberated the first city from Axis control in the War. In August, 1943, the 9th was part of the Sicily invasion force and landed on Utah Beach on June 7, 1944. As they pushed east, they were part of the St. Lo Operation and the Falaise Gap battle where the Allies almost pinched off the German retreat. During The Battle of the Bulge, they help defensive positions from Elsenborn to Kalterherberg. They crossed the Rhine at Remagen on March 7, 1945, and pushed east to the Hartz Mountains and ended the War on the Mulde River near Dessau.

Here is his story. . . .

I was born on June 10, 1925, in Ohio and lived in Ohio until I came here in 1987. It was in central Ohio and a town called Mount Vernon in Knox County, Ohio. I grew up in a village in the edge of town. My father died when I was a newborn child. He contracted diphtheria, and back in those days they weren't as well prepared to deal with the disease as they are now. So he died. I was born in Liberty Center, Ohio.

I was drafted in 1943. I was registered as a conscientious objector because I was a Seventh-day Adventist. When I was drafted I went automatically into the medical department of the Army and trained as a medic. I trained at Camp Grant, Illinois. The whole camp was set up just to train

Army medics. It is in the very northwest part of Illinois just south of the Wisconsin border.

It wasn't too long before we were sent overseas, making a few stops along the way. We were basically unattached and we were shipped to England as replacements. I shipped out of New York City, and headed for Europe. The name of the ship was the "Dominion Monarch," a converted Canadian luxury ship to a troop ship. We sailed to England, and I was at two different camps in England. One was on the southwest coast of England, just below the Bristol Channel. I have difficulty at times with the English names, even though we are English. Sometimes I felt like a misfit. I was there when the invasion of France began and after we had been in England a short time and I was shipped with a large group of others to France as replacements. Once we got there we were placed in a replacement pool and as they needed replacement soldiers they would find a place for us. Since I was a medic, I was going to be placed and a medical capacity.

I was assigned to the 9th Infantry Division. I was assigned to the 39th Regiment, 2nd Battalion, E Company. I was a member of the battalion aid station but most the time I was attached to a rifle company in the field as the medic to the company. I was with the same company pretty much the whole time, E company, but I also was with F Company for a while.

The 9th Infantry Division started out in Africa in 1942, and it patrolled the Spanish Moroccan border after the French resistance collapsed. They also were in Sicily.

When I was assigned to them, our first objective was Cherbourg, France. The Germans had built a huge underground fortress in Cherbourg. I just happened to be at the upper end of that fortress at the edge of town away from the harbor area. The Germans were in the underground fortress, it was a huge concrete area, and they had a very large steel back door to it. We were about a city block away from the back door from this fortress. An American Colonel came down to look at the huge steel door on the back of this concrete fortress and he said I want that door open.

They had an easy way to open that door, they brought a Sherman tank down and he fired a few rounds into it and it came open. The German soldiers started pouring out of there; it looked like there were hundreds of them pouring out of that open door. When they came out they were prisoners, they surrendered. When they took a step out of the door they were our prisoners.

Cherbourg was a port city but we did not get down to the waterfront. The city is on a hill and as you come into town the hill progresses down the hill through the town. I do remember running down the sidewalk in Cherbourg in a business part of town and there were storefronts with big plate glass windows. The Windows were all broken out and there were chunks of plate glass all over the sidewalk. When you're running on that and you step on it you slip, that was a hard surface to run on. I ran down the sidewalk and slipped several times and finally fell down getting a nasty cut on my arm from the glass.

The 9th Infantry Division moved on from Cherbourg after it was under U.S. control and we went up the coast. There were no other big towns up the coast that we went through. We moved from one point of the peninsula, east to the other point and by the time we reached that location the campaign was over. We captured or defeated all the Germans there, and we spent two or three days resting. They got us loaded up on trucks and moved to the next place. We went on to St. Lo and were in the Falaise Gap battle where we almost cut off the Germans.

I suppose I am the same as a lot of other veterans in that your memory has hot spots. There some things I remember very very well and other things are just a fleeting memory. We went into a rest area for about a week then we moved to another town. I don't remember the name of the town but we are back online and it wasn't too long after that that I took a piece of shrapnel in the butt and had to go to the field hospital for a while. I think that's why it's hard for me to get a grip on where I was because some things happened so fast and then I was out of it for a while with the shrapnel wound. Eventually I came back to my unit.

I was in the field hospital two different times for two different wounds. I wish I could have kept a diary of all the places I've been but it was not allowed. The thought was that if the Germans got it they would know where we were going so we were not allowed to keep diaries. But it would've been nice to have notes on all these places. The thing is that there can be a lot going on and every soldier is not in the thick of it all the time. Maybe one unit is in battle and other units aren't and it shifts back and forth that way. For that reason there are some pretty important things that might've happened that you weren't even aware of at the time. War is a weird experience.

As far as hearing about Ardennes Offensive, the Battle of the Bulge, we heard that the Germans had taken over some territory that we had already cleared earlier and that was kind of discouraging when you think about the whole thing turning around going backwards. As I recall the German advance was checked fairly soon, I don't think it was a prolonged situation. It's amazing. You can be right there. Like I was with the 9th Infantry Division, and you don't get any up to date news about anything. You had no idea what the overall picture was of what was going on, only what was going on in your little part of the battle. It is a pity to have been there and then not known much about what was happening. We didn't really even know what country we were in. In fact, I've heard soldiers say it sure would be nice to know what country we're going to die in.

I don't think I was aware that the German offensive and that the Battle of the Bulge was as big as it was. I knew they had made some breakthroughs and had reoccupied some territories that we had already been in but it was probably a much worse situation than I ever imagined at the time. I don't think we in the 9th Division lost much ground during the Battle of the Bulge. We lost some ground but not much and certainly not as bad as on the furthest tip of the offensive. After looking at a situation map showing where the German units were in front of us, I'm glad at the time I didn't know they were out there. I do remember the Hurtgen Forest.

Since I was a combat medic I never carried a weapon. We were not required to, we could if we wanted to, but I never did. And I'll tell you the truth I don't think I ever needed it. I was in some situations that were pretty tight, but there were plenty of guys there with weapons and I don't think I needed one.

The medics are usually pretty popular guys because you never knew when you needed one or how badly you would need them. I do remember how bitterly cold the winter was and my feet being frozen pretty badly.

I remember some of the air raids when there were hundreds of planes. They would start coming over and several hours later they were still coming and you had to think to yourself where are they all coming from? Of course, they're coming from England, but it was amazing how they can keep track of all those planes so they would be in the right place at the right time. But it worked. I liked to see those American raids coming over. Sometimes they're dropping bombs so close to where we were you could hear them exploding and to just hoping that they have figured out where they're all supposed to go. There were mishaps, but in any human endeavor, especially in war, there are going to be mishaps.

To my knowledge I never saw any of the Germans dress up like Americans during the Bulge although I was aware of it. It was a very, tough situation when the enemy is dressed up like you and you would not know for sure what you had to do. Even the officers in charge probably did not know what orders to give all of the time either, so it was tough on everybody.

There was one problem for the Germans in all of this. That was that the Americans had certain slang language and the Germans could learn English words but didn't necessarily know the slang. And if you're the one speaking, no matter what he was wearing on his uniform, you knew that they could not conceal their identity very well because they didn't know our slang. They may not have taken that as much in consideration as they should have.

After the Bulge, we crossed the Roer River and then on to the Rhine. We crossed the Rhine at Remagen. They had just captured the railroad bridge and the engineers got real busy laying ties so that the trucks could drive across where the rails were. They worked like beavers because they did that almost overnight, and when they were finished, the vehicles just started pouring through there. I crossed that bridge riding on the back of a half track but it wasn't too long after that the bridge suffered from such heavy use and had enough damage that it finally collapsed. I'm glad it didn't do that when I was in the middle of it.

We crossed in the daytime and every little bit a German dive bomber would come and drop a bomb. But it was my luck that none of the bombs hit the bridge while I was on it. It would hit the water splash and jolt the bridge from the concussion but they never hit the bridge. They were coming one right after another, they were desperate to take that bridge out. Some were the new German jet planes. The Messerschmitt piston engines had a certain sound. The jets looked almost the same, but the sound was altogether different with the jet engine. I did see enough of those that they had become a common thing, I don't know how many they had, and they probably wish they had a lot more than they had.

On east of the Rhine, the country was a nice country setting, but a lot of the villages were all pretty much the same, same type houses and everything and same type layout. I don't remember going through very many large city areas, but hundreds of small villages, but I did see some of those like Cherbourg. A lot of times the 1st and 9th Divisions worked closely together as they went through Germany.

One of the problems was that we didn't have maps, none of the soldiers had maps, so half the time you had no idea where we were. You try to put all this information all together later as an afterthought and after it already happened just trying to figure out where I was. I know some of our estimates of our locations weren't as accurate as they should have been, but I was always keenly interested in trying to keep track of where I was. Sometimes it worked and sometimes it did not.

There was a concentration camp that we went by and as we approached we could see the guards of course, had quit their posts and the prisoners could just wander out of the place. The prisoners were in such poor shape that they could hardly walk. At the aid station a German guy came in who had been a prisoner and somebody carried him into the station. When he got there he was so nearly starved to death that he couldn't stay on his feet and I remember some guy said to get this guy something to eat. The doctor said no. His system would not accept food right now; he didn't have enough strength in the system to digest it. If you gave him something you probably would be killing him. We had to give him injections to build him back up. It's hard to imagine anybody in that bad a shape.

I remember that we finally ended up on the Elbe River at Dessau, Germany. The Elbe River is west of Berlin not far maybe 40 or 50 miles out. The Americans were ordered to stop, and the Russians were to occupy the area beyond the Elbe River. Somebody discovered there is a prison camp with a bunch of American soldiers beyond the Elbe River. The river was big enough that had barge traffic on it, so our people got a river barge, and we took our Jeep that was equipped to put four litters on it for patients transfer and four was a pretty good load. They brought in the barge and got some big planks to run the Jeep up onto the barge to cross the Elbe River.

We were not even supposed to be over there on the other side in the first place but we evacuated a bunch of these prisoners and they were about starved to death. We got the barge and the captain who knew how to run it. We went across the river and loaded up a bunch of these wounded American prisoners there on stretchers and brought them back. We must have had 100 or more on the barge. We brought them back and processed them; you could see them very much in the beginning of dying from starvation.

It was a pity that they were in the condition that they were in but that's war. Not everything in war is pleasant.

Somebody had captured a German Officer and they brought him to the aid station. He was a doctor, a POW and he came into our aid station and told us that there were wounded American soldiers in the town hospital. And he said if we would go there then he would release them to us. Since they were Americans, we stuck our necks out pretty big to do this, we took the German officer back with us because we needed him to show us where to go.

From our aid station we went back across and into the German lines and German territory to the town to find Americans in the German hospital who were prisoners. We took off in the jeep and the German doctor sat right up on the front of the Jeep with his boots up on the front bumper, showing us the way.

The hospital was like a big high school building, a pretty good-sized school building, and it was just full of wounded. There were two German guards at the front door and we stopped and the German prisoner who was an officer walked up to the guards and the guards snapped their heels together, saluted the "Heil Hitler" salute and I thought I don't know what we're getting into.

We went in this old high school building and there was a basement floor and then regular floors and the whole place was just full of wounded soldiers laying on cots and stretchers and whatever they laid them on. This American was in the basement and the basement was full too. We went down there and when he saw us he lit up like a Roman candle. He was so happy and excited that we were there and then we told him we had to pick up the other guy first in another building and we would come back to pick him up on our way out. He just came unglued when we said we are leaving and coming back. He said we would never come back. I can imagine why he was so distraught, he was a prisoner.

So we went there and found a guy on a stretcher. He was in a different building. We just did what we needed to do to make the whole trip work together. So we went back and got the guy who was able to walk and had his arm in a sling.

He came out got on the Jeep we went and carried out the guy that was on a litter. The German doctor that took us there and showed us where to go stayed on the hood of the Jeep with his feet on the front bumper after we got everybody loaded up. He was our prisoner and he had to go back with us, he didn't have to do this and he didn't have to take us to the hospital.

I have felt extremely good about what he did. He was a doctor and had some compassion for human life because he was trained that way. And of course we were medics and we were trained that way. When we got back, he remained a POW.

I received a Silver Star for being a medic and taking care of some soldiers under very severe circumstances. I also received two purple hearts. I'm really proud of the 9th Division insignia that was given to me. We were a great division. I went back to France years later and was treated like a king. The people really appreciated what we did. I have a picture of Anna and me right at the end of the war. She and a couple of her girlfriends and I went for a walk.

I came back to America by myself, not with my unit. Of course it was a whole boatload of soldiers but we did not come back with the units. It was in the fall of 1945, probably about October. I left from Marseille, France, on a liberty ship. We docked at a port on the Atlantic in Virginia. I was glad to get home.

I wrote my mother a letter telling her I was due to come home but she did not know exactly what my schedule would be. I also had a brother who was in the station hospital and was still in Europe when I came home. Where the hospital went, he went with them. I visited my brother in Verdun France. I was on furlough and he was stationed there. So we were able to get together. I left so suddenly it was hard to tell Anna and her friends and family when I was leaving. When I left to visit my brother they thought that I had left for good but I returned, told them goodbye and left.

When I got out, I was a technical Sgt. That means that I was a technician Sgt., it was a three stripes Sgt. with the T under the stripe.

When I returned home, I learned the masonry trade and learned to lay brick and building block in Mount Vernon, Ohio. I have three children, and seven grandchildren. My son was in the Army in Vietnam in the 9th Division and the Colonel I had in WWII was now my son's division commander. I wrote him a letter and told my son was there in his command and he made a special trip to the field to visit my son. It was a wonderful thing for my son and I think the general enjoyed it too.

During the war we didn't figure we would be around long enough to tell anybody about it but it was a very interesting experience. I was slightly wounded a couple times but nothing serious. I was lucky and I came home. I have a piece of shrapnel in my butt still that bothers me every once in a while. I was very lucky!

Dale was awarded the Silver Star, Bronze Star, 2 Purple Hearts, Presidential Unit Citation, and the ETO Ribbon with 5 Battle Stars.

Malcolm during WWII.

Malcolm, 2010

Place of Birth:
Dwight, Kansas
Active Duty Date:
December 8, 1941
Unit:
102nd Evacuation Hospital supporting the 28th Division
Division Location:
Huy BE
Arrival ETO:
April 19, 1944
Rank:
First Lieutenant

MALCOLM STROM

The 102nd Evacuation Hospital was activated on March 18, 1943. After a training period in England, they landed on Omaha Beach on July 18, 1944. Following the Infantry campaign, they headed north east and were at Bastogne, Belgium, by September 29 and on to Ettlebruck, Luxembourg in November.

At the beginning of the Battle of the Bulge, they rushed to Huy, Belgium, and remained there for most of the Battle. They crossed the Roer River on February 5, 1945, and the Rhine River on March 14, 1945. In April, they moved to Gera, Germany, close to Liepzig and were there until the end of the War. They were shipped back to the States in July, 1945, and disbanded.

Here is his story. . . .

I was born March 24, 1918, here in this house. In fact, my Dad and his brothers were all born here. I went to a country school by the name of Hurino, and then to Dwight Rural High School.

I grew up here, worked on the farm and went to Kansas State at Manhattan in 1935, graduating in 1939. I was without a job for a little while. I was interested in conservation and that's what my career really was. But at the time I was in college — those were the Dust Bowl days — the government was in the process of developing a program for protection of soil and water. So I became interested in that and decided that when I graduated I wanted to do something along that line — and eventually I got on as a staff member of the Soil Conservation Service, a Bureau in the Department of Agriculture.

I did work for a while for the Union Central Life Insurance Company in their real estate and loan section. They had acquired a lot of farm land during the dry years when crops were poor and people could not pay on their mortgages. The insurance company had about 700 farms in the State of Kansas. So I worked for them in Topeka while in the meantime I was trying to get on, which I finally did in 1941. This became my career except for timeout during WWII (1941-1945) and the Korean War (1951-1952.)

I was in South Dakota when my draft number came up so I reported for duty on December 7, 1941, at Ft. Leavenworth and sworn in the next day, the day President Roosevelt declared war on Japan and Germany. As a result of that, I stayed at Ft. Leavenworth and did not go to a training camp as most did. They first put me to work interviewing the recruits, but soon I was assigned to the Medical Section where I remained until February 1942. In the meantime my commander kept urging me to apply for a commission. I finally did and was sent to Abilene, Texas for OCS (Officer Candidate School) and became a shave tail (The term 'shave tail' was coined during WW I. When a promotion was made to a second lieutenant in the field, epaulettes for the shirt were made by cutting a piece from the shirt tail.) Everybody who graduated became a member of the Medical Administrative Corps.

I got my commission along with a lot of others and not long after that was assigned to the 102nd Evacuation Hospital being developed in California. I was ordered out there to become the transportation officer for the hospital. The 102nd Evacuation Hospital was semi-mobile. We took our training in California and eventually included maneuvers in the Mohave Desert.

In April 1943, we got orders to go overseas. We didn't really know at the moment where we were going. We got on the train at Needles and went to New York as a unit. We left New York, I think the 6th of April 1943, and went to England, were there for three months, and in July left for Normandy. We became officially activated in the War on July 23, 1944.

From Normandy our unit was sent to Brest and were there about six weeks, until Brest had fallen. We were shipped all the way across France and actually camped at Bastogne for four or five days, then moved to a chateau southwest of Bastogne in November.

From there we were moved to a place called Ettlebruck at the north end of Luxembourg about 20 miles straight north of Luxembourg City, in support of the 28th Division. That is where we were on the 16th of December when the Battle of the Bulge really began. There was not very much going on at the time. Everything was pretty stable and quiet. The main battle line was only about five kilometers from Ettlebruck in a town called Diekirch just down the road a little way, and then all of a sudden things began to pop. I had a feeling that our leadership felt like we were really pretty close to falling. But not very much was going on in that whole sector, and I think that we were fooled into thinking that things were easing up.

We were pretty thin along that line. And of course the Germans knew that. They were building up in a way that we didn't realize. They slipped in there and all of a sudden they took off.

I wrote later about that Saturday night: *Lieutenant, Lieutenant, wake up, we have to go! The Germans are attacking!* That was in the middle of the night. It was around 1:00 o'clock in the morning and I was sound asleep. Sgt. Knudson, my sergeant in charge of the motor pool, had come to get me as we needed to prepare to leave. That next morning we got orders to go to a place to the north called Spa. We knew that the Germans were not very far directly north so our route getting up there was to go by way of Bastogne. We went through Bastogne and then north up through Houffalize, and I think we did not go through Malmedy as Spa was located kind of northwest from Malmedy not very far.

Before I left, an order came down to us to reverse and go back to Ettlebruck where I was still waiting to leave. We didn't have walkie-talkies or cell phones so the only thing I could do was get in the Jeep and catch the Colonel. My sergeant and I got in the Jeep and raced down the road, catch each driver and tell him to go back to Ettlebruck. We would catch one, shout to him to go back, race ahead to catch the next, and so on. The Sergeant and I got up to Spa just about the time the Colonel did.

By the time I got back to Ettlebruck it was dark and we had heavy traffic all day long. By that time the hospital unit had received further orders to go to a place called Huy on the Meuse River in Belgium. It was somewhere to the west. Because of the confusion, everyone was on his own to get there, but everybody made it. There was one vehicle with a couple of guys that didn't make it until the next morning, but finally we all got there. I arrived in Huy at 1:00 a.m. The hospital was set up in the elementary school and back in business on Monday, December 18th. The 102nd remained in Huy during the duration of the Battle of the Bulge.

Once the hospital was set up the normal job was to pick up medical supplies and food supplies. Because of the turmoil there was the urgent additional job of helping other medical units such as medical supply and hospitals get moved to safer locations. Some were having problems getting relocated, so we were assisting other medical units to get relocated. My drivers and I were on the move all the time up until just past

Christmas. I remember we sent 12 of our trucks up to Malmedy almost immediately after the 18th to help some medical units get out. The German troops were just outside of town. I know that there were a couple of trucks that got hit by gun fire, but none of the medical people were killed in that involvement as far as I know; none of ours did. It was very risky. I remember staying overnight in Malmedy and there was shelling but nobody got hurt. I stayed in the basement of a school house where there'd been a medical unit located.

Our own unit lost a lot of equipment and supplies at the site in Ettlebruck. The day after Christmas I went back to see if I could locate any of it. The town was not being held by the Germans. The bridges had been knocked out so we couldn't drive up to where we had been, but we could walk. The place itself had been pretty badly damaged. Some of our medics had stayed back with patients that could not be moved for a couple of days. They managed to get out without any problem. None of the stuff was there.

Our drivers were on the go all the time collecting medical supplies and food. There would be a depot someplace, and then move and you didn't know where it went. Drivers would have to go hunting for its new location. The same problem was true about maintenance and repair of vehicles.

We were an individual unit sometime supporting one division for a while, then move and support another. In effect, the casualties that came in would come from a little bit of all over from different outfits. Often you did not know initially where you were moving to until the last minute.

We were outside of Brest for about six weeks and while that was going on, Patton was going the other way so consequently all of the different kinds of depots were following him. I remember one time the supply officer and I and a couple of trucks went east to locate a medical depot. We had no idea. We couldn't talk to anybody that knew. I remember we went through Paris and finally found the medical depot where we could get the supplies needed.

Everything just kind of moved forward. The depots with the proper equipment, supplies and so forth would move up and set up some place and everybody would have to find them. I know that people tried to communicate but it was not easy.

For me, I really felt fortunate in the job I had. Of course I was not on the front lines somewhere. Medical people did not carry arms at all. We wore the Red Cross band on our arm which indicated our noncombat role.

We travelled quite a lot just to find things; there was a lot of independence. The drivers were all good people who had to have a lot of personal motivation because they worked on their own so much. In fact, I don't know as far as our particular unit was concerned if we ever had anybody that was AWOL.

We did have a problem after we were in Germany the War was not yet over. One night two guys raped a German girl. The girl found out who they were through a letter one of them lost. One committed suicide leaving a family in Oklahoma and the other went to prison.

The day the skies cleared and our planes were flying again was a beautiful sight. I knew we were about to move forward again.

From the Bulge, our unit actually went into Germany and stayed a short time just outside of Aachen at a place called Brand. Brand apparently had a military camp there of some sort where we stayed for about a week. The next order was to go to Bad Neunahr just west of the Remagen Bridge. Here we set up the hospital in a hotel that had been used by the Germans as a hospital. When we took it over from the Germans, they still had quite a few of their own casualties with some of ours. After we took over, some of the German staff stayed with us for a while to take care of people and we had them help clean up and get reorganized.

We crossed the Rhine River at Remagen using pontoons. After crossing we continued northeast but never got as far as the Elbe River where our troops met the Russians. We went on to Gera, about 40 miles from the Czech border, then on to the Leipzig area. Casualties were mostly from accidents and from citizens, some prisoners. We also had our people who had been prisoners and were released. We were at Gera on May 8th when we had the official news that the War was over.

I remember the camp at Nordhausen that was pretty tough. I wasn't involved with it but I saw the place. The men were still there and things hadn't been cleaned up yet. There were stacks of bodies and the ones living were emaciated and stood around kind of staring into space. They were in terrible condition.

We had heard that Nordhausen had been taken. We were a few miles back. The sergeant and I went there to see if we could find anything that would be useful. The railroad yards had been hit pretty badly from the air and there were some flat cars with new German trucks on them. We loaded up one of our trucks with tires because we were out of tires. When GIs opened tins of food they tossed the lids which became hazardous for tires. The lids would lay there on the road and would cut tires like cookie cutters and thus ruining the tires. We took the wheels and tires off of these brand new trucks and got a whole truckload of them. They were the same size as ours and worked perfectly.

My family knew we were coming home soon because the war was over, but we really thought it was likely we would go on to the Pacific. Some units did and we anticipated that possibility, so we didn't know whether we were going to come home or not. As time went on it became evident that we were coming home because things had changed in the Pacific. In the process we had people who were fairly new in Europe, had not spent much time there while others had been there from the very beginning. So there was a point system to decide who would go home first. The 102nd Evacuation Hospital and the 95th Evacuation Hospital traded people so that the 102nd kept people who had to stick around for a while, and the 95th took the people who would go home fairly soon. I joined the 95th among quite a few of the others. So we got shuffled around. I did not get home until around December 1. The War was over for a little while. My wife, Edith, didn't know what day I was arriving but she knew I was coming home.

From New York I took the train to Ft. Smith, Arkansas. Edith met me there. After receiving my discharge a recruiter signed me up for the Reserves.

A few years later I got called up because of the Korean War. I hadn't really planned on going anywhere. I had two young kids. I did not want to go. I was in the service about 18 months but first they assigned me to a training company at Ft. Meade, Maryland. I had a training platoon there for 13 weeks of training and another for eight or nine months before I was called to go to Korea. I was there about eight months in 1952.

It was an unexpected turn of events but it turned out okay for me. I had to leave the family; we were in South Dakota at the time. The kids and Edith came back to Kansas and either stayed with my folks or her folks at Baldwin.

This briefly completes the story of my time in the U.S. Army in Europe during World War II.

For his service in WWII, Malcolm was awarded the American Theater Ribbon, European, Asiatic, Middle Eastern Theatre Ribbon w/5 Bronze Stars, and Victory Medal.

Ivan in WWII.

Ivan now.

Place of Birth:
Clay Center, Kansas
Active Duty Date:
October 21, 1942
Unit:
238 General Hospital
Location:
Arlon, BE
Arrival ETO:
December 17, 1944
Rank:
Sergeant

IVAN W. WOELLHOF

I was born on November 16, 1920, and grew up in Clay Center, Kansas. I went to high school there and graduated in 1939. Harold O'Malley, one of the other Battle of the Bulge vets was a year behind me in school. When Dad was drafted over at Camp Fundson in WWI, the soldiers were dying there from the flu. Dad didn't get drafted then and that was the only thing that saved him from getting ill.

I was drafted on October 21, 1942. We moved from Clay Center to a farm, and we were farmers in Geary County. At the time kids were needed on the family farms and were given agricultural deferments. I had several deferments already and I got drafted. I met Maureen in my neighborhood but we didn't get married before I got drafted, we got married shortly after I got drafted. We had a hard time getting married because I couldn't leave the post but we finally made it all work.

After basic training I was sent to El Paso, Texas, to surgical technician school at William Beaumont. After that I went to Harmon General Hospital at Longview, Texas, for two years. Then they decided that all of the 1-A's would be shipped overseas in 1944. I was overseas for about 9 or 10 months but I got there just in time for the Battle.

While I was gone, Maureen stayed in the Manhattan area. She got an allotment of $50 per month; she was rich in those days.

It took us ten days to get over by ship. We had 10,000 troops on the ship US *West Point* and never had an escort or anything. The ship was like the *Queen Mary*, a luxury liner converted to a troop ship, RMS *Queen Elizabeth*. They said it was fast enough on its own that it didn't need an escort and could get away from the German subs on its own. We had no idea what we were getting into, just didn't know. When you got on that ship to go overseas you didn't plan on getting back, you thought it was an impossibility to get back. We had 10,000 troops on the ship and they could have dropped a bomb on us, anything.

Our unit was shipped out and we sailed directly to France, landed in Marseilles, France on December 17, 1944. There were a lot of units on the ship; ours was the 238th General Hospital. There were Doctors and nurses, our whole unit on the ship.

We landed and were moved to a staging area and it was cold, we just froze to death. Then they loaded us on to a train and Merecourt for a while then went by train to Nancy and on to Arlon, Belgium to set up.

Before we got our hospital set up at Arlon a lot of us worked in the station hospital. My main job was to give penicillin shots. I had a hundred patients that I would give penicillin to. I gave penicillin shots three times a day.

I was in the 238th General hospital. It was a fully functioning stand alone hospital and we were not attached to anybody. I remember running and jumping a ditch sometimes when we got strafed there in Arlon. When those guys came over to strafe you they made a believer out of you. We only had ditches for cover and that's where we went when the German planes came in. Sometimes we would hear the buzz bombs go over head and wonder where they were going to hit.

But I was pretty safe most of the time. I know the men in the field had it bad. I had a neighbor boy who was killed in the Battle and my wife had two cousins. My son goes to Europe on vacation in the Belgian-Germany area and visits the American Cemeteries.

They asked for volunteers to work temporarily with another medical unit. They gave me a license so I could drive an ambulance to transfer people for sick call. The station hospital was at Mericourt, France, and I would drive at night with no lights carrying people to sick call at Dijon. And that was my main job for four or five months. I have picture of all of the General Hospital people and a map.

When I first heard about the Battle we really didn't know what was going on. We heard it on the radio but we had no idea what it meant. We never really thought that we would make it back so we just went ahead and did our jobs. We just didn't know what was happening. We thought the war was almost over when we got to the German border. The Russians almost beat the Germans and had beaten them back to the German border so we thought the Germans were beaten.

After the Battle of the Bulge, we stayed right there in Arlon. I remember seeing General Eisenhower in Luxembourg City. We ended the War with the unit at Arlon. I remember seeing lots of WWI cemeteries around and I remember the huge cemeteries around Verdun.

I remember going to Trier seeing the Moselle River and seeing the vineyards along the banks of the river. On the hills along the river you could see old castles. You could sure tell where the border was because that is where the jets and buzz bombs showed up. I remember going to Brussels. The buildings were huge. How did they build them that big? I know it took over a hundreds of years to build some of them.

When the war ended, we went back the same way we came. This time we were in coal cars. Can you imagine just sitting in coal cars? We were staged at Marseilles until August and I remember going to a bull fight there. I was really surprised that they had them

We were put on the "USS *General W. F. Hase*" (AP-146) to return home. It sure was a relief when the war with Germany had ended but after the war in Germany was over we thought we would be going to the Pacific to fight Japan. They were putting together the right kind of people and units to win the war there. So we thought we were going home for leave, then on to the Pacific War. We were out to sea, maybe a couple of hours and they announced over the ship's speakers that the war in the Pacific was over. That was when we heard that the war in Japan was over.

The first thing I tried when I got home was to work for Manhattan Motors as a grease monkey. My brother in law was doing some building and he asked me to work with him. So I got into the building contractor business. I built 290 homes in this sub division along with ten or twelve townhouses and other buildings. My son went back and took pictures for me in 1999. I recognize a lot of the places in Arlon, Belgium.

Maureen and I have three boys, one lives in McPherson, one in Texas and one in Kentucky.

The thing I remember most about the Battle of the Bulge is the confusion. Nobody knew what was happening.

Ivan was awarded the ETO Ribbons with 3 Battle Stars.

Chapter 10

The Mechanics

THE UNSUNG HEROES of the war were those soldiers who brought civilian skills to the battlefield and risked their lives each day to make sure that the weapons of war worked as designed. Each of the millions of guns, trucks, tanks and other critical equipment had to be maintained and repaired by skilled craftsmen.

A typical entry in the maintenance log would be as follows: "Contact parties began working with 784th Ordnance LM Company on 9 December 1944, making direct contact with units assigned and attached to Division. Artillery pieces and fire control instruments on artillery pieces repaired with piece in position wherever possible by contact parties." Equipment was repaired in the field under the most extreme circumstances.

In addition to providing a much needed skill, the mechanics were subjected to much of the same indirect fire and revelations of the horrors of war as any others.

The Mechanics were responsible for many "field expedients," inventions that would modify war machines to deal with special, unanticipated obstacles and problems. One Paul Sheid remembers was the "hedge row buster" for giving the tanks a method of breaking through the Normandy hedge rows. "There were about six of us chosen and someone came up with an idea to make these teeth out of the beach obstacles. We worked around the clock. We set up a large tent in an orchard and we'd just pull the tank in there and fit the teeth right on it."

J. D. Sexton came upon a German Concentration camp late in the war. "These individuals relayed the horrible happenings. As I note what we saw, it makes me sick at heart as to what people can do to others."

Nobody was left untouched by the war in Europe. They can only tell us of their pain but we cannot feel it as they do because for some reason, they were destined to live it.

J. D. in WWII.

J. D. in 2010.

Place of Birth: Rural Honoraville, Alabama
Active Duty Date: December 27, 1942
Unit: 132nd Ordinance Medium Maintenance Company
Division Location: Liege, BE
Arrival ETO: September 24, 1944
Rank: T-6 Welder

J. D. SEXTON

I was born on a farm in the Honoraville, Alabama, area on May 21, 1922. Farming was hard work with only a mule and plow to rely on. When my Father passed away in 1930, leaving my Mother with six children and the beginning of the Depression, it was rough going. There were no food stamps, commodities, etc., etc., at least none that my family was aware of. The doctor who attended to my Father's illness became aware of the Civilian Conservation Corps (CCC) and suggested to my Mother that I enroll even though I was not yet 16 years of age. On the application it was indicated that I was born in 1921. This is verified on the transcript of my CCC *Ser*vice from Federal Personnel in St. Louis, MO, which also indicated my year of birth as 1921.

So, off to the CCC I went from Alabama to the State of Washington to my assignment at Cathlamet, Washington, the Grayer River Area. There a group of CCC enrollees and I went up the mountains and sawed down dead trees. This was difficult work, especially on the slopes since the trees were huge at ground level and tapered off at tops. Our noon meal was brown bag style – sandwiches with an apple. The morning and evening meal were very good. CCC enrollees were paid $30.00 monthly, of which $25.00 was automatically sent home – to the family. This monthly $25.00 helped to pay off the mortgage on our family farm. I got to keep $5.00 each month for my spending money. Some months, I didn't use up all of my $5.00 – that was a lot of money to me.

I spent a year in the State of Washington and then was transferred to a CCC Camp nearest to my home – which was Greenville, Alabama, where I spent my second year as a "CCC" Boy. At the Greenville Camp, we constructed terraces, planted pine trees and laid rocks at the end of terraces so water would run off into ditches. At the end of my 2-year enrollment I was honorable discharged.

Then I went to work at the Trailer Works in Montgomery, Alabama, for approximately a year. At this time, I learned of a Welding Course being taught at Sydney Lanier High Night School. I applied and attended the course and passed the welding course in five weeks instead of the normal six weeks course. My instructor indicated that I was qualified to weld at the shipyards in Mobile, Alabama, and gave me a certificate to that effect.

Now it was off to Mobile, Alabama, where I applied for and was hired to do welding at the Shipyard. Here, I was assigned to weld at the bottom of the ship, progressing to various stages and became a Hull Welder on the outer part of the ship. I enjoyed my welding work at the Shipyard until I received a greeting from my Uncle Sam, and on December 27, 1942, I received my Uncle Sam's greeting and was drafted into the U.S. Army.

I left home for the Army and *was* sent to Fort McClellan, Alabama, and on to California where I was processed into the U.S. Army. Since I was a welder and having welded on ships, I was sent to Camp Phillips, Kansas, where a new military Company was formed, the 132nd Ordnance Medium Maintenance Company. All those assigned to the 132nd Ordnance took our basic training at Camp Phillips, Salina, Kansas.

The 132nd Ordnance unit departed Camp Phillips on December 19, 1943, for Camp Forrest, near Tullahoma, Tennessee. We all worked in Base Shop during the rest period between phases of maneuvers. We continued work in Base Shop or repairing vehicles, armament and instruments. On May 13, 1944, we were relieved from work in Base Shop and began intensive training for overseas services.

On July 5, we departed Camp Forrest for Fort Jackson, South Carolina — on permanent duty station. Here we continued intensive training for overseas assignment. On September 4, 1944, the 132nd Ordnance MM Company departed Fort Jackson, by rail, for Camp Shanks, New York, NYPOE for permanent duty station. We arrived at Camp Shanks, NY, on September 5, 1944, and began processing for shipment. On 11 September 1944, the unit proceeded by rail from Camp Shanks, NY, to port and loaded on Ship NY 319 (HMT Scythia). On September 12, 1944, we sailed from New York Port of Embarkation for The European Theater of Operations.

The 132nd Ordnance MN Company arrived at Cherbourg harbor on September 24, 1944; however the harbor water was too rough to permit unloading of personnel and equipment from the ship. Therefore, the next day, we debarked at Cherbourg, Manche, France and moved to Valognes Staging Area to await receipt of equipment and assignment.

We received orders, October 7, 1944, assigning the unit to Third U.S. Army and attached to XVI Corps. On November 25, 1944, we departed Valognes Staging Area for Valkenburg, Holland, via motor convoy, and arrived at Laon, France, on November 26, 1944, a distance of 134 miles. The next day we arrived in Onhaye, Belgium, a distance of 90 miles.

On November 28, 1944, we arrived at Valkenburg, Holland, and preceded to Waubach, Holland, and attached to 320th Ordinance BN, 59th Ordinance Group and given maintenance assignment of 84th Infantry Division, distance of 97 miles. Total distance moved from Valognes Staging Area was 499 miles.

Contact parties began working with 784th Ordnance LM Company on December 9, 1944, making direct contact with units assigned and attached to Division. Artillery pieces and fire control instruments on artillery pieces were being repaired with the piece in position wherever possible by the contact parties.

On December 23, 1944, unit personnel became very jumpy since this was first time any of the men in unit have been in combat zone. Artillery fire, both enemy and Allied could be heard distinctly, air raids were frequent with bombs being dropped on nearby installations. At this time the 132nd Ordinance MM Company was the farthest forward of any ordnance unit in this sector, including the Division Ordnance Light Maintenance Companies. I believe it was at this time that we (132nd Ordnance MM Company) became aware of what actually was happening. A Huge War Battle was taking place.

On December 24, 1944, the unit departed Waubach, Holland, and arrived at Chapon-Seraing, Belgium, near Huy. The unit was relieved from attachment to 320th Ordinance Battalion, 59th Ordinance Group and attached to 184th Ordinance Battalion, 52nd Ordinance Group. We were relieved from 9th U.S. Army and assigned to First U.S. Army. There was no change in maintenance assignment; however the instrument section was attached to 784th Ordinance LM and worked with them.

On December 27, 1944, the unit moved from Chapon-Seraing to Regissa Station, 3 miles south of Huy, Belgium. During the stay at this station, service section of company was called upon to devise a means whereby the grenade launcher could be kept attached to the M1 rifle and did so.

The 84th Infantry Division was already at the horrible scene created by the German Nazi troops, in the Gardelegen, Germany, area when the

132nd Ordnance MM Company arrived. The 132nd Ordnance MM Company came upon a barn wherein the Nazi troops had packed individuals (old people, Jews, handicapped, ill, some used as guinea pigs) into this barn and shot phosphorus shells and grenades into the barn setting the barn and those inside on fire.

The objective of the Nazi troops was to burn the barn and all those inside to destroy any evidence of the horrible atrocities of Hilter's troops. The 84th Infantry Division already was in the process of ordering German civilians to properly bury all the victims. There were approximately 1,100 victims already put in trenches behind the barn. The victims were removed from the trenches and the German civilians were ordered by the 84th Infantry Division to properly bury each victim. There were a few individuals who were inside the barn and survived simply because they were at the bottom of the heap of bodies.

These individuals relayed the horrible happenings. As I note what we saw, it makes me sick at heart as to what people can do to others. I have original snapshots of deceased individuals lying scattered around. The thing I remember most about the Battle of the Bulge was the extreme cold temperatures and the huge amount of snow, going out into the field where the battles had been and moving from one position to another to accomplish assignments.

My unit was caught up in the Battle of the Bulge when we got to the Liege, Belgium area. I don't know just where the Germans broke through however, I believe it was in the Liege, Belgium area.

I was called out to go into the field and weld the trail which broke from the axle on a 155mm, a large artillery piece. The snow everywhere was from 10 to 18 inches deep. The artillery unit was just behind the enemy lines and when the ammo was loaded with a number 7 charge to reach the enemy, the intense cold caused the trail to crack under the load of the gun firing.

Arriving at the unit I pulled in with my welding truck and an American Military man motioned for me to get back over on the road instead of pulling out into the field. He later told me "you don't know how close you are to the front lines when you pulled into the field". He said that the damaged equipment was in a shed, ready for me to do the welding. It was jacked up with the wheels removed and, everything was set up for me to do the welding. The trail was completely twisted off the axle. The welding truck I had was loaded with acetylene and a large Hobart gasoline-powered welder. All I had to do was mash the starter and it kicked off. I had all of my welding rods, grinder and things which I needed all right out in the field on my truck.

From January 15, 1945, the 132nd Ordnance MM Company moved from Regissa Station to Durbuy, Belgium. The Instrument Section returned to the unit from 784th Ordinance LM Company. At Comblain-Au-Pont, Belgium, one platoon of Belgian Fusilier Guards was attached for guard purposes. Twelve (12) civilian employees (Belgian) were employed to assist with shop work (paid by Belgian Government).

On February 5, 1945, Belgian guards and civilian employees were relieved from the organization. On 6 February 1945, we departed Comblain-Au-Pont, Belgium and arrived at Heerlen, Holland, a distance of 58 miles. The unit was relieved from attachment to 83rd OrdinanceBN, 52nd Ordinance Group and assigned to the First U.S. Army and attached to 320th Ordinance Battalion, 59th Ordinance Group and assigned to 9th U.S. Army, the Maintenance assignment remaining with 84th Infantry Division. During this period contact parties found artillery units were having difficulty with wrenches which were furnished with the cavatized fuses and Service Section designed a wrench using a piece of steel pipe as part of the wrench. The pipe was machined on the outside and inside to work both the fuse and inside of the shell, a handle was attached by means of a split bar. Sufficient numbers of this wrench were produced to supply the artillery units of 84th Infantry Division and 95th Infantry Division.

German Airplane
Photo by JD Sexton

On February 12, 1945, eleven Netherlands (enlisted) guards were attached for guard purposes and relieved from the organization on February 27, 1945, on

February 28, 1945, to March 12, 1945, the unit moved from Heerlen, Holland, to Korrenzig, Germany, then to Dulken, Germany, and Krefeld, Germany. On April 2, 1945, we moved from Krefeld to Rhade, Germany, in conjunction with 84th Infantry Division.

On April 7, 1945, to April 15, 1945, unit moved from Rhade, Germany, to Herford, Germany, to Hanover, Germany, and Langlingen, Germany, for a total of 204 miles and on April 18, 1945, the unit moved from Langlingen, Germany, to Packbusch, Germany.

On April 11, 1945, area was strafed by lone German fighter plane at 0600 hours. On April 20, 1945, we were again strafed by lone German fighter plane at 0600 hours. Plane knocked down by AA unit approximately 4 miles. Neither causalities nor equipment damaged by either of the strafing.

On April 28, 1945, the Unit was awarded battle participation credit for Campaign "Germany" which was later redesignated "Rhineland" and on May 1, 1945, Unit was authorized battle honors for Campaign "Ardennes". On May 6, 1945, the unit moved from Packebusch, Germany, to Salzwedel, Germany, a distance of 25 miles.

On May 17, 1945, the unit moved from Salzwedel, Germany, to Sarstedt, Germany, a distance of 118 miles. At about this time, 84th Infantry Division initiated and forwarded to 9th U.S. Army a recommendation 132nd Ordinance MM Company is awarded Meritorious Service Unit plague. On May 21, 1945, unit authorized battle honors for Campaign "Central Europe" and moved from Sarstedt to Hameln, Germany, on May 25, 1945.

On May 26, 1945, the Unit alerted for redeployment through U.S.A. to the Pacific War. On May 31, 1945, unit relieved maintenance assignment of 84th Infantry Division and began processing for redeployment. On June 8, 1945, we relieved an attachment to 320th Ordinance Battalion and 59th Ordinance Group, June 9, 1945. Began movement to port. Departed Hameln and arrived Munster, Germany, a distance of 110 miles, June 10, 1945, left Munster, Germany, and arrived Liege, Belgium, a distance of 151 miles. Left Liege, Belgium, and arrived at Cambria on June 11, 1945, a distance of 112 miles and on Jun 12, 1945, left Cambria, France, and arrived at Camp Twenty Grand, near Duclair, France, a distance of 131 miles.

On June 17, 1945, left Camp Twenty Grand for Le Havre, France, and arrived at Le Havre and boarded the ship (American Export Line "SS Sea Owl") for redeployment. On June 18, 1945, the unit disembarked and was sent to Camp Herbert, Tareyton near Le Havre, France, to await repairs to the "SS Sea Owl" and on June 22, the unit embarked on the "SS Sea Owl" and ship cleared the port of Le Havre, France, for redeployment through the USA. 132nd Ordinance MM Company was the first unit on board the boat and was the only unit the men (enlisted) were not double loaded.

On July 1, 1945, the unit disembarked at Camp Shanks, NY, NYPOE, on July 2, 1945, all personnel, except for advance detachment, departed Camp Shanks for reception stations and 30 days at home for RR&R. On July 3, 1945, to July 9, 1945, Advance detachment departed Camp Shanks NY for new permanent station (Camp San Luis Obispo, California). Advance detachment arrived at Camp San Luis Obispo on July 7, 1945, and assigned to XXXVI Corps. Advance detachment left Camp San Luis Obispo, California, for 30 days at home for RR&R. On August 13, 1945, United alerted for movement, readiness dates, Equipment October 5, 1945, Personnel, October 15, 1945. On August 16, 1945, Majority of personnel rejoined unit and training began for redeployment and on August 21, 1945, unit given Port of Los Angeles, California, as port of departure.

On August 31, 1945, the Alert for movement was cancelled. September 4, 1945, Unit moved by motor convoy to Hunter Liggett Military Reservation, California, for rifle practice. On September 7, 1945, unit returned to Camp San Luis Obispo, California, on completion of range practice. Unit attached to 324th Ordinance Battalion on September 8, 1945. On 15 September 45, relieved from assignment to XXXVI Corps and assigned to VII Corps, remaining attached to 324th Ordinance BN. On October 20, 1945, unit changed from Category 2 (unit to be redeployed) to Category 4 (unit to be inactivated) and on November 10, 1945, the unit was inactivated.

I was honorably discharged from the U.S. Army and returned home to Alabama. After a short time, I returned to welding at the Mobile Shipyard. There was a dramatic change working in the shipyard after the war. I was able to get a job at the Mobile City Lines. I married the girl I met and corresponded with during my Army assignments. We came to Topeka, and I was employed at some auto dealerships and then entered the U.S. Army Reserves. I was able to get employment with the USAR Shop until retirement. I have 28-1/2 years of active duty and USAR service.

We have a son, J. Stan, an attorney, who is married to Tommye who is a Counselor. They have a son, Jay, an American tenured professor at Oxford University, England. Jay has a Ph.D., is married to Julie, and they have a daughter, Georgia. Chris, our son's daughter, received her Ph.D.

with Distinction in Developmental Psychology from Columbia University, NY. She is Principal Scientist at Medco Health Solutions, Inc. She is married to Dr. Christopher, M.D. Oncologist. They have a daughter, Sophie.

Our daughter Carlene has a Ph.D. and is a Licensed Clinical Social Worker at a VA installation. Her husband, Joe, is a retired Language Instructor. They have a daughter, Caroline, who is a Spanish teacher, is certified as ESL, has a Master's in Education and this past summer (2010) completed requirements for a Masters in Spanish at Salamanca University in Spain.

I am proud to say I have two children and two grandchildren who all have a PhD.

JD was awarded the European African Middle East Service Medal, Good Conduct Medal, and Campaign Medals for the Ardennes, Rhineland, and Central Europe.

Paul Scheid, WWII

Paul Scheid, 2010

Place of Birth:
Holton, Kansas
Active Duty Date:
March 27, 1943
Unit:
913th Ordnance Maintenance Company Assigned to 101st Airborne Division First Allied Airborne Corps
Location:
LaRoche, Belgium
Arrival ETO:
June 28, 1944
Rank:
Tech Sgt.

PAUL SCHEID

The 913th Ordnance Heavy Automotive Maintenance Company was activated on April 1, 1943, at Fort Lewis Washington. Their mission was to repair and maintain all varieties of military guns, wheeled and tracked transportation equipment and to develop solutions to unique problems outside the original equipment design and intent. Assigned to the First Allied Airborne Corps, the company landed on Omaha Beach on June 28, 1944. During the War they provided vital support to the 101st Airborne and 82nde Airborne Divisions.

Here is his story. . . .

I'm Paul L. Scheid; I was born on August 29th, 1924. I grew up in rural Holton on a farm. Dad was a farmer. After I graduated high school I went to San Francisco, California, to work as a welder in the shipyards. I worked there three months and after I turned 18, I was drafted on March 27, 1943. I knew the minute I registered, I was bound to go.

I started out as a tacker and then before I was through, I was an apprentice welder welding deck plates and things on the ships. It was all arc welding. I started out as the low man and had the worst jobs. The worst I had to weld were air ducts which are galvanized and made you sick from the gas caused by the zinc. When you would go home at night, you'd be so sick. We used to have to drink a lot of milk because it kind of counteracted the zinc in your system. They had a fan blowing somewhere, but if you were down in that ship in a hole welding, it didn't help much.

I worked on aircraft carriers. They gave us a little week-long course or something. We would go in for three hours of training and then go to work. We got $1.63 an hour.

When I got drafted, I went to the 4th Army at Ft. Lewis, Washington, for 13 weeks' basic training. I knew how to weld so they put me into a maintenance company. First they put me into an engineer outfit but then I was transferred to a brand new ordnance company — the 913th Ordnance. When I got there all they had was a cadre, and they put almost 200 draftees in it. Ninety percent of us were under 20 years old. I was eighteen. From there I went to some special training at Camp San Anita, California.

We went over as a unit to Camp Myles Standish and shipped out to Bristol, England on the USS *James Parker*, an Army transport, and landed on February 24, 1944. From Bristol we went to Scarne Cross Camp, Launceston, Cornwall, England where we were assigned to Headquarters, 1st Army. We had a little training there but not much.

Everything was a pretty well kept secret. You were either gung-ho or didn't know what to expect because you knew an invasion was coming up. We were issued all of our equipment when we got up there, all of our vehicles and such.

We had to waterproof all of our vehicles in case they dumped us off out a ways from the shore. They would drive underwater, if necessary. We put a paste just like Play Doh over all the spark plugs, distributor and carburetor. The carburetor was fitted with a hose that stuck up in the air and it was the same way on the exhaust pipe. The hose sticking up brought in air to the engine and vented the exhaust above the water.

All the vents on the transmissions were sealed off. Then we took the fan belt off and you could drive out into the water. We tested them in England by being actually dumped off an LST because we were transported in LSTs. If they'd have dumped us off out very far, the vehicles wouldn't have worked. The minute we got on shore we had to take it all off, everything had to be taken off.

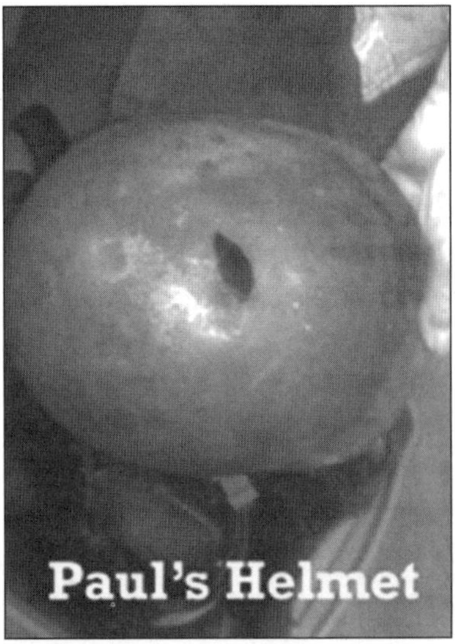

We landed on Omaha Beach on June 28, 1945, and were put in support of the 101st Airborne Division. So that's where things happened that stick in your mind. The first night in Normandy on Omaha, we had fox holes and I was supposed to get in it for the night. So I took my shoes and my helmet off. There was a command car next to me and I thought why would I want to sleep in the fox hole if I could lay on the backseat? So I got in and I laid my shoes down, put my helmet on top of them and snuggled in. During the night we were strafed by a German plane and a bullet went through my helmet and my shoes. Right beside me! I have that helmet today in the basement right now. I went ahead and wore my shoes. I guess I wore them out. I wore the helmet the rest of the War, but after awhile I just stuck it down in the bottom of my duffel bag. That's the closest I came to being shot that I know of.

Then we went right to this hedge row problem. They were losing tanks and knew they needed something. When the tanks pulled up on the hedge rows, they would normally rise up exposing their underbelly. Most anything could shoot a tank through the bottom. So they needed something to crash through the hedge rows.

For instance, there were about six of us chosen and someone came up with an idea to make these teeth out of the beach obstacles. We worked around the clock. We set up a large tent in an orchard, and we'd just pull the tank in there and fit the teeth right on it. I don't know how many we built. But after we got started, it went well.

They were like a manure scoop sticking out and we put them on the front of tanks. They were called "hedge row busters" and we welded them onto "Rhino tanks." To build the "busters", we pulled up the obstacles that looked like jacks on the beach. We sent a wrecker out and they would just pull them out, bring them back and we would cut them up. They were heavy metal, probably like either an I-beam or t-rail — about 8 inches wide. We cut them up and made them like teeth or a fork out front to cut through the hedge rows. On the front of the tank there were tie down loops that came out and that's where we hooked them on. If they hit the hedge row and could pull the front of the tank up, they could flip out quite a row. Then other tanks could go through and go on. They worked great. After the tank slipped through the hedge row, eventually they just took off the "busters". That is, if the tank made it.

Hedge Cutter made from beach obstacles

National Archives photo.

I just wonder if those hedge rows are still there. Every little field had a hedge row around it and they were not like our hedges. They built them up with dirt — maybe four to five feet thick — then the hedge was on top of that.

Another of our projects was building a bar to go up in front of a jeep. The Germans would string wires across the road to take your head off so we built the bars to cut the wire. We did things like that.

When the Battle of the Bulge started, we were in La Roche, Belgium. It was a small town about the size of Holton — about 4,000 people — and at that time we spent 65 days there. We set up quite an operation while there.

People in Belgium were so friendly and that's the way they were in La Roche. I met one family of whom I have pictures. He was the chief of police and had a little daughter, probably around 12. She followed me around like a puppy. I guess maybe it was because I might have a candy bar. I would like to make contact with them now, any of that family, if there's anyone left. But I have never tried. She gave me a little lock of her hair as a souvenir. Her name was Madeline Goffine.

We never worried about addresses, we were scattered all over the town. Any garage or any building that was big enough to work in, we did, and when we could not get inside, we worked outside. We worked on whatever equipment came in to be repaired. We had some people who worked on guns, others worked on vehicles. Some worked on heavy equipment, but I didn't do much of that because I had about 60 guys under me. I still have my little book that I used to keep track of who was working on what and what was wrong with it. It's funny the things you wind up keeping. I can sit down and look at that tub of souvenirs and every time, it brings back some memories.

We received notice that there was an attack by way of a rumor, that the Germans were infiltrating our lines dressed as Americans. On December 17th, we were ordered to set up center roadblocks with seven half tracks. We had 50 caliber machine guns and bazookas and our own rifles. We had explosives to blow up the bridges if we had to. This was all equipment that we'd worked on, or had confiscated you might say. We were not issued half tracks or bazookas or anything. We went out and just sat in a place that looked like an old dilapidated filing station. We backed into it so we weren't out in the highway, and we were along a little river. We figured we could maybe do something. Ordnance didn't do a lot of the fighting, but they did help.

We were there three days, just the six of us. We had a jeep that would send any information back and forth. They called us back to the Company on the 19th. Then we went out with a group that was supposed to find some Panzer tanks that were going to a certain spot. But we never did find any. Taking out a half track with a 50 caliber gun is like taking after a bear with a fly swatter. You might scare him, but that would be all.

At night, up in a mountain, it's so cold. We rode Army Harley Davidson's. I was not an accomplished rider but my buddy was. We saw a house and it had a dim light on. We just went in unannounced. We pulled over and took our guns with us. They were Belgian and had a fire going in the fireplace. They gave us some hot bullion and after we warmed up, we got up and jumped on the bikes and went on. It was real interesting.

I didn't know anything about the Offensive until we got run out of La Roche and ended up in Heer Agimont which is not very far. You start breathing easier and then for some reason, I guess because we weren't a combat outfit, they didn't send us back in. Oh, boy!

I have all kinds of praise for those boys that were there. We were pushed west out of La Roche. The Germans just kind of went around and we were in the group in the area that they were going through. We moved on to Ampsin. We supported the 101st and 82nd Airborne mostly, and we ended up in the First Allied Airborne Army. It was made up of the 101st, 82nd, XVII Airborne Corps, XVII Airborne Division and some British airborne units. It seemed like it covered everything.

I don't remember when the Battle was over. It was rumored right from the start that there were Germans dressed as Americans and, you know, it made you think. You would use your passwords to get back and forth. They shelled us from 7:00 o'clock in the morning until 11:30 at night on the 20th of December. That is what ran us out because they practically destroyed the town of La Roche. I never saw a German, the artillery was bad enough. Then some of us went back the next day from Namurs which wasn't so far. We took a lot of wreckers and went back and tried to salvage some of the equipment because the Germans didn't actually come in there yet. They just shelled us. It was scary but it was still better than being on the front lines. You were always looking around the corner.

We had information on what was developing during the Battle and knew when we had pushed the Germans back to the border. I can't remember a radio but the Company probably had something and we got the information just passed on to us.

We were back at Ampsin at the end of the Battle. It was under about two feet of snow and it was a nice, neat little town from what I remember of it. I was very fortunate. They billeted us in every place they could get us in and I got in with a family that was an old man and his spinster daughter. They had a little house and they had a room. They took us in and it was just like being home. That's the only family that I ever lived within their house.

We crossed the Rhine on pontoon bridges that the engineers had built in late March. We got into the little town of Erksdorf and they said we had bypassed some German soldiers who were up in the timber. I took a half track and four men up there to try to capture them. We figured they wanted to surrender if they could but we chased them all over those woods. We scared up a herd of deer in the process and we started shooting at the deer. We forgot all about the Germans and it turned out we had scared them off. I shot a wild boar that we were able to take back to our camp for dinner.

I remember the time during the Battle of the Bulge when the weather cleared and the sky looked like it did in Normandy, or early in France with the waves of planes. I tell you it was unbelievable the number of planes that would be going to bomb somewhere. We were far enough away that we could see the planes going over but not close enough to any of the actual bombing or strafing. It was just like flocks of geese or several flocks. It was fascinating.

We were at a Luftwaffe airport at Erfurt, Germany, when the War ended on May 8, 1945. We couldn't go any farther into Germany because the Russians wanted to take Berlin and they wouldn't let us in until they got it secured. They wanted to be the ones that took it, so we sat there and waited for about eleven days.

One day I saw a German plane coming in to surrender. It was a German jet plane coming in for a landing and it was the first jet I had ever seen. The pilot circled the field just at treetop level, then came around and skidded in without landing gears. There were four of us that were at the headquarters watching as it came in. We were at headquarters because some of us had been put in for Battlefield Promotions and we went for interviews. We had just come outside when that plane came in to land. We had a command car and driver so we took after the jet and we were the first ones there. We jumped up on the wing of the plane and the pilot wanted to surrender. We ended up getting his night bag just as the MPs got there and they took him prisoner. His night bag had some kind of a crème liqueur in it and we all had to have a drink of that!

I saw a lot of buzz bombs. At first they scare you and after a while you get used to them because as long as you hear them, it's all right. When it

Gallows
Unidentified Concentrtion
Camp in Germany
Phot by Paul Scheid

runs out of fuel and the motor cuts off, that is when you wonder where it is going to come down. When we were first in southern England, we didn't hear any of them, but when we got ready to ship out, they started to really come in.

I don't think we knew the Concentration Camps were there in Germany. The only one that I saw was a Polish prisoner of war camp somewhere in Germany. I have pictures of the camp and burial pits. When I took the picture, there were body parts sticking out, like a foot or a hand, but they never showed up in the picture. They were all Polish prisoners but I don't remember where we were. There were railroad track ties where they burned bodies. They would set the prisoners on those barrels and railroad ties and build a big fire and burn them. There was a German cemetery next to it. When I walked into the camp, it was scary to think how cruel it is that people would treat other people like that. You can't imagine sane people doing that.

When we were moving through Germany we had all of our equipment on trucks and we had big wreckers. You didn't really know what was going on. At the end of the War, just to occupy our time, one of the guys tore up a Fiat and made a racer out of it. You could salvage parts from a jeep but that's about all. We made a wrecker out of a jeep because it was better to take a jeep when you would go out rather than take a big truck. The jeep was faster.

The part of Berlin that we actually got into was great, lots of parks and things. But when you got down into the business section, it was devastated. I was able to take pictures of a lot of Berlin. I spent a lot of time around Brandenburg Gate. The Russians never got paid during the War, but they got it all at the end. A few of us GIs managed to sell a few cigarettes and things to them.

I was able to bring my helmet home along with a disarmed German burp gun and some bayonets. I also have a German canteen and a flag that I picked up. I don't know exactly where I got them. I have an old dueling pistol that I got somewhere. It's all in the basement in tubs and every time my great grandkids come over, they want to play with them.

We got into Berlin and we went home based on points. The oldest guys and a few that were married had some points and we didn't. We were in Berlin from July 5th until about December 20th. We had hoped all along to get home for Christmas. I came home on an aircraft carrier which was great. Going over we were sea sick all the time but the carrier was nice. When I got to New York, I was sick. I had picked up strep throat and I didn't get to have our steak supper when we landed. I was in Pennsylvania on the troop train headed home when the MPs took me off. They said I was practically out of my head, I was so sick. I wasn't going to go to the doctor because I was going home. But they put me in a hospital in Pennsylvania and I spent seven days there.

I came on home after that and of all things, when we got to Camp Chaffee Arkansas, I went into the restroom and there I met my cousin. He was on his way home, too. He lived in Valley Falls, Kansas, so we came home together. We took the bus from Camp Chaffee up to Topeka and then his parents took us home. You couldn't have that happen again in a hundred years.

I know I got home to Muscotah, Kansas, about 7:00 o'clock in the morning. My dad had a truck line and he was off somewhere getting a load of livestock. I went out and I met him where he was loading. That was the first time I had seen him in two and a half years. I got home January 6, 1946, but they waited Christmas for me.

I met my wife, Betty Cochren, playing baseball in 1946. She was living in Whiting, Kansas, and Muscotah was only five miles away. I loved to play baseball and Whiting had a baseball team. The manager was her dad so she was at every game. We just got acquainted and we got married in 1947. We had two girls and a boy, four grandkids and eight great-grandkids.

I went into the trucking business with my dad after the War. That lasted about two years and then I started farming. I farmed for over twenty years but in 1971 there was an opening for the post master in Whiting. Thirteen of us took the exam but I was lucky enough and I got it. I was post master for 29 years. I still kept the farm and farmed a much as I could after hours. Now my son-in-law and grandson help take care of it.

When I grew up we didn't own our farm. We couldn't buy it because we were poor. Everybody was. Dad worked WPA and found whatever work he could do. You didn't know you were poor because everyone was in the same boat. We were like everybody else and it's not until somebody tells you that you know you're poor.

It was difficult for us to get jobs when we came home. When I first came out, the Continental Grain Co. elevator wanted managers. The guy that ran the elevator knew me and he set it up that I could go to Kansas City and interview for the job. They wanted to send me out to Beattie or somewhere out West. I worked about a week and a half but I had never been in an office before. I couldn't stand it, it just felt like I was tied and just said, I can't do it. I've got to get outside so I passed that job up. But I worked as a carpenter and welder. I worked in Atchison in a foundry at nights and farmed during the day. I also carpentered for a man in Holton because with three kids and a house, farm payments and things, it just took everything.

You know, at those times, your wife didn't work outside of the house and farming still wasn't too lucrative there in the early '70s. We had a lot of expenses and were trying to do a lot of things plus farm. I think sometime in the '70s, my wife started working in the Holton bank. By then our kids were in college and we put all three of them through college.

Six years ago, we moved to Topeka and our son lives next door. We built houses here together. One daughter lives about five blocks away and our other daughter lives in Whiting.

Paul was awarded the ETO Medal, one silver Service star, WWII Victory Ribbon, Good Conduct Medal, Battle Campaign Stars for Normandy, Northern France, Ardennes, Rhineland and Central Europe.

Chapter 11

The Navy

SOME MIGHT ASK WHY a Navy sailor is in a book dedicated to the Battle of the Bulge. Consider for a minute, the preparation for the D-Day landing on the Normandy Beaches and who the first Americans there were: the U.S. Navy. The battle for Western Europe started with the invasion of Normandy and without its success, Allied forces might not have been in positions along the German frontier. Had it not been for the courage of sailors like John Brooks who went onto the beaches of Normandy hours before the Infantry landed, the success of the mission might have been seen differently by history.

"At 0200 in the morning, darker than heck, we went to shore on D-Day from our ship, the USS *Henrico*. That was quite an experience, wind blowing, it was raining, the tide was high, what a mess we had to go through," John said. "We had our rubber rafts and when we got close we climbed out into the water, about waist deep, and waded around."

John and the other survivors floated on their rafts, watching the assault on Omaha Beach, witnessing the carnage as though in slow motion. "I was 21 years old. We sat there on the boats and watched it all."

An interesting twist of circumstances resulted in the inclusion of another Navy story about a Kamikaze attack off the shore of Okinawa.

John Hawkins was aboard the USS *Goodhue* when its sister ship, the USS *Henrico* was hit by a Japanese bomb and John Brooks was trapped on the bow. The *Goodhue* pulled alongside the *Henrico* and brought John Brooks and others aboard. The *Goodhue* had also been hit by a Kamikaze and John Brooks was able to assist in freeing her prop from an obstruction.

Mary Ellen Mock was a Navy WAVE. She married John Mock and it is important to record her experiences.

John Brooks, WWII, U.S.

John Brooks, 2010.

Place of Birth:
Green, Kansas
on a farm
Active Duty Date:
October 1942
Unit:
USS *Henrico*
Location:
Omaha Beach D-Day
Arrival ETO:
June 6, 1944
Rank:
Machinist Mate
1st Class

JOHN BROOKS

*The USS **Henrico** County APA 45 was launched in Pascagoula, Mississippi, on March 31, 1943, as a Naval Transport but was refitted by the Navy and commissioned as an Attack Transport on July 8, 1943. The ship participated in the Normandy D-Day invasion delivering Navy demolition men to Omaha beach in the early morning hours of June 6, 1944. Assault troops of the 16th Infantry Regiment 1st Infantry Division disembarked from the USS **Henrico** in the first wave. As the fighting progressed, it took in casualties and transported them back to England.*

*From Europe, the **Henrico** participated in the invasion of Southern France, the Pacific and the invasion of Okinawa. It also participated in the Korean War, the Cuban missile crisis and the War in Vietnam. She was decommissioned on February 15, 1968.*

Here is his story. . . .

I grew up on a farm and I worked at Ft. Riley in the wintertime when I was 15 years old until I was 18. I would go back to school in the winter.

I joined the Navy in 1942 and was sent to Great Lakes Basic Training. When I got out of that, they sent me to Dearborn, Michigan, the Henry Ford Machinist School. I got out of there and they sent me to deep sea diving school. I was sent to Pier 88, New York City for diving school to help raise the SS Normandy that capsized in 1939. It caught on fire and they pumped so much water on it, it sank it or capsized it.

After schooling, we raised the Normandy and they moved it out and so my diving days were over. We were diving with the helmet and hoses and everything, 290 pounds. You weren't floating or swimming free.

They put you down with a line and you had on a suit and helmet. I have a replica.

When we got through with that, they sent me to Pier 92 in New York City, Navy Receiving Station to be reassigned. I was sent to Solomon, Maryland for small boats training and then Norfolk, Virginia, for more boat training. From there I went down to Fort Pierce, Florida, a little Navy base right off the edge of Fort Pierce to the salvage demolition school and we got through with that and they told me, "It looks like you're doing all right, we'll keep you a while and send you to underwater demolition training, CUD school." It was the precursor to the Navy SEALS but they didn't call it SEALS in those days but that's what it was. Our job was to do reconnaissance and destroy enemy defensive obstacles on beaches where we were going to make beach landings.

That was quite an ordeal. Later on they called them SEALS and the training we went through was similar to today's training. They sent me out to New York City at the docks, ready to go aboard ship at 0200 in the morning and the next morning I was pretty tired. There were 96 of us guys. The next morning I woke up around 05:30 and said, damn, so I went out topside and looked around and couldn't see any buildings. This ship was a big converted transport, converted from luxury liner to troop carrier. It would carry 20,000 troops. We were out to sea and they said we left at 04:00 o'clock this morning, two hours after we got aboard.

We went over to Scotland, to a little town by Glasgow. So at 0200 in the morning, they unloaded us on some Army trucks that came to pick us up and took us in the side of the mountains there and unloaded us around 0400 something in the morning.

There was just a little camp with a bunch of maybe eight or nine Quonset huts, one hut for the medics, one for the chow and one for the officers and then six more, enough to handle 96 of us guys. So we trained there, more and more. I don't know if you follow me or not how severe the training can get.

We knew we'd be on the invasion, our job was to prepare for invasions and we knew that we'd be facing pretty close calls. We might be captured. So we had to learn to walk quite a ways. About three to four nights a week we would walk all night, walk, trot, run, stop, rest and then have a lunch. Sun would come up then we'd go, we'd go so far, like eight or 10 mile out and then 10 miles back, have lunch, rest a couple hours and then go at it again. And we got pretty tough, you know how the Marines get in shape, they're pretty tough going. Some guys couldn't take it and some would go berserk. We ended up after about 10 months of severe training, we were pretty tough. So we got assigned to our ship, the USS *Henrico* and it was there waiting for us.

I was in Edinburgh twice and saw them secure the guards at night. That was interesting. Then I was in Glasgow four or five times in a year and a half. We didn't get much liberty in 11 months, maybe a few times, two or three times. You know, it made it kind of rough. We trained for 11 months before the invasion pretty damn tough.

We spent more training time on land because we had to learn the tactics of mines. We had a buddy system, two of us and my buddy was a black kid. We'd take exams every once in a while and tests in underwater demolitions. He and I, out of 125 questions and problems, we were the closest to getting all of them right in the class. I think mine was 122 and his was 119. The rest were around 85, 90 or 95 and 100. So I picked him as my partner. I was afraid, but he wasn't gushy, he was tough just like I thought he would be.

We'd go board ship every two or three days and we'd camp on the beaches at Northern Scotland. Then we got onto the ship at Glasgow and started down the coast of Southern Scotland and it took us a couple weeks to get from there down to England. We'd stop and walk over those hills just every which way. We worked at night and trained so we could see in the dark pretty good.

We got down to Liverpool and went down around to the English Channel. We monkeyed around there a couple months before the invasion. We'd watch Hitler's airplanes come over at night and big old spotlights would get them and they'd just spin around and our tracers were getting up there after them. But I never did see them knock any out. But you know, we weren't over 25, maybe 50 miles from London for a couple months there. German subs would come in around the shore and they got a few of our ships. That made us all kind of goosey. I was a machinist so I had to stay on board for room duty while I was aboard the ship.

I was a motor mechanic, a decent mechanic on the small boats. Our job was when we got through picking up mines and clearing the beach, we would go back to our boats. My boat was a rocket boat. It was 36 footer, just like the landing craft, and the point of the bow had a bunch of rocket crates and some guns.

We'd back out over the obstacles and start firing, firing those rockets onto the beach, about 325 of those 30-pound rockets. You got through a spot here, didn't see anything and so all hell was breaking loose and the airplanes going over, trying to bomb the hills and the Germans were shooting back with their big guns.

Here we were shooting in and the Germans had holes in the bank along the beach where they were just shooting back at our boats and I thought we'll never make it in. Well Hitler and his generals said we would never get our tanks ashore because they had obstacles on the beach. Looked like a small fence line.

Can you imagine we were trained to take these landing craft off of ships? Mine was an APA, the USS *Henrico*, personnel assault and it carried 26 boats. I was the motor mechanic and we had a coxswain, a deckhand and the officer. My duties were to make sure that the engine started. They had a flat pointed bow and a narrow windshield.

I would lie along the side of the windshield and crawl up there and coxswain would hold of my feet and unmouse the hook. It's the hook that hooked into the eye right in front of the boat and one at the back. So they would get a hold of that, pull it up.

Just soon as the boat would go down, 20 feet below the bottom of the ship or way above the mast depending upon the sea, you were trained to unhook it. Well we got by pretty good but a lot of them didn't get it. One inexperienced guy hooked wrong and here they were hanging up in the air.

The night before the invasion, there were six of us guys who went ashore on Omaha beach, the invasion beach, and waded around in the dark, trying to feel what was there and you know what we found? They had 10, 12 or 16 inch old bent up I-beams fastened down deep in cement I guess.

At 0200 in the morning, darker than heck, we went to shore on D-Day. That was quite an experience, wind blowing, it was raining, the tide was high, what a mess we had to go through. I did not make any reconnaissance trips or anything up to the beach before the invasion, but I know that bunch. They went ahead and slowly did some of that but they didn't get much information.

So in the morning in the dark, we were feeling around. The beams were six foot apart and Hitler said our tanks would never get through. Later, there would be a whole bunch of guys helping us. Two guys would carry the reel of electric wiring, about 500 feet. We didn't carry 30 pounds of TNT along with us; instead we carried 100 pounds so we had to have four or five guys and another boat, really a whole bunch of them.

We had our rubber rafts and when we got close we climbed out into the water, about waist deep, and waded around. We went up there where we were told to detonate those holes in the sand. We couldn't see any Germans up on their positions but they were awake and watching us. We were out of there by 0430, quarter 'til 5:00.

Each one of us six original guys who went on the beach the night before had five other guys carrying our TNT. We found out in our experiment there in Scotland that if you add four or five ounces of diesel fuel to the TNT, it became a detonation bomb and it increased the power of the explosive to give it about 300 percent more power. So we put 100 pounds under each one of these obstacle piers and there were 280 some piers there made of angle irons. They were crossed in every which way, like that, they're hard to describe, some straight up, some criss-crossed.

They looked like they were just simple little things but inside of those big I-beams were mines. We blew them and had good success but some blew the I-beams just a little but some of them blew clear over. These were in the ground, the top would be bent over but there was a bottom up about thigh high. So the tanks had to put one track up on the side of this beam and some of the tanks tipped over.

That big old tank would take two or three other tanks to pull them back over, pull them back up right. Here the Germans were shooting with the big guns just tearing the living hell out of those tanks and trucks coming in and the 88's were just coming out of their holes in the bank, just slaughtering our troops.

After we got the detonators in, we always wondered if we did the right thing by hooking onto the right wires. We had two guys carry the reels of cable out and put on our pontoon. The pontoon would take us out to our boat out about 500 feet. When they started shooting, we would come down on the pressure switch and you have never seen sand go so high in your life. We were out there, 500 or 600 feet and it just showered us with sand.

We left the big ship USS *Henrico*, on a smaller boat, a Higgins boat. It carried all our supplies and the officer stayed on the boat. Then we went up to like 500 or 600 yards from the coast which seemed basically on the beach and we got on a canvas raft. They didn't have rubber rafts, they were canvas, a little better than rubber. If the Germans heard us coming in, they knew better than to open up on us or they'd be torn all to hell because they would give away their positions.

It was a storm so the sea was a pretty choppy at that point. Hitler said we'd never invade because there was a terrible storm brewing. Instead of three or four foot surf, it was 15 to 18 foot surfs. Rommel said they won't come in this week so I'm going on a few days vacation, don't bother to call me. The generals decide well, cats away, the mice will play. So they took off and went every which way I guess. Nobody was left in control but there were some generals that were watching. Because the SS troops, the ones we were going to have to battle in the black uniforms, we had to contend with them. We didn't see them because there was so much Hell raised there on the beach, nobody survives either way.

When the Germans started shooting, there were so many big ships shooting out there and the airplanes would come in and bomb them. The bombers missed their drops a lot of times, got a lot of our troops, and some of our landing craft and our ships.

The ships had to drop a lot of shells before they knew how high to elevate their guns. I think our own ships killed more of our troops than Hitler did. There were 11,000 troops went in the first wave and only about 5,000 survived, 4,000 killed and there was a couple thousand wounded.

After we left the beach, we stayed in our boats and watched the invasion. We got on our rubber rafts, a whole bunch of rubber rafts, paddled back and it wasn't light yet when we got back to the boats. They started shooting at 0715, that's when we blew our charges. All hell started firing then when we blew our charges.

Out in the water, the defenses of the beach looked like crosses. Teller mines that were on poles that were sticking back up towards the beach. There were a lot of them. We tried to dynamite the big crossed beams but did not get them all. But we had to make way for the Scouts and the Rangers. And we thought we did a pretty good job despite the high tides.

I lost my partner that morning because you know, high tides, you just couldn't do a good job of picking up mines. And you didn't pick up two mines at one time or you were in trouble because they would attract each other and blow up. We had a whole armload of these wooden stakes, little orange flags on them, like the highway department. (They copied off of us.) We pushed them in the ground. We didn't have any belts on, we had string for belts and we didn't have any watches. We had lots of strings kind of lined around our waists, had a whole bunch of short strings and underneath this line. We would pull one out and tie it to the mine. It was hell.

We had a kind of plastic dog tag. We took all the shoelaces out of our shoes and tennis shoes. The mines were magnetic and we had to remove all metal from our bodies or it would detonate the mine. I don't think this partner of mine meant to do anything wrong. The tide was so high, so rough, you'd come in, you may land 50 foot up on the beach or it may suck you back underneath out to sea and drown you. Well the mine went off. I don't know what happened to him. I didn't see him anymore. The blast blew me on top of the water and the tide took me up on the beach and that's what saved me. I don't know what my partner did but it blew him to pieces I guess.

The assault troops could see the big orange flags and know where to go. We figured every 200 feet they were watching, they were trained with night vision, like we were, and they could see in the dark. So by George, after the Germans started shooting, the troops came in and they couldn't get anywhere. The boys started up toward the beach and those 88's just opened up and cleaned them out. Then they cleaned out all the boats that kept coming in. Some boats came in on high tide and they would land clear up on the beach 50 to 100 feet on the sand. They would sit there and the German 88s just opened up. It was a slaughter.

I was 21 years old. We sat there on the boats and watched it all. Soon after that, they started to bring in the airplanes. They said there were 280 bombers that came in and we thought it would just lower that hill down a couple feet but it didn't. Well, we figured the troops were next, the airplanes were going to quit bombing in that two-mile stretch of beach and we thought that the 280 planes should have done a lot of damage. When we saw the troops running up the beach there were still bombs hitting it, a couple or three bombs hit from the beach to the hill and we figured they were going to quit bombing. Just about the time the troops just got 9/10ths the way up there, they quit bombing and Hitler's 88s and 240mms opened up and just slaughtered those boys in the boats and on the beach. They came in again with more planes and finally the troops got up there on the beach in the afternoon.

I don't know where Hitler or Rommel was at that time or whether they got the message or not. But the big guns, they had, the 88s, 240s and, big guns on a railroad track on the other side of the cliff, they'd come out on a track and shoot and they said the recoil would shove this cannon back down the track. The bunkers for the artillery were 16 foot cement walls, hard to knock those out.

When our boys finally did get on the beach and up the hills, two or three miles from Omaha and Utah Beaches there were lakes or ponds where the Germans had flooded the fields. About every two or three blocks there was a dike and tank traps and the tanks had to go over the traps. They made the dikes kind of curved and the tank would come up and expose its belly when it went over a dike. The Germans could knock a tank out by hitting him underneath when it was running up on the dike.

Back there about a quarter to half a mile, they had planted thousands of trees and they had grown up 20, 30 feet high, a thick forest. And you had those big guns, in the trees. Hitler's troops on top of the hills were still in control around these ponds. On each side of the pond was timber so thick you couldn't see but you could tell where these hedgerows and dikes were. The Germans could see a Yankee tank coming over and they had artillery set and they lobbed rounds in and they got 40 some of our tanks before they got stopped. They had to stop that so they brought in several hundred more bombers and stripped that forest and before the end of the day they finally got through.

While we were watching, there was a point where we thought we weren't going to take the beach. But they just kept on coming, taking a terrible beating and our ships and airplanes twice hit them hard but still couldn't take it. The fact is we finally had so many troops we just overran everybody.

I named Normandy Beach down there, Hitler's Hill. I got that down. But that was quite an experience. I have never been back to Normandy since the War, never had the opportunity.

John's response when told that Omaha Beach looks like a resort area: You got to be kidding. Can you see the bunkers? Editor to John: There's one bunker, it's a machine gun bunker, and right on the beach in the middle of Omaha Beach and it's made of blocks. The beach kind of comes up and then it was just on a little bit of a rise, not much like a plateau.

We lost a lot of tanks that day. They'd unload the tanks from these LSTs. They were heavy and they couldn't get the LSTs up on the beach. If it were timed just right when the high tide came in, yes, you could. Otherwise they were in maybe 20 feet of water and down the tanks would go. And then sometimes when they had them on different kinds of pontoons and unload off of the ships, that didn't work at all. It was just too heavy, the tanks sank the pontoons. Oh, what a mess because you're talking about 6,000 ships out there.

They had 49 or 50 old worn out merchant ships they stripped and they sank out about a mile for a water breaker. That probably helped some. They brought in a group of floating concrete pontoons called Mulberries from England, tugs would bring two or three of them and each one was about 300 feet long. They were fastened together; can you imagine them bringing them over, crossing the English Channel 20 miles and trying to get into a beach with all these ships?

The wind and tide would whip around the ships and they had a mess. So after about 15 or 20 tanks or tugs, they would get them lined up somewhere near the beach, not very close, probably half a mile from the beach. The idea of that was to bring the merchant ship up alongside and unload them on the temporary docks. Then they would unload from the temporary docks onto boats like LST's or LCI's and take them on to the beach. That didn't work too well because the tide was too high.

I was never exposed to seeing a lot of military guys getting killed. Those of us guys in the small boats had to stick around and watch. There were 16 of us guys who made it out that morning, out of 96.

We lost most of them disarming those mines. So we had to stick around and help the Army take jeeps and stretchers and put wounded men across the jeep and run down to the water, and then put them on the landing crafts, the LST's.

We had to help the wounded; the ones that were killed were taken care of by Graves Registration, a different bunch. God, there were several hundred of us helping. And there were seven big hospital ships out there and a whole bunch of converted ships for lesser injuries. I wasn't used to seeing guys tore all to pieces and arms off and laying there screaming and hollering for help and you couldn't do anything. There was a lack of medics and you know, geez, I thought of that for 25 or 30 years every night. I know it doesn't bother me anymore. I think about it every night but it doesn't bother me anymore. But I often wondered what old Hitler thought when he found out we invaded. Ike said we're going in and we did.

We had my rocket boat so we had some K rations and C rations and some jugs of water. We were prepared to float around there for a while. We would eat off of different boats, going from one to another for our meals. Then a bunch of the boats ran across our boats, of course you're talking about 2,200 of those little boats plus a lot of other big invasion boats.

We finally got to an English cruiser. My ship went back to England to get more troops. We carried about 2,200 troops so I just stayed there. There were 751 of us, ship's company. Our ship was neat and clean and I got on this dirty old Limey ship and it was a cruiser, water and paper floating all over, filthy. We were in our dirty old clothes for four or five days so the Limeys gave us their uniforms. My shorts, one leg would be clear down to my knees and the other clear up to my waist and t-shirt sleeves clear down to my elbows. In a few days, they took us to Dartmouth, England and dumped us off in an Army camp.

Can you imagine what those Army guys thought when they saw us? What the heck are the Navy guys doing here? They didn't know what to do with us so they sent us to Weymouth to the Navy base. They didn't know what to do with us, no orders. We fooled around there a few days. Then they sent us to the White Cliffs of Dover on the Fourth of July. It took about 30 days to get this far because it was the Fourth of July.

After a few days, that place didn't know what to do with us so they sent us to Liverpool and there was the *Queen Mary* and it handled 20,000 troops. They loaded us aboard the HMS *Aquitania* and we made our way back to the States. We had a bunch of German prisoners aboard. There was one German soldier who was older and about four young Germans. The younger Germans would run up and hit you so they had to control them better.

This one officer says that's all right, guys. I was in the Merchant Marines in '39 in New York City. All you have to do is say the lights are on in the United States and it will make the Germans mad. So the German says, yes, I know the lights are on in New York boss. They were on when we sailed into New York. Some of the prisoners ended up here in Concordia, Kansas and all over the country. HMS *Aquitania* was this ship that I came back from Europe on. It was a converted luxury liner in July 1944.

We landed at Pier 92, stayed there quite a while and then finally they gave me a 30 day leave. There were six or eight of us, something like that. I headed home. I was raised up west of Randolph, up Fancy Creek area towards Greene. Five days after I got home, there were people from the Clay Center Red Cross to my house looking for a John Brooks. Now, I'm John, my dad was John. I'm John Junior. So they called me Junior around home and in the Navy they called me John. Well this lady told my Mom they were looking for a younger man. Ooh, you're looking for Junior; well he's down in Manhattan with his sister today. So Mom called, she had one of those crank up phones and says you tell your brother to get home right now. He's got to be in Boston, his ship's going to leave Boston tomorrow night. That's how they were tracking me down.

My Mom took me to the Manhattan railroad depot. There were several hundred soldiers lying around. After a while we got on a train, went to New York City. It took me 23 hours to get to New York and by that time, my ship should have been leaving. I still had to grab another train from there to Boston, another six or eight, hours on the train.

I got up there and went aboard, and I knew I'd be challenged; I didn't have any liberty card or anything. There were another two or three guys standing there trying to get on. By that time our Officer, Lieutenant Junior Grade File, came by and welcomed us and asked us if we were having trouble? Yes, they said we're AWOL. He went up to this officer and he told him we were not AWOL. They took our names and we had to go to Captain's Mast next day. The Captain was there and Lt. File was there and he told them that we were not AWOL. They've been through hell and back again. Captain laughed and says he would clear this up. He knew.

From there we went out to the Pacific and participated in a few invasions, I was still on the USS *Henrico*. We went back to the Philippines and retrained on the 2nd of April we invaded Okinawa.

Our last invasion was Okinawa and that was a nasty one. We lost a lot of our troops. That evening I had to come back to ship from the island about 4:00 in the afternoon and stand 4:00 to 8:00 boiler room duty. About dusk, the old ship kind of leaned over and I said what the heck, the lights went out, must have hit another ship. I was with an officer and another guy and we were standing by the smoke stack. A Japanese suicide plane hit us, one bomb went off and blew the bridge up and then another bomb hit, it went down through the seven bulkheads two inches thick, past us guys, three or four feet from us, through the smoke stack, out the port side, about six feet above the water. Then it went off, blew all the boats off the side.

The Jap plane hit the Officer's mess, on the bridge, and since this was mealtime, (evening) most of the officers were in the room eating and were killed. We also took an armor-piercing bomb on one side. It went down through two LCVP boats and on down through two steel decks to explode at about or below the incinerator room. The electrical was on the same deck we were on and the generators were just right ahead of us. We had what we called big battle lights, big flashlights. They turned them on and found us and pulled us out of the twisted steel and ruble and we escaped out through the bow up through the barber shop. We got out that way.

Then by George, we had to stay above deck that night or above the bow of the ship because it was all burnt. We didn't sink but it was burnt. People say a ship won't burn but it will because that oil is blowing everywhere.

We stayed in the bow and the next morning there was a boat that came along and took us to the USS *Goodhue*. The *Goodhue* had a cable wrapped around the propeller. I went and helped those guys get the cable off the prop. I put on the diving gear and went down and untangled the cable around the prop. We got that undone so they kept us. I wanted to get back to my ship and get my souvenirs and clothes but they told me I didn't need them. I had made a whole bunch of souvenirs in the machine shop.

The Goodhue got hit again after I got on it. It hit the boom, spun around, didn't do much damage but it bent the boom down over the side of the ship.

We were a couple months training in the Philippines, and then we were heading for Japan. We were 100 miles out from Yokohama when they dropped the big A bomb.

We went through a big typhoon and that was something terrible. The two Atomic Bombs brought the Japanese to their senses and they surrendered and the war was over. We loaded up a bunch of American prisoners of war and brought them to the Philippines and we came home and I got discharged.

August 11, 1947, Wilma Werning and I were married. We had three sons, Ralph, Harold and Danny. We and Ralph have cattle together on his ranch. Harold retired from the telephone company and Danny works at K-Hill Engine Service, a business we owned and managed for 43 years.

John tells his war experiences through the models he builds. *Author photo*

Beach Battalion No. 6 and the USS *GOODHUE* APA 107

By John Hawkins

John Hawkins in WWII.

The best I recall we were Beach Battalion No. 6 and trained at Oceanside California across the railroad tracks from the main entrance to Marine Camp Pendleton. Our Quonset huts were set up between the tracks and the ocean. The hut I was assigned was right at the edge of the bluff above the beach. We could hear the breakers rolling and crashing all night long.

It was the job of the beach party to land in the fourth to seventh wave of an assault of an enemy beach and set up beach markers to designate where supplies were to be brought in for the troops. There were large banners mounted on two poles and we were to get them set up so the boat crews who were bringing in the supplies could see them. A banner with a funnel on it was for fuel, a red cross was for medical supplies and I can't recall another very important one for, ammunition. There were banners for food supplies, drinking water and others I can't recall.

The Beach Party was made up of men of many skills. The largest group was the Hydrographic group. It was made up of Boatswain Mates, Coxswains, Seaman 1st and 2nd class. Their job was to keep the boats from broaching while unloading at the beach and direct unloading of the boats. They, under the Beach master, were in charge of everything up to the high water mark of the beach. The Army or Marines took over there. We had Signalmen. They used radios but did a lot of flag waving (semaphore) also to communicate with the incoming boats. We had a Doctor and a number of Corpsmen to handle wounded on the beach. We had boat motor macs to help with boat motor problems. There were two Carpenter Mates and one Ship Fitter and one Electricians Mate. There may have been others that I don't recall. The Ship Fitter and Electrician were also the demolition men. We were trained to blow up anything that was in the way of getting the troops and supplies onto the beach. Lovejoy was the Ship Fitter's name and I was the Electrician. We trained on the demolition grounds at Camp Pendleton. Marines trained us in the use of many explosives. Lovejoy was a First Class Ship Fitter and I was striking for Electricians Mate. Lovejoy carried a submachine gun and I carried an old 1903 Springfield rifle. In addition to our regular back packs, canteen and fox hole shovel, I had a large tool box to do electrical work on boats as needed, and two haversacks of tetratol explosives to carry onto the beach. Lovejoy carried the caps and fuse reels plus his large toolbox. We were to dig a foxhole first for the demolition supplies then for ourselves. We were to dig in at a distance from each other so a mortar wouldn't get both of us. I helped the Hydrographic group keep the boats from broaching and unloading supplies. I never had to blow anything up on a landing and I don't recall having to do any electrical repairs to a boat during a landing. I did have to blow up a boat skid that was floating among ships that had been hit by kamikaze planes, off of Okinawa, the first of April 1945. (This is a story by its self)

After participating in the taking of Kerama Retto (April 1st), we were standing by with reserves for Okinawa when we were attacked by a flight of Nips (April 2nd), three of whom our gunners shot down. But the third suicided into our ship, hitting one of the masts and exploding in mid-air over the after decks and gun positions. Material damage was light; but casualties, both killed and wounded, were heavy. Our Ship, USS *Goodhue*.

APA 107 was one of three ships of our Transport Group hit the evening of April 2, 1945, off of Okinawa. The APA 120 and PA 45 from our Transport Group 51 were both hit that evening. Our ship lost 27 men with about 130 wounded and I believe suffered less than the other two. The APA 120 took a Jap plane in the side at the water line and it went into the boiler room and exploded. She did not sink and was towed into our refuge harbor off Okinawa. The PA 45 took a Jap plane in the Officer's mess, on the bridge, and since this was mealtime, (evening) most of the officers were in the room eating and were killed. The 45 also took an armor-piercing bomb on one side. It went down through two LCVP boats and on down through two steel decks to explode at about or below the incinerator room. I never heard the number of casualties for either of the other ships. There were other crippled ships with us for a couple of weeks. We had thirty-two days and eighty-two air raids at Okinawa. Our combat troops were part of the 77th Army Division. We had picked them up off the beach at Leyte (Philippines) in February for training in preparation for the Kerama Retto-Okinawa operation.

On April 17, we sent our troops in on Ie Shima where Ernie Pyle was killed. One of our boats returned him to a ship but I believe he was already deceased. It was a flag from our ship that was raised on Ie Shima. I was not asked to make the assault so I did not get to go. One Beach Party member (a signalman from the Flag Group on our ship) lost the tip of a finger to shrapnel. As I recall, the troops and Beach Party were returned to the ship after one day or two. We latter landed our troops on North Okinawa on April 20. I did not have to go on this assault. This was spectacular to watch from our station out beyond the bombardment ships. Three battle ships were steaming along one behind the other just beyond us toward the beach. Next were cruisers and then a line of destroyers all firing onto the beach. There were three LCMR's spaced about 1/2 mile apart, going into the beach and firing rows of rockets all the way in. Then they turned around and as they moved away from the beach they moved over, they fired their 5 inch forty gun onto the beach until they turned around and started another run sending in rockets all the way. They would reload the rocket launchers as they came out from the beach. This must have softened up a lot of beach. All the time there were airplanes in the air, in fours and they were in four leaf, clover leaf, patterns and one would be diving at all times with the other three waiting their turn. They dropped bombs fired rockets and strafed certain targets back away from the beach. After considerable bombardment, the assault boats took the troops in behind a large number of AMTRACK tanks. In spite of all the bombardment there was still a stiff fight after the troops got on the beach and started inland. The men of the Army 77th Division were mostly older and I suspect most of them who survived the Okinawa operation are gone now (Aug. 2001).

We had a reunion with our 77th Division Army men Sept. 27, 1945, at Cebu, Philippines. They took a large group of our ship's crew on a picnic into the interior. We rode in their DUCKS (large boat like vehicles that they used to transport material and supplies from the ship to the beach.) We had shared a lot of loading and unloading those DUCKS at Leyte and at the three landings or more at Okinawa. It was a great day for the USS *Goodhue*'s Boat crew and Beach Party sailors.

We picked up the Americal Army Division at Cebu Sept. 1 and delivered it to Yokohama for occupation duty. We passed the Battleship *Missouri* where the Armistice was signed.

We picked up RAMP's (x-prisoners of war the Japs held.) I can't recall how many nations the prisoners represented. We took them to Manila where they went to hospitals to be checked and treated. They were all just skin and bones but VERY HAPPY to be free and with Americans.

I never had to make a landing under fire (Thank God). We put troops on several small islands around Okinawa and the day we took Ie Shima (Where Ernie Pyle was killed.) I was not asked to be a part of the Beach Party that made the landing. I recall our Beach Party was landed on an island in the Philippines where there were still some Japs and we had to dig in and wait. We spent the day on the beach in our fox holes. They would not let us climb a coconut tree for something to eat so we were hungry and since it rained all afternoon we were wet. They sent the boats back for us after dark and when I climbed up the cargo net and over the rail the barrel of my 1903 Springfield rifle had water pour out of it. We went armed and with full packs and explosives and tools. We didn't get shot at but we didn't offer very good targets either. Troops never did come in. It may have been a training exercise but we didn't think so and was scared, hungry and wet from about noon until we were back on the ship. We were on the beach at Lingayen Gulf training with elements of the 33rd Army Division when a jeep came speeding down the beach announcing the war was over and within minutes the many ships lying off the beach started shooting star shells and smoke shells etc. into the sky. That afternoon the ships send lots of beer onto the beach and we were all allowed two beers. That was our last Beach Party landing and exercise. (Aug. 15, 1945)

Taking a second load of troops from Cebu, Philippines, to Japan was canceled (Sept. 26) and we were sent back to manila to pick up Navy discharge's and return them to San Francisco but were diverted to San Pedro where we were greeted by a large and grateful turn out of civilians, bands etc.

We then loaded the officer's cars onto the ship and sailed for the Panama Canal. We passed through and had liberty on the Atlantic side and it was the vilest, thieving place I have ever been. The little kids would gang up around you and before you knew what was going on they would steal everything out of you jumper pocket and anything else they could grab. They passed it to others who out ran us and the Shore Patrol. They got an expensive fountain pen from my buddy.

We sailed up to Norfolk, Virginia, for de-commissioning. I was selected with one other Electricians Mate to stay on the ship until it was turned over to some non-Navy owner. There was a very small group of us left on the ship for about a month until all of the Navy spare parts etc. could be removed. I went by train to Chicago and then on the Santa Fe to my home in Missouri. Was home a couple of weeks and went back to Great Lakes Naval Training Center for discharge which only took a few days and I was on my way home free. May of 1946. I was in for two years, two months and 17 days. It was the greatest experience of my life. Some tough times and the trauma of seeing ship mates blown up and dying is still a burden, but there were many beautiful, wonderful things I was able to see and do. After having been with army infantry men who had fought on the front lines, I have a great respect and appreciation for them and the miserable conditions they served under. I'm glad I chose the Navy, glad I was in the Beach Party, glad I was on a ship and glad I was able to cover so much of the world and survive.

John H. Hawkins Em/2c, Beach Battalion 6, USS *Goodhue* APA 107. (Nov.1944 to April 1946) Nov. 12, 2001JHH

MARY ELLEN MOCK

John and Mary Ellen Brown were married on April 3, 1993. Mary Ellen is also a WWII Navy veteran.

While teaching in the Eureka Public Schools, I enlisted in the United States Navy (Waves) in March, 1944 during World War II. I reported for duty July 25, 1944, at Hunter College, New York City. (During the war it was called The USS Hunter) for six weeks boot training. When this was completed, I was assigned to Mail School where all aspects of the Postal Service had to be mastered. Upon graduating from Mail School, those with grades in the upper ten percent had a choice of postal stations.

From the list of vacancies, another WAVE friend and I chose to be stationed at the Fleet Post Office in San Francisco, California. My first assignment there was working in the Fleet Records Section. This required keeping the addresses of all Naval personnel stationed on ships, the Pacific War Zones and those stationed on the west coast current so the service men and women's mail could be dispatched to them as quickly as possible. Next to family, mail was the next priority.

I was assigned all the addresses cards that began with the letter "H." Other duties performed when so ordered were working in the V-mail section in the Fleet Post Office pitching mail, and at the time of discharge was working back in Fleet Records as a supervisor checking the address card reports to see if the workers were changing the addresses accurately and if the "Missing in Action" and "Deceased" mail was being handled properly and promptly especially as the war ended.

On December 2, 1945, I married former Tech-Sgt. Michael Sissman. We met at the B-29 Great Bend Air Force Base when I was teaching there and helping in the USO and we were married in San Francisco. He had just been discharged from the Army Air Force after six years in the service, two of which were in the China Burma-India (CBI) war zone as a crew chief on B-29 bombers.

March 4, 1946, I was discharged from the Navy with the rank of Mailman Second Class and continued to live in San Francisco until my husband who was employed by TWA, was transferred to the Los Angeles International Airport. We established a home in Hawthorne, California.

I returned to Kansas in 1951 when my husband's health failed. He was hospitalized at Fitzsimons Army hospital, Denver for over three years. I used my GI Bill, went back to college for several years until I had a Masters Degree and upgraded my teaching credentials.

I had one little daughter often going to summer school classes with me. I retired after teaching first grade and high school special education for 35 years, 20 years in the Eureka Public Schools. I am a member of the Eureka Lioness club. Eureka Music Club, Senior Vice Commander of DAV Auxiliary #36, Emporia and many other similar positions.

I was a widow for 8 years. John and I were quite active in the Disabled American Veterans and Auxiliary. At times our meetings and social related activities were in Emporia. We car pooled trips with another couple so were riding together frequently. I held DAVA Department national Chaplain's office which required a lot of traveling at times. John would be my chauffeur getting me to and meeting me at the Wichita airport. After four years, he finally ended his bachelor days and asked me to marry him!

Index

Page	NAME	UNIT	ARRIVAL TO ETO	UNIT ARRIVAL TO ETO
ix	**FOREWORD**			
1	**Chapter 1 — THE NICKEL**			
11	**Chapter 2 — UNCLE BILL**			
11	WILLIAM F. "BILL" DRIVER	1st Infantry Division	Feb. 4, 1945	June 6, 1944
15	**Chapter 3 — BATH TO BATTLE**			
18	**Chapter 4 — THE PRISONERS**			
21	JOHN MOCK	106th Infantry Division	Dec. 11, 1944	Dec. 11, 1944
33	JULIAN SIEBERT	26th Infantry Division	Nov. 1944	Sept. 17, 1944
43	BILL STAHL	106th Infantry Division	Dec. 11, 1944	Dec. 11, 1944
51	**Chapter 5 — THE INFANTRY**			
55	RAYMOND BROWN	95th Infantry Division	Sept. 15, 1944	Sept. 15, 1944
65	ART COTTRELL	90th Infantry Division	June 6, 1944	June 6, 1944
75	HOWARD GOODWIN	102nd Infantry Division	Sept. 30, 1944	Sept. 30, 1944
87	HORACE HIGH	75th Infantry Division	Jan. 12, 1945	Dec. 13, 1944
99	ART HOLTMAN	XVIII Airborne Corps	Aug. 25, 1944	Aug. 27, 1944
103	LENHARDT HOMEIR	2nd Infantry Division	Dec. 12, 1944	June 7, 1944
113	DICK JEPSEN	30th Infantry Division	July 14, 1944	June 11, 1944
121	ROBERT "BOB" KNIGHT	84th Infantry Division	Nov. 16, 1944	Oct. 1, 1944
133	VIRGIL MYERS	80th Infantry Division	Sept. 28, 1944	Aug. 3, 1944
147	CHARLES NEALE	XVII Airborne Corps	Dec. 23, 1944	Dec. 25, 1944
151	DON NIXON	82 Airborne Division	Jan. 3, 1945	June 6, 1944
157	ROBERT PEARSON	26th Infantry Division	Sept. 30, 1944	Sept. 17, 1944
163	FRANK RHODES	509th Parachute Battalion	Aug. 15, 1944	Aug. 15, 1944
173	DON RICHARDS	80th Infantry Division	Aug. 1944	Aug. 5, 1944
179	JIM SHARP	1st Infantry Division	Jan. 10, 1945	June 6, 1944

Page	NAME	UNIT	ARRIVAL TO ETO	UNIT ARRIVAL TO ETO
189	**Chapter 6 — THE TANKERS**			
191	ORVAL ABEL	2nd Armored Division	June 28, 1945	June 28, 1945
197	JACK GRAGERT	8th Armored Division	Jan. 1945	Sept. 7, 1944
205	HARLAN HENRY	7th Armored Division	Dec. 1, 1944	Aug. 24, 1944
213	KENNY LUIGS	9th Armored Division	Oct. 23, 1944	Oct. 23, 1944
227	CARL SHELL	9th Armored Division	Oct. 23, 1944	Oct. 23, 1944
237	HERMAN W. WESTMEYER	11th Armored Division	Dec. 17, 1944	Dec. 15, 1944
249	CARL WHITE	749th Tank Battalion	Dec. 1, 1944	June 29, 1944
257	**Chapter 7 — THE ARTILLERYMEN**			
259	ELMER BLANKENHAGEN	666 Field Artillery	Dec. 20, 1944	Dec. 20, 1944
267	DON HUSE	76th Field Artillery	Aug. 11, 1944	Aug. 10, 1944
277	HAROLD O'MALLEY	87th Infantry Division	Dec. 2, 1944	Dec. 2, 1944
283	WALLACE JEFFREY	327th Fighter Control Group	June 7, 1944	June 7, 1944
295	CARROL JOY	406 Fighter Group	Aug. 1, 1944	Aug. 1, 1944
305	**Chapter 9 — THE MEDICS**			
307	FORREST ADAMS	29th Infantry Division	June 6, 1944	June 6, 1944
319	DR. JAMES McCONCHIE	77th Medical Evacuation	July 7, 1944	July 7, 1944
329	DALE SMITH	9th Infantry Division	June 10, 1944	June 7, 1944
337	MALCOLM STROM	102nd Evacuation Hospital	April 18, 1944	July 18, 1944
325	IVAN W. WOELLHOF	238th General Hospital	Dec. 17, 1944	
347	**Chapter 10 — THE MECHANICS**			
349	J. D. SEXTON	132nd Ordnance Co.	Sept. 12, 1944	Sept. 12, 1944
357	PAUL SCHEID	913nd Ordnance Co.	June 28, 1944	June 28, 1944
367	**Chapter 11 — THE NAVY**			
369	JOHN BROOKS	USS *Henrico*	June 6, 1944	June 5, 1944
381	JOHN HAWKINS	USS *Goodhue*		
387	MARY ELLEN MOCK	Fleet Post Office Pacific		